Sustainable Fish Production and Processing

Sustainable Fish Production and Processing

Edited by

Charis M. Galanakis

Research & Innovation Department, Galanakis Laboratories, Chania, Greece;
Food Waste Recovery Group, ISEKI Food Association, Vienna, Austria

ELSEVIER

ACADEMIC PRESS
An imprint of Elsevier

Academic Press is an imprint of Elsevier
125 London Wall, London EC2Y 5AS, United Kingdom
525 B Street, Suite 1650, San Diego, CA 92101, United States
50 Hampshire Street, 5th Floor, Cambridge, MA 02139, United States
The Boulevard, Langford Lane, Kidlington, Oxford OX5 1GB, United Kingdom

British Library Cataloguing-in-Publication Data
A catalogue record for this book is available from the British Library

Library of Congress Cataloging-in-Publication Data
A catalog record for this book is available from the Library of Congress

ISBN: 978-0-12-824296-4

For Information on all Academic Press publications
visit our website at https://www.elsevier.com/books-and-journals

Publisher: Nikki Levy
Acquisitions Editor: Anna Valutkevich
Editorial Project Manager: Samantha Allard
Production Project Manager: Selvaraj Raviraj
Cover Designer: Vicky Pearson Esser

Typeset by MPS Limited, Chennai, India

Contents

List of contributors

Heba H. Abdel-Mohsen
National Institute of Oceanography and Fisheries (NIOF), Fish Nutrition Laboratory, Aquaculture Division, Cairo, Egypt

Kelvin Adrah
Food and Nutritional Sciences Program, North Carolina Agricultural and Technical State University, Greensboro, NC, United States

Shehu Latunji Akintola
Fisheries Department, Lagos State University, Lagos, Nigeria

Ines Andretta
Universidade Federal do Rio Grande do Sul, Porto Alegre, Rio Grande do Sul, Brazil

Christoph Josef Backi
Department of Chemical Engineering, Norwegian University of Science and Technology (NTNU), Trondheim, Norway

Ishita Banerjee
Department of Biochemistry and Molecular Biology, Pondicherry University, Puducherry, India

Mayank Bhushan
Department of Nanotechnology, North-Eastern Hill University, Shillong, India

Charles Brennan
Department of Wine, Food and Molecular Biosciences, Lincoln University, Christchurch, New Zealand; Riddet Research Institute, Palmerston North, New Zealand; School of Food Science and Engineering, South China University of Technology, Guangzhou, P.R. China

Margaret Brennan
Department of Wine, Food and Molecular Biosciences, Lincoln University, Christchurch, New Zealand

Raquel Carrilho
Centre of Marine Sciences (CCMAR), University of Algarve, Faro, Portugal; Faculty of Sciences and Technology, University of Algarve, Faro, Portugal

Marco Cerqueira
Centre of Marine Sciences (CCMAR), University of Algarve, Faro, Portugal

Paula Gabriela da Silva Pires
Universidade Federal do Rio Grande do Sul, Porto Alegre, Rio Grande do Sul, Brazil

Cláudia Raposo de Magalhães
Centre of Marine Sciences (CCMAR), University of Algarve, Faro, Portugal; Faculty of Sciences and Technology, University of Algarve, Faro, Portugal

Naglezi de Menezes Lovatto
Universidade Federal de Santa Maria, Santa Maria, Rio Grande do Sul, Brazil

Ajay S. Desai
Department of Wine, Food and Molecular Biosciences, Lincoln University, Christchurch, New Zealand; Riddet Research Institute, Palmerston North, New Zealand; Dr. Balasaheb Sawant Konkan Krishi Vidyapeeth, Dapoli, India

Swapnamoy Dutta
Department of Green Energy Technology, Pondicherry University, Puducherry, India

Isa Olalekan Elegbede
Environmental Planning, Cottbus-Senftenberg Brandenburg University of Technology, Cottbus, Germany

Kafayat Adetoun Fakoya
Fisheries Department, Lagos State University, Lagos, Nigeria

Ana Paula Farinha
Centre of Marine Sciences (CCMAR), University of Algarve, Faro, Portugal

S.S. Gangan
Dr. Balasaheb Sawant Konkan Krishi Vidyapeeth, Dapoli, India

Krishna Kumar Jaiswal
Department of Chemistry, Uttaranchal University, Dehradun, India; Institute for Water and Wastewater Technology, Durban University of Technology, Durban, South Africa

Toheeb Jolaosho (Lekan)
Fisheries Department, Lagos State University, Lagos, Nigeria

Marcos Kipper
Elanco Animal Health, São Paulo, São Paulo, Brazil

Bruno Bianchi Loureiro
Universidade Federal de Santa Maria, Santa Maria, Rio Grande do Sul, Brazil

Alfred Maluwa
Malawi University of Science and Technology, Limbe, Malawi

Sahar Fahmy Mehanna
Fish Population Dynamics Lab, Fisheries Division, National Institute of Oceanography and Fisheries, Egypt

Maria Sara Cabrera Mendéz
Universidade Federal do Rio Grande do Sul, Porto Alegre, Rio Grande do Sul, Brazil

Márcio Moreira
Centre of Marine Sciences (CCMAR), University of Algarve, Faro, Portugal; National Institute for the Sea and Atmosphere (IPMA), EPPO – Aquaculture Research Station, Olhão, Portugal; Faculty of Sciences and Technology, University of Algarve, Faro, Portugal

Esther Odunsi (Odunayo)
Department of Aquaculture and Fisheries Management, Federal University of Agriculture, Abeokuta, Nigeria

Pedro M. Rodrigues
Centre of Marine Sciences (CCMAR), University of Algarve, Faro, Portugal; Faculty of Sciences and Technology, University of Algarve, Faro, Portugal

Norhan E. Saleh
National Institute of Oceanography and Fisheries (NIOF), Fish Nutrition Laboratory, Aquaculture Division, Cairo, Egypt

Denise Schrama
Centre of Marine Sciences (CCMAR), University of Algarve, Faro, Portugal; Faculty of Sciences and Technology, University of Algarve, Faro, Portugal

Reza Tahergorabi
Food and Nutritional Sciences Program, North Carolina Agricultural and Technical State University, Greensboro, NC, United States

Mayookha V.P.
Department of Food Science and Technology, Pondicherry University, Puducherry, India

Elham A. Wassef
National Institute of Oceanography and Fisheries (NIOF), Fish Nutrition Laboratory, Aquaculture Division, Cairo, Egypt

Preface

Over the last decades, global aquaculture presented substantial growth due to the increased need for human consumption. This continuous production growth is still expected to come over the next decades leading to the depletion of aquatic animal food. As resources become more restricted and demand grows, sustainability becomes a major issue in the fish and marine industries. At the same time, aquaculture is a very competitive market. Thus a global awareness is raised in spite of using scientific knowledge and emerging technologies to obtain a more sustainable production. This fact has enhanced the importance of issues like genomics, genetics, and breeding in aquaculture. In contrast, it has turned the fish industry's interest to investigate the handling of their by-products with another perspective, for example, by adopting more profitable options.

Sustainability relies on the approach that we should meet the present needs without compromising the future and conditions of the next generations. To respond to sustainability goals, the maximum utilization of aquatic materials is required to integrate activities throughout all the production-to-consumption stages. To maximize the conversion of raw materials into consumer goods, we need to optimize production and processing, improve postcatching practices, and increase the by-products' valorization. Subsequently, there is a need for a new guide covering the latest developments in this particular direction.

Food Waste Recovery Group (www.foodwasterecovery.group) has developed a portfolio of numerous activities, including consultation reports, e-courses, webinars, workshops, reference modules, publications, and multiple books in the broad fields of food, nutrition, bioresources, and environment. Following this trend, the current book aspires to fill in this gap existing in the existing literature by providing specific solutions for industrial sustainability in spite of fish processing, production, and waste management, as well as discussing current trends. The present book covers the importance of aspects concerning the sustainable production and processing of fish at all relevant stages. The ultimate goal is to optimize the current aquaculture production systems by providing a guide to all fish and aquaculture scientists, technologists, professionals, and producers, trying to optimize industrial performance and reduce environmental impact.

The book consists of 2 parts and 10 chapters. **Part A** deals with sustainable fish and seafood production. Chapter 1, Genomics, Genetics, and Breeding in Aquaculture, discusses the various techniques in fish genetics, breeding, and genomics; structural to functional genomics; and their application during marker-assisted selection. Aquaculture can supplement the dwindling fish catches from the wild sources to feed the ever-growing human population. The biology of aquatic organisms, which shows high genetic variability and high fecundity coupled with a short generation interval, has facilitated rapid progress with genetic improvement programs. The vast genetic variation in the aquaculture species can be exploited to increase and sustain aquaculture production using a combination of breeding strategies.

Chapter 2, Proteomics for Quality and Safety in Fishery Products, discusses the main applications of proteomics-based strategies in fish welfare, quality, and safety assessment toward sustainable aquaculture. Fish quality, safety, and sustainability are highly interconnected, involving many stakeholders, from consumers to policymakers/legislators responsible for the implementation of rules, in a continuous feedback loop where environmental concerns are getting a central role. Postharvest activities remain critical to eliminating hunger and midpoints between field harvest and consumption in the fisheries value chain. Chapter 3, Postharvest Practices in Small-Scale Fisheries, presents different postharvest practices of fisheries together with relevant recommendations for small-scale fisheries. Policy changes are needed to improve sustainability in postharvest practices and ensure food and nutritional security, especially in the developing countries that lack cold chains, storage facilities, and advanced logistics systems.

The sustainable management and use of aquatic resources and adoption of ecosystem approaches can enhance economic growth, increase food security, improve nutrition, and reduce poverty. Chapter 4, Egyptian Marine Fisheries and Its Sustainability, discusses the sustainability of the seafood sector using Egypt's industry as a case study where fish production from both natural resources and aquaculture plays a vital role in food security, livelihood, and the creation of employment opportunities in rural areas. The chapter describes the current status of Egyptian marine fisheries, addresses the main challenges faced during its development, and provides practical solutions for its sustainability. Chapter 5, Life Cycle Impact of Industrial Aquaculture Systems, deals with the life cycle impact of industrial aquaculture systems. The capture fishery production has overexploited the wild-fish stocks, and the global demand for aquatic products is still increasing. Industrial aquaculture systems, such as marine net-pen, have a life cycle impact, most of which is due to the energy needed to produce feed and the marine biotic resource overexploitation.

Part B deals with fish and seafood processing and valorization of by-products. In Chapter 6, Industrial Thawing of Fish Blocks, thawing methods are presented in terms of their functional principles, technical specifications, and to some degree, also quality preserving factors. Two main methods exist: heat/energy to the frozen product only via the boundary layers and generate heat inside the product. Both have their advantages and disadvantages, but the methods providing heat via the product's boundary layers are typically used in large-scale industrial applications. Chapter 7, Utilization of Fish

Waste as a Value-Added Ingredient: Sources and Bioactive Properties of Fish Protein Hydrolysate, deals with the sources, functional properties, and valorization opportunities of fish protein hydrolysates. Those are the breakdown product of enzymatic hydrolysis into smaller peptides. Due to the vast amount of fish waste material generated and potential sources of high protein content with balanced amino acids, fish protein hydrolysates have a better alternative. Bioactive peptides obtained from fish waste material such as skin, fin, viscera, and scale with antioxidative properties may be a potential alternative for the synthetic antioxidants, which exhibit the potential benefits in human nutrition.

Chapter 8, Ready-to-Eat Products Elaborated With Mechanically Separated Fish Meat From Waste Processing, discusses the mechanical separation of fish meat to develop and manufacture ready-to-eat products for human consumption and evaluate their chemical stability. The mechanical separation of fish meat from waste processing presents a sustainable and alternative way of manufacturing ready-to-eat products for human consumption, considering its social, economic, and environmental benefits. Chapter 9, Sustainable Fish and Seafood Production and Processing, discusses the opportunity of valorizing seafood by-products for the formation of aquaculture fish feeds. These aquatic by-products include unsalable undersized fish and shrimp and those parts of fish that are not consumed as human food. The unwanted bioparts are rich in protein, lipids, minerals, and vitamins and thus can be utilized as animal feeds.

The ability to extract, refine, and commercialize fish oils faces a challenge in processing facilities where infrastructure is limited and plants are located remotely. Under these conditions, extraction of oil from fish by-products for use as an in-house or regional fuel may be both economically and environmentally a more sustainable approach. Chapter 10, Lipid Extraction From Fish Processing Residues for Sustainable Biofuel Production, presents the developments in fish oil extraction methodologies, including physical, chemical, biochemical, and biological, for biofuel production. The characteristic properties of biodiesel have been described concerning international standards. Prospects and industrial development have a significant impact on a sustainable environment.

Conclusively, the book addresses food and aquaculture scientists and technologists, and professionals working in the fish industry seeking to improve their by-products' management by actively utilizing efficient waste applications. University libraries and institutes could use it as a textbook and ancillary reading in graduates and postgraduate level multidiscipline courses dealing with fish processing technology, aquaculture science, food technology, environmental, and bioresource technology.

I want to thank and acknowledge all authors for their valuable collaboration in bringing together different topics of sustainable fish production and processing in one essential reference. Their acceptance of my invitation, editorial guidelines, and timeline schedule is highly appreciated. I would also like to thank the book manager Lena Sparks, the acquisition editor Patricia Osborn, and Elsevier's publication team for helping during the editing and production process. Finally, a message for all the readers: This collaborative scientific effort contains hundreds of thousands of words that may have some gaps or errors. Comments and criticism are always welcome, so please do not hesitate to contact me to discuss any relevant issue.

Charis M. Galanakis[1,2]

[1]*Research & Innovation Department, Galanakis Laboratories, Chania, Greece*

[2]*Food Waste Recovery Group, ISEKI Food Association, Vienna, Austria*

PART | 1

Sustainable Fish and Seafood Production

Chapter | one

Genomics, Genetics, and Breeding in Aquaculture

Alfred Maluwa
Malawi University of Science and Technology, Limbe, Malawi

1.1 INTRODUCTION

The chapter on aquaculture genomics, genetics, and breeding describes the techniques and strategies that have been developed to improve the genetic performance of aquaculture species. In addition, detailed applications of the techniques with genetic improvement programs on the diverse aquaculture species are presented. Aquaculture is important because it is supplementing fish production from capture fisheries, where the catches are dwindling. The global per capita fish consumption has increased, despite the growing human population due to fish production that has been contributed by aquaculture (FAO, 2020). However, there is a need for novel and innovative ways to improve the production efficiency that can sustain the profitability of aquaculture. To achieve this, genetic improvement programs have been initiated in many countries with the aim of improving further the efficiency of aquaculture production. The genetic improvement programs have also focused on the quality of aquatic organisms that are produced, the economics of production in terms of profitability, hence sustainability and the development of viable markets. Researchers in aquaculture genetics have studied and applied animal breeding principles, genetics, and genomics to improve growth, age to maturation, disease resistance, and tolerance to diverse environmental conditions.

This chapter starts with the justification of aquaculture by stating its importance as a food source and also for other food products. The chapter further describes the genesis of genetic improvement for aquaculture species. Progress that has been achieved is also described specifically in relation to the exploitation of the additive genetic components of the aquaculture organisms. Specific examples have been presented concerning the work with rainbow trout, salmonids, tilapias, catfishes, and marine species of aquaculture

Sustainable Fish Production and Processing. DOI: https://doi.org/10.1016/B978-0-12-824296-4.00003-7

importance. Efforts to exploit heterosis and the limitations and challenges in exploiting hybrid vigor for aquaculture development have also been discussed. Other genetic improvement strategies, such as chromosome set manipulation, gene transfer, and different molecular technics, have also been presented. Although the application of chromosome set manipulation, gene transfer, and molecular technics have not resulted in increased commercial aquaculture production, these technologies have formed the foundation for the scientific work that has contributed to the advancement in genomics research in aquaculture.

Genomics has now become popular for the discovery and exploitation of important gene(s) that code for important aquaculture traits. The major work in genomics in aquaculture has concentrated on the identification and exploitation of important traits such as growth, age to maturation, disease resistance, and extreme environmental tolerance such as cold weather. This chapter further discusses the genesis of aquaculture genetics, from the discovery of the deoxyribonucleic acid, which contains the genetic code and the various gene sequencing techniques that followed functional genomics. The application of genomics with quantitative trait loci, expressed quantitative trait loci, gene regulation, and conservation at the levels of genome, transcription, proteome, epigenome, and system biology have also been described. The combination of genetics, breeding, and genome techniques is providing useful information which is being applied to improve and sustain aquaculture production.

1.2 IMPORTANCE OF AQUACULTURE AND AQUATIC ORGANISMS

Fish, including other teleost organisms such as mollusks and crustaceans, which are collectively called aquaculture species, are important food species worldwide. In 2018 global aquaculture production was estimated at 179 million tonnes, with an estimated value of US $401 billion (FAO, 2020). Of this production, 156 million tonnes were consumed directly, and 22 million tonnes were used as an ingredient during feed manufacturing processes. Aquaculture contributed 82 million tonnes, valued at about US $250 billion (FAO, 2020). The global human population is increasing rapidly at around 1.6% per annum. Consequently, there are now 7.8 billion people on the globe (World Population Barometer, 2020). The high human population translates to high demand for food, which includes fish. For example, global fish consumption for a period spanning from 1961 to 2017, increased at an annual rate of 3.1% (FAO, 2020). This rate was almost twice the annual world population growth rate and also higher than that of the other sources of protein, such as meat, dairy, and milk, which increased by 2.1% per annum during the same period (FAO, 2020).

Aquaculture has been practiced since ancient times, from 3500 BCE in China with the farming of common carp, which is a native species to the country (Parker, 2000; Fagan, 2017). The practice has now spread globally, giving a history of thousands of years (Bentsen and Gjerde, 1994). Due to the ever-increasing human population the demand for fish and fish products is also increasing all the time (FAO, 2020). Consequently, there is a need to improve the scale of aquaculture production and the efficiency and sustainability of fish production from aquaculture (Jose, 2015). To sustain aquaculture production the major challenge has been how to exploit the genes that are responsible for aquaculture performance traits in the cultured species. In addition, the expression of the genes that are responsible for aquaculture traits is also sensitive to the environment in which the species are raised.

As a starting point a genetic improvement on the aquaculture performance of the fish species initially concentrated on important traits which are prevention and/or control of infectious disease outbreak, control of sexual maturation (as most teleost's growth slows down after attaining sexual maturation), finding appropriate and sustainable feed for carnivorous aquaculture species, and tolerance to diverse and changing environments (Bentsen and Gjerde, 1994). The genetic constitution of organisms and their interaction with the environment are responsible for the performance of the species in aquaculture. Hence a need arose for the selection of families or individuals within families that would resist or tolerate adverse environmental conditions to optimize the yields.

1.2.1 The Genesis of Genetic Improvement for Aquaculture Fish Species

The first approach in the attempts to improve the aquaculture performance of the cultured fish species was through genetic selection. Genetic selection research was conducted following the example of breeding work in domesticated mammals and birds (livestock). The genetic improvement research with livestock was based on the animal model (Bentsen and Gjerde, 1994). During the early 1990s, in livestock, the productivity was 3–5 times higher than that of their wild populations (Eknath et al., 1993), and this improved performance was attributed to the genetic selection. Consequently the breeding for desirable aquaculture traits has lagged behind that of livestock. In 2005 more than 90% of the aquaculture production was coming from wild stocks, that were genetically unimproved (Gjedrem, 2005). Before the onset of breeding programs for the aquaculture species, efforts concentrated on mating what was perceived phenotypically best males with phenotypically best females. The problem with this method was that the phenotype was also due to the environment in which the individuals were raised and not necessarily due to their genetic constitution. To prevent inbreeding, after several generations of mating, the best phenotypically broodstock was mated with wild stocks to increase the genetic variation in the farmed broodstock (Eknath et al., 1993).

In the late 1990s, research on fish breeding concentrated on studies with farmed fish species that would exhibit superior genetic performance compared to their wild progenitors (Bentsen and Gjerde, 1994). The work on selective breeding in aquaculture species, therefore, focused on the exploitation of the additive genetic component in the species being cultured. More emphasis was made on screening the available wild stocks for superior strains but this effort did not yield the desired results because these techniques did not produce the fish that was demanded by the industry. Crossbreeding programs were undertaken to exploit heterosis or hybrid vigor, however, with most aquaculture species, especially fish, there is moderate to low heterosis for the aquaculture performance traits (Gjerde and Refstie, 1984; Dunham, 1987; Wohlfarth, 1993). Furthermore, where heterosis was substantial or high, it could not be used in the later generations, because heterosis does not accumulate for exploitation during the latter generations. Hybrid vigor is enjoyed only in the first generation and is lost in future generations. Hence the need to recreate heterosis in every generation was a major setback in the exploitation of hybrid vigor in aquaculture (Bentsen and Gjerde, 1994). In the 1990s selective breeding to exploit the additive genetic components of aquaculture species was considered more appropriate and efficient (Bentsen and Gjerde, 1994). Since then, there has been steady progress in improving aquaculture traits in the cultured fish species. The high fecundity in fish and the short generation interval are some of the factors that have facilitated fast progress in fish breeding programs (Gjedrem, 2005).

A lot of progress on breeding for aquaculture development has been achieved during the late 1990s and early 2000s. According to Gjedrem (2005), successful breeding programs in aquaculture have been developed for salmonids, carps, rainbow trout, catfishes, and tilapia. In these breeding programs, genetic gains per generation for growth have ranged from 8.4% to 20.0% (Gjedrem, 2005; Ponzoni et al., 2005; Maluwa and Gjerde, 2007). In *Oreochromis niloticus* which is the most extensively studied tilapia, a realized response to selection for harvest body weight relative to the base population of 101% has been achieved within five generations of selection (Bentsen et al., 2003). For *Oreochromis shiranus* a realized response to selection of 13.2% per generation has been reported (Maluwa and Gjerde, 2007).

The background of research in fish genetics has been extensively reviewed by Dunham (2004) and Gjedrem (2005). The review showed that earlier researchers in fish genetics concentrated mostly on fish growth, age to maturation, tolerance to a wide range of environmental conditions (plasticity), and disease resistance (McAndrew and Napier, 2010). There has been fast progress with fish breeding programs because of the high fecundity in fish that provides many gametes and offspring, thereby enabling the implementation of breeding designs that effectively provide better estimates to enable the separation of genetic and environmental effects. Consequently, fish breeding research has allowed high selection pressure, and the resultant improved fish populations have been passed on to the aquaculture industry for commercial production. Many fish species are still close to their wild ancestors, as a result, there are high levels of genetic variation which is being exploited to improve the aquaculture performance of the species (McAndrew and Napier, 2010). A number of strategies to improve the performance of aquaculture species include selective breeding, hybridization, chromosome set manipulation, gene transfer, and molecular techniques.

1.2.2 Exploitation of the Additive Genetic Component in Aquaculture

Selective breeding work to exploit the additive genetic component of aquaculture species has initially concentrated on common and major carps (Ann and Thien, 1993; Bakos and Gorda, 1994; Gjerde et al., 2002; Basavaraju et al., 2003), rainbow trout (Kause et al., 2002; Pante et al., 2002; Mambrini et al., 2004); salmonids (Refstie, 1990; Martinez et al., 1999; Bentsen and Olesen, 2000; Neira et al., 2005), tilapias (Uraiwan and Doyle, 1986; Pullin et al., 1991; Macaranas et al., 1997; Bentsen et al., 1998; Bolivar and Newkirk, 2002; Ponzoni et al., 2005; Rutten et al., 2005; Maluwa, 2005), catfishes (Dunham, 1987; Waldbieser and Wolters, 1999), and ornamental fish species. According to Kirpichnikov (1981), selective breeding work with carps was also initiated in Central and Eastern Europe. Other breeding programs to further develop the carps were implemented in Vietnam and India (Penman et al., 2005). The work with the carps has shown that both crossbreeding and pedigree-based selection can still produce fish with superior performance than their wild or unselected counterparts (Vandeputte, 2009).

1.2.2.1 ADDITIVE GENETIC COMPONENT EXPLOITATION IN RAINBOW TROUT

In the United States, rainbow trout has been domesticated for a long time (Gall and Haung, 1988). In Norway, selective breeding work of the species concentrated on selection for seawater production (Gjedrem, 1992). In Denmark (Henryon et al., 2002) and Finland (Kause et al., 2003) the rainbow trout has also been subjected to selective breeding to improve the aquaculture performance. Despite these efforts, there has not been a significant positive impact of the improved rainbow trout populations for the development of the aquaculture industry.

1.2.2.2 ADDITIVE GENETIC COMPONENT EXPLOITATION IN SALMONIDS

Breeding programs on Norwegian salmon were pioneered with funding from the Norwegian government but are now being implemented by private companies such as Aquagen and Salmobreed (Gjedrem, 2005). In this breeding program, individual and family selection was assessed for a wide range of commercial traits such as growth, sexual maturation, and body conformation. As farming became more intense a need to select for disease-resistant traits was felt. Hence recent studies on genetic improvement in the Norwegian salmon have concentrated on disease resistance. The most notorious diseases in the species have been the infectious pancreatic necrosis virus and aeromonas infectious salmon anemia. Hence, selective breeding studies have paid particular attention to the selection of families or individuals within families that would resist these current diseases and the other diseases that would emerge in the future.

The selective breeding for body weight in the Norwegian salmon resulted in estimated genetic gains of 8%–10% per generation (Gjoen and Bentsen, 1997). Similar breeding programs for salmonids were initiated in Scotland by Landcatch Natural selection and in South America, in Chile by Aquachile and Stofnfiskur. Most farmed eggs were obtained from these scientific breeding programs. These genetic improvement studies significantly improved the performance of the selected fish compared to their parental generations. Currently the genetic selection work is now being complemented by genomic technologies, including the quantitative trait loci and marker-assisted selection approaches. These technologies are being applied to produce genetically improved fish strains that are also resistant to viruses (Houston et al., 2008; Moen et al., 2009).

1.2.2.3 ADDITIVE GENETIC COMPONENT EXPLOITATION IN TILAPIA

Tilapias originated from Africa and have now become the second most important aquaculture species, only superseded by carps (FAO, 2020). The Nile tilapia is the most widely studied tilapia species. Other selective breeding programs have been implemented with *O. shiranus* (Maluwa, 2005). Early work on selective breeding programs with tilapias was hindered by the degraded nature of strains that were being used (McAndrew and Napier, 2010). The major pioneer work with selective breeding work in tilapia was the genetically improved farm tilapia program, which was first implemented in the Philippines in 1987. The genetically improved farm tilapia fish was shared with other eight southeast Asian countries under a project that was funded by the United Nations Development Program through the then International Center for the Living Aquatic Resources Management and now the World Fish Center. The participating countries were Bangladesh, Fiji, India, Indonesia, Malaysia, Thailand, and Vietnam. All the Asian countries, except India, have been working on *O. niloticus*, whose founder population came from the genetically improved farmed tilapia project in the Philippines. In India the targeted species were the Indian major carps (Rohu and Rohita). Four African countries (Egypt, Cote d'Ivoire, Ghana, and Malawi) were selected to participate in this project. However, to conserve tilapia biodiversity in Africa, it was decided that the genetically improved farm tilapia should not be exported back to Africa, rather, the African countries should be capacitated to initiate their own breeding programs, using their indigenous fish species. Egypt, Cote d'Ivoire, and Ghana started their tilapia breeding programs using their local strains of Nile tilapia. However, Malawi initiated a breeding program using an endemic tilapia, *O. shiranus*. The Asian and African countries formed a network, known as the International Network on Genetics in Aquaculture (Acosta and Gupta, 2010). The genetically improved farm tilapia strain has become a foundation for most breeding programs among the Asian countries in the International Network for Genetics in Aquaculture. According to Eknath et al. (1993), other tilapia breeding programs that used *O. niloticus* have also used the genetically improved farmed tilapia strain as a base population. In Europe, tilapia breeding programs developed new synthetic populations of *O. niloticus* using a number of selected lines (Rutten et al., 2005).

Despite the existence of the tilapia breeding programs, most of the tilapia production in aquaculture is not coming from the genetically improved strains (Beverage and McAndrew, 2000). Significant quantities of commercial tilapia production are coming from all-male monosex culture because tilapias exhibit sex dimorphism for growth, in which males grow faster than females. This phenomenon is known as stunting or runting in the species, in which the individuals when raised in ponds and other aquaculture rearing units, attain sexual maturity early, thereby compromising growth (Maluwa and Gjerde, 2006; Fuentes-Silva et al., 2013; Lin et al., 2015). To produce the large number of all-male fingerlings that are required to stock the aquaculture units the fish has been subjected to sex reversal techniques (Wohlfarth and Hulata, 1983). In tilapias, feeding the fry at the first feeding stage for 28 days with a diet that contains a male hormone (17α-methyl testosterone) incorporated at 60 mg/kg of feed (Wassermann and Afonso, 2002; Wohlfarth and Hulata, 1983) has produced progeny that comprises 80%–95% males. Another method has been the crossing of the tilapias. Inter- and intraspecific hybridization of tilapias produces offspring that is predominantly male (Wohlfarth and Wedekind, 1991; Cnaani et al., 2008; Volff and Schartl, 2001; Snake et al., 2020). The use of genetically super-male tilapias having YY chromosomes on the 23rd pair (Mair and Little, 1991) also produce all-male progeny when mated with normal heterozygous XY females. The all-male tilapia production has been responsible for the boom of the tilapia aquaculture industry because, in addition to attaining higher growth rates than with mixed-sex culture, all-male cultures produce uniform sized fish at harvest (Bartley et al., 2001; El-Zaeem and Salam, 2013) which fetches more money for the farmers than using mixed-sex culture (Kamaruzzaman et al., 2009). Among the sex reversal techniques in tilapia the nonhormonal methods are being preferred over the hormonal methods due to health and environmental concerns (Snake et al., 2020).

1.2.2.4 ADDITIVE GENETIC COMPONENT EXPLOITATION OF HIGH-VALUE MARINE SPECIES

Recently, in the early 2000s, selective breeding programs have been applied to other marine high-value aquaculture species such as the European sea bass, gilthead sea bream, and turbot (McAndrew and Napier, 2010). Other genetic improvement programs were initiated with cod, halibut, and tuna (McAndrew and Napier, 2010). Consequently, by 2005, as high as 80% of the European sea bass production was from commercial populations that had undergone genetic selection. At the commercial level, one large-scale family-based selective breeding program on sea bream had been established in 2002 (McAndrew and Napier, 2010).

1.2.3 Fish Genetic Improvement Through Hybridization

Hybridization is possible in many fish species, and it is easy to collect gametes from most fish species. There are two types of hybridization in fish; intraspecific hybridization, which is the crossing of different subspecies within a species and interspecific hybridization, which is the crossing of fish from the same genus but different species. The contribution of hybrids in aquaculture is described by Bartley et al. (2001). For example, the interspecific hybridization *of O. niloticus* and *Oreochromis aureus* has been used in Israel because the cross produces predominantly male populations that also grew equally well during the cold season (Hulata et al., 1993). In the United States, intraspecific hybridization of sunshine bass (*Morone chrysops* × *Morone saxatilis*) produced hybrids that grew faster and exhibited better overall culture characteristics than either of the parental species under commercial aquaculture conditions (Smith, 1988). Details of tilapia hybridization techniques have been described by a number of authors such as Wohlfarth and Wedekind (1991), Cnaani et al. (2008), Volff and Schartl (2001), and Snake et al. (2020). However, despite the reported advantages of hybrid vigor, there has been very little commercial production of hybrids because the vigor and other advantages of hybrids are not cumulative, hence the heterosis has to be recreated in every future generation because there is introgression of performance of the hybrids back to the parental species with successive breeding of the fish.

1.2.4 Fish Genetic Improvement Through Chromosome Set Manipulation

The chromosomes of aquatic species, including fish, can be manipulated to produce offspring with desirable traits for aquaculture. For example, if fertilized eggs are subjected to heat, strong irradiation with UV or Gamma rays, or shock pressure during the second meiotic division, a triploid embryo containing a three chromosome set is produced. Protocols are available for the production of triploids in 30 different fish and shellfish species (Dunham, 2004). Triploid fish are sterile and the method has been used, especially with rainbow trout, to produce large-sized fish from all-female strains (Bye and Lincoln, 1986). The technique has also been successfully used with channel catfish (Wolters et al., 1991) and common carp (Basavaraju et al., 2002). In other fish species, triploidy has also been used especially with species that exhibit stunting, such as tilapias. In Oysters the technique has been used to improve growth rate and flesh quality (Guo et al., 1996). The other application of triploidy concerns the conservation of fish biodiversity, as with triploids, there is no risk of the farmed strains to interbreed with other local fish populations (Kozfkay et al., 2006). The technique has also been used to prevent exotic species to get established in the introduced ecosystem. A good example species for this is the grass carp, which was introduced in ponds to control weeds. Triploids were preferred because of their sterility because they cannot get established in the introduced environment.

Transgenic fish pose a threat when they are introduced into the environment. To avoid the breeding of transgenic fish with the local wild fish species, transgenic triploids have been used and this method has prevented potential adverse impacts of the escaped transgenic fish (Mair et al., 2007). A simpler way of producing triploids is shocking fertilized eggs at first mitotic division to produce a tetraploid embryo, that is, fish with four sets of chromosomes. Male tetraploid produces diploid sperm which when crossed with normal female fish, produces a triploid (Guo et al., 1996). This method produces triploids directly and avoids the shocking of large numbers of eggs. Most of the research works on direct triploid production have been extensively conducted with rainbow trout (Myers et al., 1986).

Another application of chromosome set manipulation in fish has been parthenogenesis, which is a technique that is used to produce offspring from one parent, for example, maternal or paternal origins only (Komen and Thorgaard, 2007). Gamma or ultraviolet rays destroy the nuclear deoxyribonucleic acid in eggs or sperm. The treated gamete is then fused with untreated sperm or egg to produce a haploid embryo, which undergoes a shock by inhibiting the second meiotic division to produce a diploid set of chromosomes in the fish. The gynogenetic process of inhibiting the second meiotic division is known as meiotic gynogenesis, and the resultant fish are referred to as meiotic gynogenetic fish. This process produces all-female progeny if the X or W chromosome was obtained from a female. When the first mitotic division is inhibited in fish whose Y or Z male chromosome was used, the process is known as parthenogenesis. The fish that are produced using parthenogenesis are known as mitotic gynogenetic or androgenetic fish and the progeny of parthenogenesis comprises all-male fish. The gynogenetic fish generate unique genotypes, consequently, this technique has been used to analyze complex tilapia traits. Gynogenesis has also been used to generate new strains of fish and has facilitated the advancement of sequencing of genes that code for important traits in aquaculture. This has been possible because the haploid embryos contain sufficient deoxyribonucleic acid which has been used for gene mapping (Kocher et al., 1998). The other important application of gynogenesis is that it facilitates the generation of unnatural genotypes such as the super YY male tilapia broodstock. The tilapia super males when mated with normal XX females produce all-male (XY) tilapia offspring (Myers et al., 1986). In silver barb (*Puntius gonionotus*) a large number of all-female progeny have been produced through gynogenesis (Pongthana et al., 1999). Further work with meiotic gynogenetic fish includes sex reversal of the silver barb to produce all-male populations (Pongthana et al., 1999), thus the application of parthenogenesis has produced a large number of broodstock for commercial production of all-female fry in silver barb. Regarding maternal gynogenesis, turbot is another fish species where meiotic gynogenesis has been applied to produce all-female populations (Cal et al., 2006).

The application of gynogenetic techniques through the production of double haploid (diploid) individuals from either meiotic gynogenetic or androgenetic background has been used to generate clonal or isogenic lines of fish. This is achieved by subjecting the haploids to the second round of parthenogenesis (Streisiner et al., 1981). In tilapias the double haploid technique has been applied to produce clonal lines (Muller-Belecke and Horstegen-Schwark, 1995; Hussain et al., 1998; Sarder et al., 1999). The wide application of gynogenesis is further shown by the production of double haploids in other species such as rainbow trout (Scheerer et al., 1986; Thorgaard et al., 1990; Quillet et al., 2007) and common carp (Bongers et al., 1997). Cloned lines of fish have been used in fish health studies and have contributed to important discoveries and breakthroughs in the area of fish diseases and the discovery of vaccines against certain known fish diseases in aquaculture. Recently, with advances in genomics, clonal lines have been a useful resource in whole-genome sequencing studies.

1.2.5 Gene Transfer Technologies in Fish

Gene transfer is the technology that has been used to donate desirable genes from one organism (donor) to another (recipient). In aquaculture, transgenic fish have been produced in goldfish using the human growth hormone gene (Zhu et al., 1985). A good example is a work by Penman et al. (1991) that produced transgenic rainbow trout. In channel catfish, transgenic work was reported by Dunham et al. (1987) and Brem et al. (1988) conducted transgenic research with Nile Tilapia. Despite the production of transgenic fish in these species the aquaculture performance of the transgenics is not as significant as expected and the results deviated with wide margins from those expected. However, though the success rate of transgenic technology has been low, from about 5% to 30% increase in growth during the first year, unlike hybrids, the improved performance in growth was also enjoyed in the subsequent generations (Devlin et al., 1995). The other important positive result with gene transfer was that a gene construct for one species, for example, that of one salmonid species, gave similar results on a different salmonid species (Du et al., 1992; Devlin et al., 1995; Devlin, 1997; Cook et al., 2000). Some fish species responded better to gene transfer technology than others. For example, it is reported that transgenic salmonids, exhibited a very positive response to the growth hormone (Gonzalez-Villasenor and Powers, 1992; Dunham and Devlin, 1998).

Contrary results with the growth hormone transgenic technology were obtained with warm-water fish such as tilapias and carps. With these warm-water species the gene transfer technique was not very successful. It was, therefore, suggested that these warm-water fish species that naturally exhibit high growth rates throughout the year under optimal water temperatures were not as reliant on growth hormone regulation as the cold water fish species. The results further showed that regulation of the growth hormone was more effective on wild than selected strains. For example, Devlin et al. (2001) found a 17-fold improvement of transgenics over their wild sibs in rainbow trout but achieved only 4.4% gain when the gene was transferred to selected domestic populations.

The major setback with transgenic fish was that the fish was not readily accepted by the general public. Most consumers feel it is unethical to eat fish that carries a human growth gene. Furthermore, transgenics are considered to pose high environmental risks if they escaped into the natural water bodies. The research work in transgenics also met a number of technical challenges, which include the low integration of the candidate gene into the target receptor fish. Consequently, there has been limited commercialization of transgenic fish (Mair et al., 2007). However, the work on transgenic fish formed the basis of the modern-day genomics work in aquaculture.

1.2.6 Molecular Techniques in Aquaculture

Molecular biology is the study of micromolecules and the macromolecular mechanisms in the cells of living organisms. Molecular biology has provided the requisite knowledge and solved the puzzle regarding the molecular nature of the gene, the mechanisms that are involving during gene replication, how genes mutate, and also how genes express themselves in the form of traits (Tebery et al., 2020). Advances in the field of molecular biology have been the result of research and contributions of results from different scientists in the field of genetics, physics, and structural chemistry. The motivation for these groups of scientists to undertake their studies was to resolve a common problem, which was the nature of inheritance. This phenomenon remained a long-standing mystery up to the 18th century (Muller, 1926). During the early times, such as the 12th century, molecular biologists and scientists in the field of genetics were guided by and adopted the Mendelian laws of gene segregation and independent assortment. Thus these Mendelian laws formed the basis of their investigations. However, there was still a gap in knowledge regarding the actual mechanisms of gene replication, gene mutation, and gene expression (Morgan, 1926). The mystery of the nature of inheritance in living organisms was resolved with the advances in genetics when genetic markers were discovered. The genetic markers are the fragments of deoxyribonucleic acid that are associated with a given location within a genome. These markers, also known as molecular markers, are now used in molecular biology and biotechnology to identify a particular sequence of deoxyribonucleic acid from the target or donor organisms.

There are many types of genetic markers in use today but each has limitations as well as strengths. The genetic markers are classified into three different categories, which are the first, the second, and new generation markers

(Mehaswaran, 2004). These types of markers are also used to identify dominance and codominance within a genome, which helps to identify alleles (i.e., forms in which a gene expresses itself) that are homozygous or heterozygous for a given gene (Contreras-Soto et al., 2017). Codominant markers are of particular importance because they are used to identify more than one allele which enables researchers to follow particular traits through mapping techniques and thus enabling amplification of particular gene sequences within the genome for comparison analysis. Markers are very useful in genetics research because they identify genetic linkages between identifiable locations within a chromosome, and they can be replicated when a need to verify their presence arises. Molecular markers can also identify genetic small changes within the population being mapped, thus enabling researchers to identify segregation of traits, their identity, and distinction among the population being studied (Mehaswaran, 2014).

The identification of particular locations on a chromosome enables the creation of physical maps. Some organisms have two alleles, that is, are biallele for a trait while some have many, that is, are polyallelic for a given trait. One of the importance of genetic markers is that they aid in the identification of the number of alleles that are responsible for a given trait in an organism (Mehaswaran, 2014). Due to the strengths and limitations of the different markers the choice of markers should be made after careful consideration. For example, a random amplified polymorphic deoxyribonucleic acid marker identifies only one band of distinction and may also be sensitive to reproducible results depending on the conditions in which the sample is produced (Contreras-Soto et al., 2017). Furthermore the use of random amplified polymorphic deoxyribonucleic acid marker is made on the assumption that two samples share the same locus when the samples are produced (Mehaswaran, 2014). The other factor that is considered is that different markers require different amounts of deoxyribonucleic acid; for example, random amplified polymorphic deoxyribonucleic nucleic acid markers require only 0.2 μg of deoxyribose nucleic acid, while restriction fragment length polymorphism marker may require 10 μg of deoxyribonucleic acid extracted from it to produce credible results (Contreras-Soto et al., 2017). Single nucleotide polymorphism markers have turned out to be a potential tool in breeding programs involving both plants and animals (Contreras-Soto et al., 2017).

The application of molecular markers in aquatic organisms has been reviewed by Liu and Cordes (2004). Markers are defined as Type 1, that is, the actual genes of known function, and Type 2, that is, the anonymous deoxyribonucleic acid segments. The most commonly used markers in aquatic organisms have been Type 2, especially the microsatellite loci, which are a small tandem repeat of deoxyribonucleic acid sequences. The microsatellite loci have been generated in salmonids, tilapia, sea bass, sea bream, and carp (Tautz, 1989). In aquatic organisms the microsatellites have been applied to define stock parentage (Norris et al., 2000). This method has been of particular importance in marine species because it is difficult to develop a single-rearing family. Consequently, studies involving mass spawning of two aquatic species; sea bream (Brown et al., 2005) and cod (Herlin et al., 2008) and the artificial spawning (using manual spawning techniques) of halibut (Jackson et al., 1998) have used microsatellite markers. These markers have guided the percentage of individuals to be selected as broodstock for the next generation. As a result the microsatellites have assisted to avoid intensifying selection pressure on a few individuals or families in the parental generation. The marker-assisted selection has shown advantages over the traditional selective breeding in fish in that there is no need to physically tag the fish for identification purposes during mass grow-out during progeny testing. Tagging introduces mortalities due to tagging stress and wounding of fish in a breeding program. The other advantage of molecular markers is that they remove the maternal full-sib effect—an environmental factor that arises due to separate rearing of fry as full-sib families when being raised to tagging size. Although the tagged fish are tested for growth in a common grow-out pond, the maternal full-sib effect due to initial separate fry rearing is prevalent at harvest. Hence, if not properly accounted for in traditional selective breeding programs, the maternal full-sib effect has the potential to cofound additive genetic effects. The maternal full-sib effect has been reported to be very high in *O. shiranus* an endemic tilapia in Malawi (Maluwa and Gjerde, 2007). In addition, the marker-assisted selection can reduce the cost of genetic improvement programs through selective breeding by removing the cost of maintaining pedigrees and breeding nucleus backup populations (McAndrew and Napier, 2010). Genetic markers have many other applications in managing wild and farmed aquatic species. For example, the genetic markers are used to identify the origin of escaped farm stocks in Atlantic salmon, rainbow trout, and Atlantic cod (Glover, 2010). Markers have also been used to define potential introgression of genes from farm escapees into wild populations of salmonids (Skaala et al., 2006).

1.3 GENOMICS

1.3.1 Introduction to Genomics

Genomics is the study of the whole genetic code of living organisms. As a study, genomics also includes the investigations of gene interactions amongst themselves and with the environment in which the organism inhabits (WHO, 2002). Genes are coded information on a strand of deoxyribonucleic acid that gives instructions to the body to produce certain types of amino acids that are needed by the body according to the coded information. The chemical compound that carries the coded information is a molecule known as deoxyribonucleic acid and this molecule contains the instructions in the form of genes that are needed to develop and direct the activities of nearly all living organisms (Calladine and Drew, 1992). There are two twisting strands that are paired in a deoxyribonucleic acid molecule. Due to the presence of the twisting, paired strands, the deoxyribonucleic acid molecule is called a double helix. The double helix is held together by four chemical compounds, which form the base of the molecule, called nucleotide bases. The bases comprise the genetic "alphabet" or code (Sinden, 1994). The bases in the deoxyribonucleic acid are adenine (A), thymine (T), guanine (G), and cytosine (C). There are two strands that carry the bases in the deoxyribonucleic acid molecule that are situated opposite to each other and they pair specifically and exactly, thus, an A always pairs with a T; and a C always pairs with a G. The order of the As, Ts, Cs, and Gs determines the meaning of the information or instruction, known as the genes that are encoded in that part of the deoxyribonucleic acid molecule (Bates and Maxwell, 1993). Most of the coded information in the deoxyribonucleic acid molecule is used to produce amino acids which in turn synthesizes proteins. The proteins are building blocks for body cells and tissues of organisms. With the four-letter code, deoxyribonucleic acid contains information that is needed to build the entire body of an organism (Bates and Maxwell, 1993; Sinden, 1994).

1.3.2 Genes and Gene Functions

A gene is a unit of deoxyribonucleic acid that carries the instructions for making a specific protein or set of proteins, depending upon the coded information in the base pairs. According to the human genome project, in humans, for example, the number of genes that code, on average, for three proteins, ranges from 20,000 to 25,000 (Calladine and Drew, 1992). In the body of all living organisms, every single cell contains a complete copy of many base pairs, for example, approximately 3 billion deoxyribonucleic acid–base pairs, or letters, makeup the entire human genome (Calladine and Drew, 1992).

A strand of deoxyribonucleic acid is referred to as a chromosome, and most living organisms have 23 pairs of chromosomes. The chromosomes are located in the nucleus of living organisms (Sinden, 1994). Genes direct the production of proteins through enzymes and messenger ribonucleic acid molecules. The enzyme's function is to copy the information in a gene's deoxyribonucleic acid into a molecule called messenger ribonucleic acid (Sinden, 1994). The messenger ribonucleic acid moves out of the nucleus to the cell's cytoplasm with the coded information, where it is read or decoded by another small molecular cell called a ribosome. According to the information that is decoded in this process, small molecules called amino acids are produced and placed in the right order to form a specific protein (Calladine and Drew, 1992). The amino acids are building blocks for proteins that make up the body structures of organisms such as organs and tissues. The proteins also control chemical reactions and carry signals between cells (Sinden, 1994).

1.3.3 Deoxyribonucleic Acid Sequencing

Deoxyribonucleic acid sequencing is the determination of the exact order of the bases in a strand of deoxyribonucleic acid (Sanger et al., 1977). The base compounds in a strand always pair specifically, therefore, when one base is identified, it leads to the identification of the other base in the pair. Hence, there is no need to report both bases of the base pair (Kaiser et al., 2003). Deoxyribonucleic acid is synthesized by an enzyme that is known as deoxyribonucleic acid polymerase. The most common type of sequencing used today is called sequencing by synthesis. With this method the deoxyribonucleic acid polymerase is used to replicate and hence generate a new strand of deoxyribonucleic acid from the original strand of interest. In the sequencing reaction a fluorescent label that acts as a chemical tag incorporates individual nucleotides into an enzyme that in turn incorporates it into the new deoxyribonucleic acid strand (Frank et al., 2013). The nucleotide emits a fluorescent signal which is detected and the signals are different for each of the four nucleotides. Thus particular nucleotide being investigated is identified based on the signal it emits and hence the signal that has been received. This method is efficient and can generate "reads" of 125 nucleotides in a roll translating into billions of reads at a time (Kaiser et al., 2003).

The sequence of overlapping segments is also read in order to assemble the sequence of all the bases in a large piece of deoxyribonucleic acid molecule. The process of reading overlapping segments allows the assembling of longer sequences from shorter pieces (Culver and Labow, 2002). Accuracy is ensured by reading the sequence in the overlapping segments several times (Culver and Labow, 2002). Through the sequencing by synthesis method, deoxyribonucleic acid has been sequenced to search for genetic variations and mutations of interest in plants, animals, and humans.

Several researchers such as National Human Genome Research Institute (2010), Klug et al. (2012), and Culver and Labow (2002) have sequenced and analyzed genomes through the use of high-throughput deoxyribonucleic acid sequencing and bioinformatics. This technique is used to assemble and analyze the function and structure of entire genomes. Recent advances in genomics have made it possible to now understand even the most complex biological systems in delicate parts such as the brain (Kadakkuzha and Puthanveettil, 2013). Genomics also includes the study of epistasis (effect of one gene on another gene within the same genome). This is the interaction of genes within a genome, which is termed intragenomic interaction. Other applications of genomics also deal with pleiotropy, a situation that arises when one gene is affecting more than one trait. Thus, genomics covers a wide area of studies because it also includes studies on heterosis or hybrid vigor, and other interactions between loci and alleles within the genome (Pevsner, 2009).

1.3.4 The Genesis of Gene Mapping

The word "genre" comes from the Greek word "gen" meaning "become," "create," "creation," and "birth" and has its subsequent variants; genealogy, genesis, genic, genomere, genotype, genus. The word genome is attributed to Hans Winkler (1920), and the term genomics was first used by Tom

Roderick in 1986 (Yadav, 2007). Rosalind Franklin and James Watson discovered the helical structure of deoxyribonucleic acid in 1953 and these results were confirmed by Francis Crick in 1955 when they reported the sequence of amino acid of insulin (Yadav, 2007). Consequently, most molecular biologists concentrated their work on nucleic acid sequencing (Ankeny, 2003). In 1964 Robert Holley and colleagues published their work on the ribonucleotide sequence of alanine transfer ribonucleic acid (Holley et al., 1965). Marshall Nirenberg and Philip Leder were the first to report the triplet nature of the genetic code. In their work, they determined the sequence of 54 of the 64 codons (Nirenberg et al., 1965). In 1972 a team of researchers led by Walter Fiers discovered the sequence of the Ms2 coat protein gene (Min Jou et al., 1972). This work was later expanded and the complete sequence of the Ms2 coat protein was determined. Therefore a complete nucleotide—sequence of a bacteriophage Ms2 ribonucleic acid was determined. It was discovered that the genome of the Ms2 ribonucleic acid encodes for four genes in 3569 base pairs (Fiers et al., 1976). In the Simian virus the Ms2 ribonucleic acid encodes for 40 genes (Fiers et al., 1978). Later, in 1977 Frederick Sanger and colleagues were the first geneticists to develop deoxyribonucleic acid sequencing techniques, which resulted in the establishment of the current comprehensive genome sequencing projects (Sanger et al., 1977).

A comprehensive sequencing procedure using deoxyribonucleic acid polymerase as an enzyme was published in 1975, by Frederick Sanger and Alan Coulson (Sanger, 1980). The procedure was called the Plus and Minus technique. There were two closely related methods that generated short oligonucleotide gel which was called polyacrylamide gel electrophoresis. The observation on the gel was done with the aid of autoradiography. The plus and minus procedure was able to sequence up to 80 nucleotides at one given time and was able to sequence most of the 5386 nucleotides of the single-stranded bacteriophage ΦX174 (Sanger et al., 1977). The plus and minus technique made a huge positive impact in sequencing work because it produced the first fully sequenced deoxyribonucleic acid–based genome (Sanger et al., 1977).

The Plus and Minus method, despite its huge positive impact, was refined and resulted in the development of the chain-termination, or Sanger method. This method formed the basis of the current techniques of genome sequencing using deoxyribonucleic acid. The analysis of the sequencing data from the Sanger method involved mapping, data storage, and bioinformatics (Kaiser et al., 2003; Sanger et al., 1977). Walter Gilbert and Allan Maxam independently developed the Maxam–Gilbert method (also known as chemical method) of deoxyribonucleic acid sequencing in the same year (1977). This method involved preferential cleavage of deoxyribonucleic acid at known bases (Maxam and Gilbert, 1977; Darden and Tabery, 2010). The development of these techniques in genome sequencing led to rapid intensification in the scope and speed of completion of sequencing projects. Some examples of the complete genome sequencing projects and their results are summarized in Table 1.1.

TABLE 1.1 Complete Genomes That Have Been Sequenced

Organism	Year Sequenced	Measurement	Reference
Human mitochondria	1981	16,568 base pairs	Anderson et al. (1981)
Plant chloroplast	1986	–	Shinozaki et al. (1986), Ohyama et al. (1986)
Microbe, chromosome III	1992	315 kilo base pairs	Oliver et al. (1992)
Microbe, *Haemophilus influenzae*	1995	1.8 mega base pairs	Fleischmann et al. (1995)
Microbe, *Saccharomyces cerevisiae*	1996	12.1 mega base pairs	Goffeau et al. (1996)

As of 2011 complete sequences were available for 2719 virus, 1115 archaea, and bacteria (30,310 and 36 eukaryotes of which 50% are fungi) (Darden and Tabery, 2010).

1.3.5 Genome Annotation

The sequence of genes along the deoxyribonucleic acid molecule and the biological function(s) that are regulated at any given gene sequence is determined by conducting additional analysis (Pevsner, 2009). The process of attaining biological information from given sequences is called genome annotation. According to Stein (2001), during genome annotation, the main steps that are followed include identification of portions of the genome that does not code for any protein; this step is followed by gene prediction, which involves the identification of the elements on the genome; finally, biological information is attached to the identified elements in the genome. These three steps can be undertaken automatically in silico, that is, "using computer simulation" or manually. When these steps are performed manually, the process is termed "curation" and involves human expertise, and follow-up experiments are conducted to possibly verify the results (Brent, 2008). Ideally the automated and manual approaches are done concurrently and they complement each other during the gene sequencing process. Traditionally the basic level of annotation is through the use of a computer program known as the basic local alignment search tool, which is an algorithm that is used to compare the primary information from biological sequencing. The information can be the amino acid sequences of proteins, which usually come from two sources; the nucleotides of deoxyribonucleic acid or those of ribonucleic acid (Stein, 2001). The tool also provides the comparison of a subject protein or nucleotide sequence (called a query) with a library or database of lines and identifies library sequences that resemble the query sequence above a certain threshold. Different types of basic local alignment search tools are available according to the query sequences and the target databases (Stein, 2001). For example, following the discovery of a previously unknown gene in one organism will enable a typical basic local alignment search tool to search for the genome of another organism. The results may show whether or not the other organism carries a similar gene; basic local alignment search tools, therefore, identifies sequences in an organism genome that resemble a known gene based on the similarity of the two sequences (Stein, 2001).

Recent advances in gene sequencing have provided additional information to the annotation platform, which allows manual identification and removal of errors and or discrepancies between genes especially when the genes have initially been assigned the same annotation (Brent, 2008). For some databases, different methods have been used to validate the results that are obtained from gene annotation. These methods include genome context information, similarity scores, experimental data, and data integration with other resources. In other databases such as Ensembl, there is use of curated data sources and a range of software tools to automatically annotate a genome under investigation (Brent, 2008). Genome annotation can be structural, which consists of identifying genomic elements primarily from the open reading frames and their localization. Alternatively, genome annotation can be done through observation of the gene structure and also functions, which consists of attaching biological information to genomic elements (Iseli et al., 1999). Regarding open reading frames, they provide two ways of extracting information. One of the methods is to use the expressed sequence tag scans, which extract coding regions from low-quality expressed sequence tags and correct frameshift errors (Iseli et al., 1999). The second method involves using the open reading frame predictor, which is a program that identifies proteins-coding sequences from expressed sequence tags through predicting the most probable coding regions from all the six translation frames (Min et al., 2005).

1.3.6 Deoxyribonucleic Acid Hybridization-Based Approaches

A number of technologies have been developed and are used to quantify gene expression. One of the methods involves hybridizing the deoxyribonucleic acid and it is therefore referred to as the hybridization-based method. The other method involves gene sequencing and is referred to as the sequence-based approach (Liu et al., 2020). Hybridization-based approaches involve the incubation of complementary deoxyribonucleic acid with customized microarrays. The complementary deoxyribonucleic acid is labeled by fluorescent light. On the other hand, unlike microarray methods, sequence-based approaches are used to determine gene expression levels directly through sequencing of the complementary deoxyribonucleic acid or through the use of expressed sequence tag libraries (Harbers and Caninci, 2005). Tag-based methods have been developed to improve the throughput, one of such methods is the serial analysis of gene expression (Velculescu et al., 1995; Harbers and Caninci, 2005). The other method is the massively parallel signature sequencing (Brenner, 2001). The tag-based sequencing approaches are high throughput and can provide precise digital gene expression levels. Recently, high throughput of deoxyribonucleic acid sequencing technologies have been developed and provide additional strengths for gene expression profiling. The method which is called ribonucleic acid—sequencing has advantages over the previous sequence-based approaches and has improved the way in which transcriptomes are analyzed for both gene expression profiling. Most of the literature in functional genomics involve gene expression profiling through sequencing-based approaches, such as expressed sequence tag analysis, serial analysis of gene expression, massively parallel signature sequencing, and ribonucleic acid—sequencing and hybridization-based microarray analysis.

1.3.7 GO Annotations

A GO, by definition, is a process that regulates the frequency, rate, and extent of a biological process. As such the GO annotations are used to identify the description of gene products in cells of organisms. During GO annotation research, gene functions are classified into three categories, which are biological processes, cellular components, and molecular functions. GO annotation process links consensus sequences to the GO terms where they are assigned the most likely function (Conesa et al., 2005). Using the process that is known as the GO enrichment analysis, the most relevant processes that are associated with a certain biological pathway are clustered together and hence the genes that are associated with that biological pathway are identified.

1.3.8 Expressed Sequence Tag Analysis

The expressed sequence tag analysis is an efficient approach for gene discovery and gene identification and has been extensively used in fish. For example, from 2001 to 2007, Catfish expressed sequence tags have increased from 10,000 to 44,000, and putative gene numbers have increased from 5905 to 25,000 (Li and Waldbieser, 2006). In Pacific Oysters (*Crassostrea gigas*), 40,845 high quality expressed sequence tags represented 29,745 unique transcribed sequences (Fleury et al., 2009). In gilthead seabream (*Sparus aurata*), 30,000 expressed sequence tags represented 18,196 putative unigenes (Louro et al., 2010). According to Li et al. (2011) there are currently, 180 aquaculture species with more than 100 expressed sequence tags in the database of expressed sequence tags. The other application of expressed sequence tags in aquaculture is that they provide a basis for comparison of gene expression profiling in different tissues and conditions. A good example is the work that was conducted by Kondo et al. (2011) in rainbow trout (*Oncorhynchus mykiss*) who sequenced 30,000 expressed sequence tags from the fish's adipose tissue. These expressed sequence tags were used to search for adipokine-related genes in rainbow trout. During the search, it was discovered that none of the tags encoded for adipokine gene despite the important roles the gene plays in the adipocytes of mammals. These results made researchers start using other methods, such as the protein chain reaction. With the protein chain reaction method the results confirmed those that were obtained from the expressed sequence tag analysis. An important breakthrough of this work was the discovery that rainbow trout adiponectin transcripts were found in low concentration and hence weakly detected in adipose tissue, however, the transcripts were highly concentrated and therefore strongly detected in muscle tissues. Based on these results, it was concluded that there seemed to be a difference in the energy metabolism between fish and mammals (Kondo et al., 2011). Following this work, Chini et al. (2008), developed normalized complementary deoxyribonucleic acid libraries from the liver, ovary, and testis of bluefish tuna (*Thunnus thynnus*). Similarly, Zou et al. (2011) constructed normalized complementary deoxyribonucleic acid libraries from testis, ovary, and mixed organs of mud crab (*Scylla paramamosain*). One of the major breakthroughs of the research with the expressed sequence tag analyses is that it led to the identification of sex-specific transcripted genes.

More specific assessment of gene characteristics in aquaculture has been achieved with the construction of microarray chips that contain many (usually in thousands) expressed sequence tags. These expressed sequence tags are derived from the whole fish genome or more focused on subsets of genes that are specific to a tissue or biological function. Most of the work in this field has been undertaken with salmonids (Rise et al., 2007). Studies in salmonids included disease-related responses such as the response to *Piscirickettsia salmonis* (Rise et al., 2004) and *Aeromonas salmonicida* (Edvardsen et al., 2011). Responses of salmonids to amoebic gill disease were studied by Morrison et al. (2006). In addition, MacKenzie and Jentoft (2016) investigated the immune response to a lipopolysaccharide challenge in the species. Some studies on expressed sequence tags were conducted to investigate live bacteria vaccines (Martin et al., 2006) and also deoxyribonucleic acid vaccinations salmonids (Purcell et al., 2011). In aquaculture, other studies with expressed sequence tags include investigating the genes for traits of interest such as response to growth in transgenic salmon (Rise et al., 2007). The expression sequence tags were also used to sequence genes that are responsible for adverse environmental conditions. Examples are studies that were conducted to sequence the genes that are responsible for stress that is associated with handling (Krasnov et al., 2005) and temperature regulation (Vornanen et al., 2005). Expressed sequence tags were also used to determine the genes that are responsible for the highly unsaturated fatty acids lipid metabolism Parkinson and Blaxter (2009).

It is evident that expressed sequence tags have been extensively applied with aquaculture species. Further work with other aquaculture species has provided sequence information that has been used for microarray development. In the area of fish health a large-scale oligonucleotide microarray platform containing 20,000 features has been developed and has facilitated the study of the immune response of the Atlantic cod spleen (Booman et al., 2011). In another aquaculture species, for example, the gilthead seabream (*Sparus aurata*), oligo microarray for the species was also developed based on the expressed sequence tags which were used to identify many (1050) differentially expressed genes (Ferraresso et al., 2008). Despite the importance in transcriptome characterization, expressed sequence tags are relatively expensive compared to the recently developed approaches (Liu et al., 2020), however, they still have great importance and value to serve as a reference for ribonucleic acid—sequence analysis and hence expressed sequence tags are essential for high-quality reference-guided assembly of the next-generation sequencer-generated short reads (Liu et al., 2020).

1.3.9 Microarray Analysis

Microarray has been described as the most powerful tool in gene sequencing because it enables the analysis of whole gene expression in individual cells or tissues and different conditions (Schena et al., 1995). The principle behind microarray is that it is used to place dense gene target sequences in small areas. It also facilitates hybridization of the deoxyribonucleic acid. Probes, which are the deoxyribonucleic acid sequences are anchored or spotted onto the solid surface of a chip in large numbers. When these probes have been fluorescence-labeled, they are hybridized with the features on the microarray. Microarrays, therefore, facilitate three important processes, which are nucleic acid hybridization, high-density signal detection, and high-resolution laser scanning (Peatman and Liu, 2007). This process is undertaken by high-density spotting robots and photolithography that allow each feature to be placed accurately on the slide in high densities. Signals are clearly detected because of the fluorescent-labeled probes and this is in contrast to the traditional radiation labeling, where the signals are not very clearly detected. In addition, the high-resolution laser scanner allows accurate fluorescent signal quantification. There are two types of arrays, that is, the spotted or printed arrays and the in situ arrays. Both of these microarrays have been applied with aquaculture species. The work by Schena et al. (1995) used the spotted arrays, while Mathavan et al. (2005) and Peatman et al. (2007) used the in situ arrays.

1.4 STRUCTURAL AND FUNCTIONAL GENOMICS IN AQUACULTURE

Genomics as a discipline has generated a large amount of biological information (Li et al., 2017). The protocols and techniques of genomics have also been used to describe many aquaculture species. Increased understanding of the structure, organization, and evolution of genomics has paved the way for the study of the functions of genes in the genomes of many organisms, including the species that are used in aquaculture. Although initial research had concentrated on structural genomics, currently, the emphasis is on gene functions. As a result, functional genomics is an expansion of the gene structural studies that represent a new phase of genome analysis. In genomics, there is now a transition from the mapping and sequencing of genomes to studying gene functions. Despite the change of focus from structural to functional genomics, structural genomics is still playing a crucial role in genomics because it is the source of information that is used to study the functions of the sequenced-coded genes. When studying the gene functions, high throughput or large-scale experimental methodologies are used, which are combined with statistical computational analysis of the raw data that is generated. Functional genomics has enabled the expansion of the scope of research in genetics, for example, instead of studying a single gene or proteins, or a few genes, functional genomics has facilitated genome-wide scale studies that involve investigations on functions of all genes or proteins at once (Liu et al., 2020).

The comprehensive nature of functional genomics has resulted in the generation of huge datasets. Consequently, there is a need to develop additional capacities in data analysis to obtain biologically relevant information. To achieve this, expertise in computer science and mathematics is crucial and has been integrated into the functional genomics. Traditionally, structural genomics has relied mostly on the expertise in computational biology because it mostly deals with genome data generation and management. In contrast, functional genomics involves the development of datasets from which to obtain valuable biological information. Hence, functional genomics has led to the understanding of the behavior of biological systems (Hieter and Bogusti, 1997).

To date the functions of many eukaryotic organisms, which have genes in the range of 20,000–40,000 genes have been determined (Liu et al., 2020). Before the development of the techniques in gene functions, historically the functions of genes were determined by observing phenotypic mutation. The process followed the genetic mapping of the mutated gene phenotype and eventually geneticists were able to trace the gene controlling the trait. This method is called the forward genetics approach (Liu et al., 2020). It is a straightforward approach; however, mutations, whether induced or natural are rare and mutated phenotypes are often difficult to observe. Consequently, reverse genetics was developed in which a candidate gene is inactivated, and then observations are made on the change in phenotype. Alternatively, new gene products are added, and then the products are determined.

1.4.1 Application of Functional Genomics in Aquaculture

In aquaculture, there has been extensive use and application of the genetic resources and technologies that have been developed in genomics, including the functional genomics. These genetic resources include complementary expressed sequence tool, transcription sequences, whole-genome sequences, and microarray. In zebrafish, microarray (which is hitherto an efficient technology that is used to identify the patterns of gene expression), both low density or high-density complementary deoxyribonucleic acid arrays and oligo arrays have been used (Ton et al., 2002; Matharan et al., 2005). These microarrays have also been used to study the functions of genes in salmonids (von Schalburg et al., 2005; Koop et al., 2008). Other aquaculture species whose gene functions have similarly been studied by microarrays include catfishes, as reported by Li and Waldbieser (2006) and Peatman et al. (2007), Atlantic cod (Booman et al., 2011; Edvardsen et al., 2011), Shrimp (Aoki et al., 2011; Leelatanawit et al., 2011), and Oyster (Wang et al., 2017; Dheilly et al., 2011).

Microarrays in aquaculture species have mainly focused on the early development stages of fish and fish immunity or disease resistance. During the early development the interest of researchers has been to determine how genes interact and change at the expression level during the developmental process. In the zebrafish the work by Ton et al. (2002) used microarray to study the fish's development stage (embryogenesis) and constructed a zebrafish complementary deoxyribonucleic acid array. The results showed a dynamic change in levels of gene expression that is responsible for the fish's development processes. In Atlantic cod, according to Driveness et al. (2011), complementary deoxyribonucleic acid microarray, represented by 7000 genes in the species, was utilized to study the temporal activity of a transcriptome also during early cod embryogenic development (Driveness et al., 2011). Studies that focused on immunity or disease resistance in fish have been conducted using microarrays mostly to search and identify genes that are responsible for the immune system and disease resistance in the species. One example of successful work was reported by Meijer et al. (2005) in zebrafish, in which oligo array containing 16K features was utilized in the analysis of a host transcription response to *Mycobacterium marinum* infection.

1.4.2 Quantitative Trait Loci Analysis in Aquaculture

In aquaculture, quantitative trait loci analysis has also been used, and the results have revealed statistically significant linkage between phenotypes and genotypes, thereby providing explanation for genetic basis of variation in complex traits (Falconer and Mackay, 1996; Lynch and Walsh, 1998). In a sense, quantitative trait loci analysis can be viewed as an incomplete genome-wide association study analysis with limited number of markers that does not cover the entire genome. If one quantitative trait locus is found, there may be more quantitative trait loci in the genome that can be discovered. More importantly, when the closely linked markers are absent in a given genomic region that contains significant quantitative trait loci for the traits, both the most significant quantitative trait loci for the traits and the most significant genes responsible for the trait can be missed (Liu and Cordes, 2004). Many quantitative trait loci studies have been conducted with aquaculture species, and most of the studies have been conducted with the salmonids (Liu et al., 2017).

A large number of the studies using quantitative trait loci in fish have shown that messenger ribonucleic acid levels for many genes are heritable, thus amenable to genetic analysis (Brem et al., 2002; Cheung et al., 2003; Schadt et al., 2003). In the past years, genetic and expression approaches have been brought together, in what has been termed "genetical genomics" to study the genetic basis of gene expression (Jansen and Nap, 2001), or expression quantitative trait loci. The quantitative trait loci have been extensively used in aquaculture.

1.4.3 Expression Quantitative Trait Loci Analysis in Aquaculture

Expression quantitative trait loci analysis technique combines traditional linkage analysis with the profiling of global gene expression. Using genetically related individuals, both the gene expression data and deoxyribonucleic acid markers are sampled. The sample contains many possible genes in the candidate species and the gene expression level is treated as a separate quantitative trait. The most studied traits with aquaculture species are body weight and disease resistance due to their importance in the industry. The expression quantitative trait loci involve the identification of the genomic loci that affect the levels of each transcript using the conventional quantitative trait loci analysis. When the results of the analysis are significant for a quantitative trait loci, the implication is that there are different genotypes at the polymorphic marker locus. In addition, it implies that the genotypes are associated with different expression levels. The advantage with expression quantitative trait loci over other methods is that it utilizes data from both assembled genome sequence and genome-wide deoxyribonucleic acid microarray. Hence, this approach is more accurate, powerful, and simpler to implement, as a result, the method is used to study any quantitative trait (Wayne and McIntyre, 2002). The expression quantitative trait loci have been made more applicable and more visibly helpful by the discovery of high-throughput array-based methods, which are methods that are used to measure the messenger ribonucleic acid abundance. Consequently, from the early 2000s to date, many expression quantitative trait loci studies have been conducted with plants, animals, and humans (Schadt et al., 2003; Morley et al., 2004; Euwart et al., 2005; Bystrykh et al., 2005; Chesler et al., 2005; Hubner et al., 2005; Lan et al., 2006).

1.4.4 Ribonucleic Acid Sequencing in Aquaculture

Ribonucleic acid-sequencer provides a more accurate assessment of gene expression, and hence this technology is being extended to studies in expression quantitative trait loci analysis (Majewski and Pastinen, 2011). Ribonucleic acid-sequencer is providing a platform that is independent and objectively standard compared with the microarray approach and has confirmed and further clarified previous microarray results (Montgomery et al., 2010;

Pickrell et al., 2010). When genomic regions show a high proportion of expression quantitative trait loci, it means that the areas under study have genes that code for common transcription regulators. These regulators are responsible for the control of (and determine the synthesis and resultant products from) important biological pathways (Cogburn et al., 2007). One of the most significant breakthroughs with the expression quantitative trait loci is that it facilitates the identification of trans-acting loci. This is very important because, in reality, expression regulators are difficult to identify, even in instances where the whole genome has been sequenced (Li et al., 2017). During data manipulation, there are two main ways in which expression quantitative trait loci data is analyzed. The first method involves the use of standard gene mapping software packages and the second utilizes custom codes such as quantitative trait loci mapping system. The latter is more preferred because it is faster than the former method (Wang et al., 2017; Wu et al., 2014). The expression quantitative trait loci method, despite its many advantages and benefits in genomics, has some limitations for use in aquaculture. The greatest limitation is the high cost that is associated with this method. As a result, the method is very expensive and requires huge projects with substantial budgetary resources for its successful implementation.

The first version of the Atlantic code (*Gadus morhua*) genome was developed by Torresen et al. (2016a). The work involved genome sequencing using high throughput sequencing technology. The genome was the first example of the teleost genome to be obtained using a pure high throughput sequencing technology strategy. It was also the first vertebrate genome that was generated exclusively using Roche/454 sequencing technology. The Roche/454 sequencing technology was preferred because it has a longer read length and a wider range of suitable software that utilizes the generated data at a time. The sequencing work involved assembly and annotation of the first Atlantic cod genome (Torresen et al., 2016b). This first genome assembly has proven to be an important resource because it is hitherto extensively used in a wide range of studies. Further advancements have been made within the latest and more advanced genomic tools to generate this first version of the Atlantic cod genome (Torresen et al., 2016b).

The rainbow trout genome was analyzed in terms of structure. The high throughput sequencing technology is a valuable resource to improve understanding of the biology of the species in aquaculture. The application of this work led to the identification of genes and molecular pathways involved in Atlantic cod growth and development. Another area in which steady progress has been achieved is on the responses of Atlantic salmon to internal (inside the body) changes such as nutrition, infections by different pathogens, and other stimuli to the immune system. Progress has also been achieved with the aquaculture species on external changes, such as environmental stressors, which include temperature, stress, and pollutants (Torresen et al., 2016b). Furthermore the high throughput sequencing method resolved a long outstanding puzzle of the immune system of the Atlantic cod. In this fish the immune system does not have components of an adaptive immune system as in other fish species. Before the discovery of the high throughput sequencing method, it was thought that the Atlantic cod immune system was an essential feature of all jawed vertebrates (Torresen et al., 2016b).

Genetic linkage maps have been developed which enable the identification of quantitative trait loci or genome sections containing genes that influence important traits. There is now more quantitative trait loci data that has been generated for species such as Atlantic salmon, tilapia, and common carp. In fish, there is a large difference in recombination rate between sexes (Hayes et al., 2007). Consequently, there has been fast progress with the use of quantitative trait loci to identify the genes that are responsible for resistance to infectious pancreatic necrosis. Hence the quantitative trait loci for these genes has been successfully incorporated into commercial Atlantic salmon breeding program (Houston et al., 2008). A major limitation with the application of quantitative trait loci in aquaculture is that most of the quantitative trait loci that have been described for aquaculture species are for the traits that are easy to measure. Limited progress has been achieved with traits that have low heritability and are difficult to measure. In addition, there is also little progress with traits that need to be identified early before maturity or traits such as flesh quality, which include fresh color, fillet conformation, and adiposity (Hayes et al., 2007). Despite these limitations, the use of marker-assisted selection technology is being extensively used in aquaculture, and there is now a large number of candidate genes that are being used in aquaculture which have been derived from terrestrial livestock (De-Santis and Jerry, 2007).

1.4.5 Gene Regulation and Conservation

In living organisms, especially in animals, the genes that are responsible for complex biological regulatory processes, in most cases, exhibit high expression levels over generations, that is, they are evolutionary stable. Genomic studies with model fish species such as the zebrafish and the pufferfish show that many gene families of fish consist of more members compared to mammals (Wittbrdt et al., 1998). This finding is related to the evolution of the fish species. The greater number of genes in fish are thought to have been caused by additional ancient genome duplication which appeared in the lineage that led to the modern ray-finned fishes but not along the lineage leading to tetrapods (Wittbrdt et al., 1998). There is an indication that the mammals and tetrapods shared a common ancestor in the past, that is, some 360 million years ago, and that some individual duplicated members of the gene family were later lost in fish (Chen and Rajawsky, 2007). It is further reported that some genes in mammals produce different traits from their homologues in fish. These results may imply that the degree of how genes have been related to evolutionary does not always provide an accurate predictor of their evolutionary conservation. In addition, evolutionary, related genes do not also provide accurate information to predict the similarity of their function. There are phenotypical differences between fish and mammals. The fish are phenotypically simpler than mammals. When considering gene regulation and conservation, the evidence suggests the possibility that evolution might have taken alternative genomics paths to gene regulation. It is, therefore, suggested that the fish have quickly responded to selection pressure, adapted, and speciated quickly due to their more complex genomic architecture than that of mammals.

1.4.5.1 GENE REGULATION AND CONSERVATION AT GENOME LEVEL

In vertebrates, genes are regulated and conserved at the genome level. This is shown by the observation that although the vertebrates share general body plans and organs, possess the same and related sets of genes, and rely on similar physiological mechanisms, they show great diversity in morphology, habitat, and behavior. This phenomenon is attributed to the conservation and regulation of genes (Chan et al., 2009). Genome-level genetic regulation and protection have many applications in aquaculture. The advances in genomic studies brought about by the human genome project have brought rapid advances in fish as well. The genome revolution has the potential to impact aquaculture and fisheries production and has implications for the future management of fish genetic resources. Aquaculture has the potential to supplement the dwindling fish catches from the wild but has also the potential to release fish that has undesirable genes into the environment and thereby adversely affecting fish biodiversity in the wild through hybridization. To date, analytical genetic technologies that are most relevant to fisheries and aquaculture have been described in this chapter and include deoxyribonucleic acid markers, genome mapping, microarrays, and sequencing. The deoxyribonucleic acid markers are also useful in genetic linkage mapping; they are also important in the analysis of genetic resources, strain differentiation, species differentiation, parentage identification, and preservation of genetic diversity, and conservation of integrity.

The application of genomic studies to aquaculture and fisheries will continue to contribute to resolving problems such as diseases, environmental impacts, and low-profit margins. Therefore there are many areas in which genomics has potential application to impact aquaculture and fisheries. The areas include marker-assisted selection, environmental manipulation to enhance productivity, development of novel programs for environmental monitoring, development of effective vaccines and vaccine delivery technology, monitoring antibiotic resistance, diagnosis of fish diseases, accurate identification of fish species, conservation of fish genetic resources, protection of endangered fish species, development and application of fish transgenic technologies.

1.4.5.2 GENE CONSERVATION AND REGULATION AT TRANSCRIPTION LEVEL

Transcription is the first of several steps of deoxyribonucleic acid–based gene expression in which a particular segment of the deoxyribonucleic acid is copied into the ribonucleic acid, especially the messenger ribonucleic acid. The enzyme responsible for this process is called ribonucleic acid polymerase. Both deoxyribonucleic acid and ribonucleic acid arc nucleic acids, which use base pairs of nucleotides to complement each other. During transcription a deoxyribonucleic acid sequence is read by a ribonucleic acid polymerase, which produces a complementary, antiparallel ribonucleic acid strand called a primary transcript. According to Berg et al. (2006) the stretch of deoxyribonucleic acid transcribed into a ribonucleic acid molecule is called a transcription unit and encodes for at least one gene. If the gene encodes a protein, the transcription produces messenger ribonucleic acid, which in turn serves as a template for the protein's synthesis through translation (Berg et al., 2006). Alternatively, the transcribed gene may encode for noncoding ribonucleic acid such as microribonucleic acid, ribosomal ribonucleic acid, transfer ribonucleic acid, or enzymatic ribonucleic acid molecules called ribozymes. The gene conservation that occurs at transcription level, is done through the alternative route, where the transcribed gene encodes for noncoding ribonucleic acid (Moore, 2005). This conservation is possible because for both transcription factors and microribonucleic acid, the regulators are well conserved over large evolutionary distances; however, the targeted products have evolved much more quickly, showing that a large-scale rewiring of regulatory networks has taken place in the course of evolution (Levine and Tjian, 2003). Extensive work on transcription in aquaculture is reported in a review by Chandhini and Kumar (2018). Some of the reported work involved diseases and immunity, reproduction and development, growth and nutrition, toxicology, and stress. Transcription is also applied to generate genetic resources, during differential gene expression, analysis of molecular pathways, and development of molecular markers (Chandhini and Kumar, 2018).

1.4.5.3 GENE CONSERVATION AND REGULATION AT PROTEOME LEVEL

A proteome is an entire set of proteins that are expressed from a given genome. The expression of proteome occurs at various levels, such as at genome, cell, tissue, or organism levels. The proteomes are expressed at any given time and under a given cell type of the organisms. In eukaryotic organisms the major regulatory posttransitional modifier of proteins is considered to be the protein phosphorylation on serine, threonine, and tyrosine. A study that described a proteome is reported by Mece et al. (2007). This study was conducted to investigate the evolutionary conservation of bacterial phosphoproteins and phosphorylation sites across the phylogenetic tree in bacteria. The results show that in the bacteria, there were significantly higher conserved phosphoproteins and phosphorylated residues than the nonphosphorylated residues (Mece et al., 2007). In addition, the study found a number of potential phosphorylation sites conserved from archaea to humans. In aquaculture, proteomics has been established as a powerful tool and unbiased instrument in environmental monitoring and risk assessment (Rodrigues et al., 2017). Proteomics, therefore, in aquaculture have been applied to address welfare, nutrition, health, environmental quality, and safety. The use of high throughput proteomics and metaproteomics has resulted in the rapid integration into aquaculture research with the goal of contributing to high-quality food production processes and environmental sustainability (Rodrigues et al., 2017).

1.4.5.4 GENE CONSERVATION AND REGULATION AT EPIGENOME LEVEL

Epigenome consists of records of chemical changes that occur in the deoxyribonucleic acid—a molecule that contains the genetic code of all organisms. The deoxyribonucleic acid molecule has a system of recording chemical changes that take place within itself and the records are stored in the epigenome. Within the epigenome the records of the chemical changes are stored in the form of histone proteins. The organisms are able to transmit the histone proteins from one generation to another. When offspring inherit the histone proteins from their parents, the process is known as transgenerational stranded epigenetic inheritance. Almouzni et al. (2014), following the human genome project, reported on the links between genetic, epigenetic, and non-genetic factors throughout the lifespan and across generations and their role in disease susceptibility and disease progression. Best et al. (2018) proposed key areas where epigenetics can be applied in both fish and shellfish, which include environmental manipulation to produce the desired phenotypes and epigenetic selection, either alone or in combination with genetic selection. These strategies have the potential to increase the productivity of aquaculture species (Granada et al., 2017; Best et al., 2018).

1.4.5.5 GENE CONSERVATION AND REGULATION USING SYSTEM BIOLOGY

The complex biological system has been modeled and the data that is collected is fitted into the models by performing analyzes using computational and mathematical techniques. System biology is therefore an interdisciplinary field of study in biology that focuses on complex interactions that take place within biological systems, using a holistic approach to biological research. System biology is made up of large numbers of components that interact in a large number of combinations. The features and interactions of a system are studied through a systems biology approach. Zebrafish have been used as a vertebrate model for research in areas such as drug development, disease modeling, and other biological exploration (Mushtaq et al., 2013). Using the system biology a study by Mushtaq et al. (2013) in zebrafish has shown that there is significant conservation of genetics and other cellular networks in the species. This process is similarly significant when data from other vertebrate models, including humans is analyzed (Mushtaq et al., 2013).

1.5 CONCLUSION

In aquaculture, extensive studies to improve the performance of the cultured fish species have been conducted. The research that has been conducted on genetic improvement of the fish species, includes exploitation of the additive genetic component, heterosis, chromosome set manipulation, gene transfer technologies, molecular techniques, and application of genomics to conduct marker-assisted selection. Aquaculture species have high fecundity in addition to external fertilization. For most fish species, they are genetically close to the wild ancestors. These reproductive features have provided powerful resources for genetic studies of complex traits, that is, those with low heritability, such as disease resistance. The enormous diversity in the aquaculture species, coupled with low heterosis for most important traits, has necessitated the establishment of species-specific breeding programs to exploit the additive genetic components. The reported genetic biodiversity in fish also presents an opportunity for choosing suitable model species to inform the biology of candidate farmed species. The review in this chapter shows that the aquaculture industry has actively applied genetic improvement strategies and technologies to improve fish production. Genomics is a significant opportunity, and the use of genomic tools can utilize more effectively the genetic variation for economically important aquaculture traits. The application of genomics techniques can replace the keeping of pedigree data and eliminate the maternal full-sib effect in a breeding program. Therefore the application of genomics can facilitate accurate estimation of the genetic components. However, the major constraints with genomics currently include the high cost and limited technical human expertise in the aquaculture industry. The genomics tools, therefore, should continue to be used to complement other fish genetic improvement strategies and techniques that have been developed for the aquaculture species. When genomics complement the other genetic improvement strategies, there is increased efficiency. Other strategies, for example, gene transfer and chromosome set manipulation have provided a solid foundation for genomics. Within the genomics field, structural genomics has provided the required information for functional genomics. Therefore genetic variation in aquaculture species can be effectively utilized using a combination of the available strategies to improve further and sustain the aquaculture production.

ACKNOWLEDGMENTS

The chapter on genomics, genetics, and breeding in aquaculture is contained in the book entitled "Sustainable Fish and Sea Food Production at Processing." Editors of genomics, genetics, and breeding at the Elsevier Academic Press are acknowledged for their assistance and guidance. Management and faculty at the Malawi University of Science and Technology are recognized for their support and encouragement.

REFERENCES

Acosta, B., and Gupta, M.V., (2010). Genetic improvement of Farmed Tilapias Project: Impact and Lessons Learnt. In: Silva, S., and Davy, B., (Editors). Success Stories in Asian Aquaculture, ISBN 978-90-481-3085-6.

Almouzni, G., Altucci, L., Widschwendter, M., 2014. Relationship between genome and epigenome – challenges and requirements for future research. BMC Genomics 15, 487.

Anderson, S., Bankier, A.T., Barrel, B.G., de Bruijn, M.H., Coulson, A.R., Drouin, J., 1981. Sequence and organization of the human mitochondrial genome. Nature 290 (5806), 457–465. Available from: https://doi.org/10.1038/290457a0. PMID 7219534. S2CID 4355527.

Ankeny, R.A., 2003. Sequencing the genome from nematode to human: changing methods, changing science. Endeavour 27 (2), 87–92. Available from: https://doi.org/10.1016/s0160-9327(03)00061-9. PMID 12798815.

Ann, A.Q., Thien, T.M., 1993. Preliminary results of silver carp breeding program in Vietnam. Naga, the ICLARM Quarterly 16 (1), 21–23.

Aoki, T., Wang, H.C., Unajak, S., Santos, M.D., Kondo, H., Hirono, I., 2011. Microarray analyses of shrimp immune responses. Marine Biotechnology (NY) 13, 629–638.

Bakos, J., Gorda, S., 1994. Breeding programs of Hungarian carp with special regard to intra-specific hybridization, Proceedings of the 5th World Congress on Genetics Applied to Livestock Productions, Guelph, Canada, 7–12 August, vol. 19. pp. 383–386.

Bartley, D.M., Rana, K., Immink, A.J., 2001. The use of interspecific hybrids in aquaculture and fisheries. Reviews in Fish Biology and Fisheries 10, 235–337.

Basavaraju, Y., Mair, G.C., Mohan Kumar, H.M., Pradeep Kumar, S., Keshavappa, G.Y., Penman, D.J., 2002. An evaluation of triploidy as a potential solution to the problem of precocious sexual maturation in common carp *Cyprinus carpio*, in Karnataka, India. Aquaculture 204, 407–418.

Basavaraju, Y., Penman, D.J., Mair, G.C., 2003. Stock evaluation and development of a breeding program for common carp (*Cyprinus carpio*) in Karnataka, India: progress of a research project. Naga, the ICLARM Quarterly 26 (2), 30–32.

Bates, A.D., Maxwell, A., 1993. DNA Topology. Oxford University Press, IRL Press, In Focus Series, ISBN 0-19-963349-5.

Bentsen, H.B., Gjerde, B., 1994. Design of fish breeding programs, Proceedings of the 5th World Congress on Genetics Applied to Livestock Production, Guelph, Canada, vol. 19. pp. 353–359.

Bentsen, H.B., Olesen, I., 2000. Designing aquaculture mass selection programs to avoid high inbreeding rates. Proceedings of the 7th World Congress on Genetics in Aquaculture, Townsville, Australia.

Bentsen, H.B., Eknath, A.E., Palada-de Vera, M.S., Danting, J.C., Reyes, H.L., Dianisio, E.E., et al., 1998. Genetic improvement of farmed tilapias: growth performance in a complete diallel cross experiment with eight strains of *Oreochromis niloticus*. Aquaculture 160, 145–173.

Bentsen, H.B., Eknath, E.A., Rye, M., Thodesen, J., Gjerde, B., 2003. Genetic improvement of farmed tilapias: response to selection for growth performance in the GIFT project. Genetics in Aquaculture VIII, Puerto Varas, November 9–15.

Berg, J., Tymoczko, J.L., Stryer, L., 2006. Biochemistry, sixth ed. W.H. Freeman, San Francisco, CA, ISBN 0-7167-8724-5.

Best, C., Ikert, H., Kostyniuk, D., Craig, P., Navarro-Martin, L., Marandel, L., et al., 2018. Epigenetics in teleost fish: from molecular mechanisms to physiological phenotypes. Comparative Biochemistry and Physiology Part B: Biochemistry and Molecular Biology 224, 210–224.

Beverage, M.C., McAndrew, B., 2000. Tilapia: Biology and Exploitation. Kluwer Academic Publishers, Dordrecht.

Bolivar, R.B., Newkirk, G.F., 2002. Response to within family selection for body weight in Nile tilapia (*Oreochromis niloticus*) using a single trait animal model. Aquaculture 204, 371–381.

Bongers, A.B.J., Bovenhuis, H., Van Stokkom, A.C., Wiegertjes, G.F., Zandieh-Doulabi, B., Komen, J., et al., 1997. Distribution of genetic variance in gynogenetic or androgenetic families. Aquaculture 153, 225–238.

Booman, M., Borza, T., Feng, C.Y., Hori, T.S., Higgins, B., Culf, A., et al., 2011. Development and experimental validation of a 20K Atlantic Cod (*Gadus morhua*) oligonucleotide microarray based on a collection of over 150,000 ESTs. Marine Biotechnology 13, 733–750.

Brem, G., Brenig, B., Horstgen-Schwark, G., Winnacker, E.L., 1988. Gene transfer in tilapia (*Oreochromis niloticus*). Aquaculture 68, 209–219.

Brem, R.B., Yvert, G., Clinton, R., Kruglyak, L., 2002. Genetic dissection of transcriptional regulation in budding yeast. Sciencc 296, 752–755.

Brenner, S.E., 2001. A tour of structural genomics (PDF). Nature Reviews Genetics 2 (10), 801–809. Available from: https://doi.org/10.1038/35093574. PMID 11584296. S2CID 5656447.

Brent, M.R., 2008. Steady progress and recent breakthroughs in the accuracy of automated genome annotation (PDF). Nature Reviews Genetics 9 (1), 62–73. Available from: https://doi.org/10.1038/nrg2220. PMID 18087260. S2CID 20412451. Archived from the original(PDF) on 2013-05-29. Retrieved 2013-01-04.

Brown, R.C., Woolliams, J.A., Mcandrew, B.J., 2005. Factors influencing effective population size in commercial populations of gilthead seabream, *Sparus aurata*. Aquaculture 247, 219–225.

Bye, V.J., Lincoln, R.F., 1986. Commercial methods for the control of sexual maturation in rainbow trout (*Salmo gairdneri* R.). Aquaculture 57, 299–309.

Bystrykh, L., Weersing, E., Dontje, B., Sutton, S., Pletcher, M.T., Wiltshire, T., et al., 2005. Uncovering regulatory pathways that affect hematopoietic stem cell function using 'genetical genomics'. Nature Genetics 37, 225–232.

Cal, R.M., Vidal, S., Martinez, P., Alvarez-Blazqquez, B., Gomez, C., Pifferrer, F., 2006. Growth and gonadal development of gynogenetic diploid *Scophthalmus maximus*. Journal of Fish Biology 68, 401–413.

Calladine, C.R., Drew, H.R., 1992. Understanding DNA: The Molecule and How It Works. Academic Press, New York, ISBN 0-12-155086-9.

Chan, E.T., Quon, G.T., Chua, G., Babak, T., Trochesset, M., Zirgibl, A., et al., 2009. Conservation of core gene expression in vertebrate tissues. Journal of Biology 8, 33.

Chandhini, S., Kumar, V.J., 2018. Transcription in aquaculture: current status and applications. Reviews in Aquaculture 11 (1). Available from: https://doi.org/10.1111/raq.12298.

Chen, K., Rajawsky, N., 2007. The evolution of gene regulation by transcription factors in microRNAs. Nature Reviews Genetics 8, 93–103.

Chesler, E.J., Lu, L., Shou, S.M., Qu, Y.H., Gu, J., Wang, J.T., et al., 2005. Complex trait analysis of gene expression uncovers polygenic and pleiotropic networks that modulate nervous system function. Nature Genetics 37, 233–242.

Cheung, V.G., Conlin, L.K., Weber, T.M., Arcaro, M., Jen, K.Y., Morley, M., et al., 2003. Natural variation in human gene expression assessed in lymphoblastoid cells. Nature Genetics 33, 422–425.

Chini, V., Cattaneo, A.G., Rossi, F., Bernardini, G., Terova, G., Saroglia, M., et al., 2008. Genes expressed in Blue Fin Tuna (*Thunnus thynnus*) liver and gonads. Gene 410, 207–213.

Cnaani, A., Lee, B.Y., Zilberman, N., OzoufoCostaz, C., Hulata, G., Ron, M., et al., 2008. Genetics of sex determination in tilapiine species. Sexual Development 2 (1), 43–54.

Cogburn, L.A., Porter, T.E., Duclos, M.J., Simon, J., Burgess, S.C., Zhu, J.J., et al., 2007. Functional genomics of the chicken—a model organism. Poultry Science 86, 2059–2094.

Conesa, A., Gotz, S., Garcia-Gomez, J.M., Terol, J., Talon, M., Robles, M., 2005. Blast2GO: a universal tool for annotation, visualization and analysis in functional genomics research. Bioinformatics 21, 3674–3676.

Contreras-Soto, R.I., Mora, F., de Olivereira, M.A.R., Higashi, W., Scapim, C.A., Schuster, I., 2017. A genome wide association study for agronomic traits in soybeans using SNP markers and SNP based haplotype analysis. PLoS ONE 2, 1–22. Available from: https://doi.org/10.1371/journal.pone.0171105. PMC 5289539 PMID 28152092.

Cook, J.T., Mcniven, M.A., Richardson, G.F., Sutterlin, A.M., 2000. Growth rate, body composition and feed digestibility/conversion of growth enhanced transgenic Atlantic salmon (*Salmo salar*). Aquaculture 188, 15–32.

Culver, K.W., Labow, M.A., 2002. Genomics. In: Robinson, R. (Ed.), Genetics. Macmillan Science Library. Macmillan Reference USA, New York, ISBN 978-0-02-865606-9.

Darden, L., Tabery, J., 2010. Molecular biology. In: Zalta, E.N. (Ed.), The Stanford Encyclopedia of Philosophy. (Fall 2010 edition).

De-Santis, C., Jerry, D.R., 2007. Growth genes in finfish—where should we be looking? Aquaculture 272 (1–4), 22–38.

Devlin, R.H., 1997. Transgenic salmonids. In: Houdebine, L.M. (Ed.), Transgenic Animals: Generation and Use. Harwood Academic Publishers, Amsterdam, pp. 105–117.

Devlin, R.H., Yesaki, T.Y., Donaldson, E.M., Du, S.-J., Hew, C.L., 1995. Production of germline transgenic Pacific salmonids with dramatically increased growth performance. Canadian Journal of Fisheries and Aquatic Sciences 52, 1376–1384.

Devlin, R.H., Biagi, C.A., Yesaki, T.Y., Smailus, D.E., Byatt, J.C., 2001. Growth of domesticated transgenic fish. Nature 409, 781–782.

Dheilly, N.M., Lelong, C., Huvet, A., Favrel, P., 2011. Development of a pacific oyster (*Crassostrea gigas*) 31,918-feature microarray: identification of reference genes and tissue enriched expression patterns. MBC Genomics 12, 468.

Drivenes, O., Taranger, G.L., Edvardsen, R.B., 2011. Gene expression profiling of Atlantic Cod (*Gadus morhua*) embryo genesis using microarray. Marine Biotechnology 14, 167–176.

Du, S.J., Gong, Z., Fletcher, G.L., Shears, M.A., King, M.J., Idler, D.R., et al., 1992. Growth enhancement in transgenic Atlantic salmon by the use of an 'all fish' chimeric growth hormone gene construct. Nature Biotechnology 10, 176–181.

Dunham, R.A., 1987. American catfish breeding programs. In: Tiews, K. (Ed.), Selection, Hybridization and Genetic Engineering in Aquaculture, vol. II. Schriften der Bundesforschungsanstalt fur Fischerei, Hamburg, pp. 407–416.

Dunham, R.A., 2004. Aquaculture and Fisheries Biotechnology: Genetic Approaches. CABI, Wallingford.

Dunham, R.A., Devlin, R., 1998. Comparison of traditional breeding and transgenesis in farmed fish with implications for growth enhancement and fitness. In: Murray, J.D., Anderson, G.B., Oberbauer, A.M., McGloughlin, M.M. (Eds.), Transgenic Animals in Agriculture. CABI, Wallingford, pp. 209–229.

Dunham, R.A., Eash, J., Askins, J., Townes, T.M., 1987. Transfer of metallothionein-human growth hormone fusion gene into channel catfish. Transactions of the American Fisheries Society 116, 87–91.

Edvardsen, R.B., Malde, K., Mitterlholzer, C., Taranger, G.L., Nilsen, F., 2011. EST resources and establishment and validation of a 16k CDNA microarray from Atlantic cod (*Gadus morhua*). Comparative Biochemistry and Physiology Part D: Genomics and Proteomics 6, 23–30.

Eknath, A.E., Tayamen, M.M., Palada-de Vera, M.S., Danting, J.C., Reyes, R.A., Dionisio, E.E., et al., 1993. Genetic improvement of Farmed Tilapias: the growth performance of eight strain of *Oreochromis niloticus* tested in different farm environments. Aquaculture 111, 171–188.

El-Zaeem, S.Y., Salam, G.M., 2013. Production of genetically male tilapia through interspecific hybridization between *Oreochromis niloticus* and *O. aureus*. Iranian Journal of Fisheries Sciences 12 (4), 802–812.

Euwart, K.V., Belonger, J.C., William, J., Karakaah, T., Penny, J., Tsri, S.C., et al., 2005. Identification of genes differentially expressed in Atlantic salmon (*Salmo salar*) in response to infection by *Aeromonas salmonicida* using CDNA microarray technology. Development and Comparative Immunology 29, 333–347.

Fagan, E., 2017. Fishing: How the Sea Fed Civilization. Yale University Press, New Haven, CT, ISBN 9780300231885.

Falconer, D.S., Mackay, T.F.C., 1996. Introduction to Quantitative Genetics, fourth ed. Prentice-Hall, Longman Group Limited, Harlow.

FAO, 2020. The State of World Fisheries and Aquaculture 2020. FAO. Available from: http://www.fao.org.

Ferraresso, S.I., Vitulo, N., Mininni, A.N., Romuald, C., Cardazzo, B., Negrisolo, E., et al., 2008. Development and validation of a gene expression oligo microarray for the gilthead sea bream (*Sparus aurata*). MBC Genomics 9, 580.

Fiers, W., Contreras, R., Duerinck, F., Haegeman, G., Iserentant, D., Merregaert, J., et al., 1976. Complete nucleotide sequence of bacteriophage MS2 RNA: primary and secondary structure of the replicate gene. Nature 260 (5551), 500.

Fiers, W., Contreras, R., Haegeman, G., Rogiers, R., Van de Voorde, A., Van Heuverswyn, H., et al., 1978. Complete nucleotide sequence of SV40 DNA. Nature 273 (5658), 113–120. Available from: https://doi.org/10.1038/273113a0. PMID 205802. S2CID 1634424.

Fleischmann, R.D., Adams, M.D., White, O., Clayton, R.A., Kirkness, E.F., Kerlavage, A.R., 1995. Whole genome random sequencing and assembly of *Haemophilus influenzae* Rd. Science 269 (5223), 496–512. Available from: https://doi.org/10.1126/science.7542800. PMID 7542800. S2CID 10423613.

Fleury, E., Huvet, A., Lelong, C., deLorgerilj, J., Boulo, V., Gueguen, Y., et al., 2009. Generation and analysis of a 29745 unique expressed sequence tags from pacific oyster (*Crassostrea gigas*) assembled into a publicly accessible database: the Gigas database. BMC Genomics 10, 341.

Frank, M., Prenzler, A., Eils, R., Graf von der Schulenburg, J.M., 2013. Genome sequencing: a systematic review of health economic evidence. Health Economics Review 3, 29.

Fuentes-Silva, C., Soto-Zarazua, G., Torres-Pacheco, I., Flores-Rangel, A., 2013. Male tilapia production techniques: a mini-review African. Journal of Biotechnology 12 (36), 5496–5502.

Gall, G.A., Haung, N., 1988. Heritability and selection schemes for rainbow trout: female reproductive performance. Aquaculture 73, 57–66.

Gjedrem, T., 1992. Breeding plans for rainbow trout. Aquaculture 100, 73–83.

Gjedrem, T., 2005. Selection and Breeding Programs in Aquaculture. Springer, Berlin. Available from: https://doi.org/10.1007/1-4020-3342-7, ebook ISBN 978-1-4020-3342-1, 364 pp.

Gjerde, B., Refstie, T., 1984. Complete diallel cross between five stocks of Atlantic salmon. Livestock Production Science 11, 207–226.

Gjerde, B., Reddy, P.V., Mahapatra, K.D., Saha, J.N., Jana, R.K., Meher, P.M., et al., 2002. Growth and survival in two complete diallel crosses with five stocks of Rohu carp (*Labeo rohita*). Aquaculture 209, 103–115.

Gjoen, H.M., Bentsen, H.B., 1997. Past, present and future of genetic improvement in salmon aquaculture. ICES Journal of Marine Science 54, 1009–1014.

Glover, K.A., 2010. Forensic identification of fish farm escapees: the Norwegian experience. Aquaculture Environment Interactions 1, 1–10.

Goffeau, A., Barrel, B.G., Bussey, H., Davis, R.W., Dujon, B., Feldmann, H., et al., 1996. Life with 6000 genes. Science 274 (5287), 546, 563–567. PMID 8849441. S2CID 211123134. Available from: https://doi.org/10.1126/science.274.5287.546.

Gonzalez-Villasenor, L.I., Powers, D.A., 1992. Transfer, expression and inheritance of salmonid growth hormone in channel catfish, *Ictalurus punctatus*, and effects on performance traits. Molecular Marine Biology and Biotechnology 1, 380–389.

Granada, L., Lemos, M.F., Cabral, H.N., Bossier, P., Sara, C.N., 2017. Epigenetics in Aquaculture – the last frontier. Reviews in Aquaculture 10 (4), 994–1013. Available from: https://doi.org/10.1111/raq.12219.

Guo, X., DeBrosse, G.A., Allen, S.K., 1996. All triploid pacific oysters (*Crassostrea gigas Thunberg*) produced by mating tetraploids and diploids. Aquaculture 142, 149–161.

Hans Winkler (1920). Vebreifung und Ursache der Parthenogenesis in Pflanzen - und Tierreiche Verlag, Fischer, Jena.

Harbers, M., Caninci, P., 2005. Tag based approaches for transcriptome research and genome annotation. Nature Methods 2, 495–502.

Hayes, B., Laerdahl, J.K., Lien, S., Moen, T., Berg, P., Hindar, K., et al., 2007. An extensive resource of single nucleotide polymorphism markers associated with Atlantic salmon (*Salmo salar*) expressed sequences. Aquaculture 265, 82–90.

Henryon, M., Jokumsen, A., Berg, P., Lund, L., Pedersen, P.B., Olesen, N.J., et al., 2002. Genetic variation for growth rate, feed conversion efficiency, and disease resistance exists within a farmed population of rainbow trout. Aquaculture 209, 59–76.

Herlin, M., Delghandi, M., Wesmajervi, M., Taggart, J.B., Mcandrew, B.J., Penman, D. J., 2008. Analysis of the parental contribution to a group of fry from a single day of spawning from a commercial Atlantic cod (*Gadus morhua*) breeding tank. Aquaculture 274, 218–224.

Hieter, P., Bogusti, M., 1997. Functional genomics: it's all how you read it. Science 278, 601–602.

Holley, R.W., Apgar, J., Everett, G.A., Madison, J.T., Marquisee, M., Merrill, S.H., et al., 1965. Structure of a ribonucleic acid. Science 147 (3664), 1462.

Houston, R.D., Haley, C.S., Hamilton, A., Guy, D.R., Tinch, A.E., Taggart, J.B., et al., 2008. Major quantitative trait loci affect resistance to infectious pancreatic necrosis in Atlantic salmon (*Salmo salar*). Genetics 178, 1109–1115.

Hubner, N., Wallace, C.A., Zimdahl, H., Petretto, E., Schulz, H., Maciver, F., et al., 2005. Integrated transcriptional profiling and linkage analysis for identification of genes underlying disease. Nature Genetics 37, 243–253.

Hulata, G., Wohlfarth, G.W., Karplus, L., Schroeder, G.L., Harpaz, S., Halevy, A., et al., 1993. Evaluation of *Oreochromis niloticus* × *Oreochromis aureus*, hybrid progeny of different geographical isolates, reared under varying management regimes. Aquaculture 115, 253–271.

Hussain, M.G., Penman, D.J., McAndrew, B.J., 1998. Production of heterozygous and homozygous clones in Nile tilapia. Aquaculture International 6, 197–205.

Iseli, C., Jongeneel, C.V., Bucher, P., 1999. ESTScan: a program for detecting, evaluating, and reconstructing potential coding regions in EST sequences. Proceedings of International Conference on Intelligent Systems for Molecular Biology. pp. 138–148.

Jackson, T.R., Ferguson, M.M., Dansman, R.G., Fishback, A.G., Ihssen, P.E., O'Connell, M., et al., 1998. Identification of two QTL influencing upper temperature tolerance in three rainbow trout (*Oncorhynchus mykiss*) half sib families. Heredity 80, 143–151.

Jansen, R.C., Nap, J.P., 2001. Genetical genomics: the added value for segregation. Trends in Genetics 17, 388–391.

Jose, A.M., 2015. Movement of regulatory RNA between animal cells. Genesis 53, 395–416.

Kadakkuzha, B.M., Puthanveettil, S.V., 2013. Genomics and proteomics in solving brain complexity. Molecular bioSystems 9 (7), 1807–1821. Available from: https://doi.org/10.1039/C3MB25391K. PMC 6425491. PMID 23615871.

Kaiser, O., Bartels, D., Bekel, T., Goesmann, A., Kespohl, S., Pühler, A., et al., 2003. Whole genome shotgun sequencing guided by bioinformatics pipelines—an optimized approach for an established technique. Journal of Biotechnology 106 (2–3), 121–133. Available from: https://doi.org/10.1016/j.jbiotec.2003.08.008. PMID 14651855.

Kamaruzzaman, N., Nguyen, N.H., Hamzah, A., Ponzoni, R.W., 2009. Growth performance of mixed sex, hormonally sex reversed and progeny of YY male tilapia of the GIFT Strain, Oreochromis niloticus. Aquaculture Research 40 (6), 720–728.

Kause, A., Ritola, O., Mathysaari, E., Eskelinen, U., 2002. Coupling body weight and its composition: a quantitative genetic analysis in rainbow trout. Aquaculture 211, 65–79.

Kause, A., Ritola, O., Paananen, T., Mantysaari, E., Eskelinen, U., 2003. Selection against early maturity in large rainbow trout (*Oncorhynchus mykiss*): the quantitative genetics of sexual dimorphism and genotype by environment interactions. Aquaculture 228, 53–68.

Kirpichnikov, V.S., 1981. Genetics Bases of Fish Selection. Springer-Verlag, Berlin.

Klug, W.S., Cummings, M.R., Spencer, C.A., Palladino, M.A., 2012. Concepts of Genetics, tenth ed. Pearson Education, San Francisco, CA, ISBN 978-0-321-72412-0.

Kocher, T.D., Lee, W.J., Sobeleweska, H., Penman, D.J., McAndrew, B.J., 1998. A genetic linkage map of a cichlid fish, the tilapia (*Oreochromis niloticus*). Genetics 148, 1225–1235.

Komen, H., Thorgaard, G.H., 2007. Androgenesis, gynogenesis and the production of clones in fishes: a review. Aquaculture 269, 150–173.

Kondo, H., Suga, R., Suda, S., Nozaki, R., Hirono, I., Nagasaka, R., et al., 2011. EST analysis on adipose tissue of rainbow trout *Oncorhynchus mykiss* and tissue distribution of adiponectin. Gene 485, 40–45.

Koop, B.F., von Schalburg, K.R., Leong, J., Walker, N., Lieph, R., Cooper, G.A., et al., 2008. A salmonid EST genomic study: genes, duplications, phylogeny and microarrays. BMC Genomics 9, 545.

Kozfkay, J.R., Dillon, J.C., Schill, D.J., 2006. Routine use of sterile fish in salmonids sport fisheries: are we there? Fisheries 31, 392–401.

Krasnov, A., Koskinen, H., Pehkonen, P., Rexroad III, C.E., Afanasyev, S., Mölsä, H., 2005. Gene expression in the brain and kidney of rainbow trout in response to handling stress. BMC Genomics 6, 3.

Lan, H., Chen, M., Flowers, J.B., Yandell, B.S., Stapleton, D.S., Mata, C.M., et al., 2006. Combined expression trait correlations and expression quantitative trait locus mapping. PLoS Genetics 2, 51–61.

Leelatanawit, R., Uawisetwathana, U., Klinbunga, S., Karoonuthaisiri, N., 2011. A cDNA microarray, UniShrimpChip, for identification of genes relevant to testicular development in the black tiger shrimp (*Penaeus monodon*). BMC Molecular Biology 12, 15.

Levine, M., Tjian, R., 2003. Transcription regulation and animal diversity. Nature 424, 147–151.

Li, R.W., Waldbieser, G.C., 2006. Production and utilization of a high-density oligonucleotide microarray in channel catfish, *Ictalurus punctatus*. BMC Genomics 7, 134.

Li, J., Boroevich, K.A., Koop, B.F., Davison, W.S., 2011. Comparative genomics identifies candidate genes for infectious salmon anemia (ISA) resistance in Atlantic salmon (*Salmo salar*). Marine Biotechnology (NY) 13 (2), 232–241.

Li, Y., San, X., Hu, X., Xau, X., Zhang, J., Guo, X., et al., 2017. Scallop genome reveals molecular adaptations to semi-sessile life and neurotoxins. Nature Communications 8, 1721.

Lin, C.E., Safari, A., Agyakwah, S.K., Attipoe, F.Y., El-Nagger, G.O., Hamzah, A., et al., 2015. Differences in sexual size dimorphism among farmed tilapia species and strains undergoing genetic improvement for body weight. Aquaculture Reports 1 (2015), 20–27.

Liu, Z.J., Cordes, J., 2004. DNA marker technology and their applications in aquaculture genetics. Aquaculture 238, 1–37.

Liu, F., Li, Y., Yu, H., Zhang, L., Hu, J., Bo, Z., et al., 2020. MolluscDB: an integrated functional and evolutionary genomics database for the hyper-diverse animal phylum Mollusca. Nucleic Acids Research 49 (D1), D988–D997. Available from: https://doi.org/10.1093/nar/gkaa918.

Louro, B., Passos, A.L.S., Souche, E.L., Tsigenopoulos, C., Beck, A., Lagnel, J., et al., 2010. Gilthead sea bream (*Sparus auratus*) and European sea bass (*Dicentrarchus labrax*) expressed sequence tags: characterization, tissue specific expression and gene markers. Marine Genomics 3, 179–191.

Lynch, M., Walsh, J.B., 1998. Genetics and Analysis of Quantitative Traits. Sinauer Associates Inc., Sunderland, MA, 98 pp.

Macaranas, J.M., Mather, P.B., Lal, S.N., Vereivalu, T., Lagibalavu, M., Capra, M.F., 1997. Genotype and environment: a comparative evaluation of for tilapia stocks in Fiji. Aquaculture 150, 11–24.

MacKenzie, S., Jentoft, S. (Eds.), 2016. Genomics in Aquaculture. Academic Press, San Diego, CA.

Mair, G.C., Little, D.C., 1991. Population control in farmed tilapias. NAGA, The ICLARM Quarterly 14, 8–13.

Mair, G.C., Nam, Y.K., Solar, I.I., 2007. Risk management: reducing risk through confinement of transgenic fish. In: Kapuscinski, A.R., Hayes, K.R., Li, S., Dana, G. (Eds.), Environmental Risk Assessment of Genetically Modified Organisms, vol. 3. CABI, Wallingford, pp. 209–238, Methodologies for Transgenic Fish.

Majewski, J., Pastinen, T., 2011. The study of eQTL variations by RNA-seq: from SNPs to phenotypes. Trends in Genetics 27, 72–79.

Maluwa, A.O., 2005. Genetic Selection for Growth of a Malawian Indigenous Tilapia, *Oreochromis shiranus* (Doctor Scientiarum thesis). Norwegian University of Life Sciences (UMB), Department of Animal and Aquacultural Sciences. ISSN 0820-3220; ISBN 82-575-0668-0.

Maluwa, A.O., Gjerde, B., 2006. Estimates of the strain additive and maternal genetic and heterosis effects for harvest body weight of *Oreochromis shiranus* F2 generation. Aquaculture 259, 38–46.

Maluwa, A.O., Gjerde, B., 2007. Response to selection for body weight of *Oreochromis shiranus*. Aquaculture 273, 33–41.

Mambrini, M., Sanchez, M.P., Chevassus, B., Labbe, L., Quillet, E., Boujard, T., 2004. Selection for growth increases feed intake and affects feeding behavior of brown trout. Livestock Production Science 88, 85–98.

Martinez, V., Neira, R., Gall, G.A.E., 1999. Estimation of genetic parameters from pedigreed populations: lessons from analysis of alevin weight in Coho salmon (*Oncorhynchus kisutch*). Aquaculture 180, 223–236.

Martin, D.C., Baets, K., Walter, K.L., MacDonald, V.E., Wlodarski, M.J., Gozani, O., et al., 2006. The Yng1P plant homeodomain finger is a methyl-histone binding module that recognizes lysine 4-methylated histone H3. Molecular and Cellular Biology 26 (21), 7871–7879.

Mathavan, S., Lee, S.G., Mak, A., Miller, L.D., Murthy, K.R., Govindarajan, K.R., et al., 2005. Transcriptome analysis of zebrafish embryogenesis using microarrays. PLoS Genetics 1, 260–276.

Maxam, A.M., Gilbert, W., 1977. A new method for sequencing DNA. Proceedings of the National Academy of Sciences of the United States of America 74 (2), 560.

McAndrew, B., Napier, J., 2010. Foresight project on global food and farming futures. Application of genetics and genomics to aquaculture development: current and future directions. Journal of Agricultural Science 149 (Suppl. S1), 143–151. Available from: https://doi.org/10.1017/S0021859610001152.

Mece, B., Gnad, F., Soufi, B., Kumar, C., Olsen, J.V., Mijakovic, I., et al., 2007. Phosphoproteome analysis of *E. coli* reveals evolutionary conservation of bacterial ser/thr/tyr phosphorylation. Molecular and Cellular Proteomics 7 (2), 299–307. Available from: https://doi.org/10.1074/mcp.M700311-MCP200.

Mehaswaran, M., 2004. Molecular Markers; History, Future and Application. Advanced Biotech.

Mehaswaran, M., 2014. Molecular Markers; History, Future and Application. Advanced Biotech.

Meijer, A.H., Verbeek, F.J., Salas-Vidal, E., Corredor-Adamez, M., Bussman, J., van der Sar, A.M., et al., 2005. Transcriptome profiling of adult zebrafish at the late stage of chronic tuberculosis due to *Mycobacterium marinum* infection. Molecular Immunology 42, 1185–1203.

Min Jou, W., Haegeman, G., Ysebaert, M., 1972. Nucleotide sequence of the gene coding for the bacteriophage MS2 coat protein. Nature 237 (5350), 82–88. https://doi.org/10.1038/237082a0. PMID 4555447.S2CID 4153893.

Min, X.J., Butler, G., Storms, R., Tsang, A., 2005. OrfPredictor: predicting protein-coding regions in EST derived sequences. Nucleic Acids Research 33, W677–W680.

Moen, T., Baranski, M., Sonesson, A.K., Kjoglum, S., 2009. Confirmation and fine-mapping of major QTL for resistance to infectious pancreatic necrosis in Atlantic salmon (*Salmo salar*): population level associations between markers and trait. BMC Genomics 10, 368. Available from: https://doi.org/10.1186/1471-2164-10-368.

Montgomery, S.B., Sammeth, M., Gutierrez-Arcelus, M., Lach, R.P., Ingle, C., Nisbett, J., et al., 2010. Transcriptome genetics using second generation sequencing in a Caucasian population. Nature 464, 773–777.

Moore, M.J., 2005. From birth to death: the complex lives of eukaryotic mRNAs. Science 309, 1514–1518.

Morgan, T.H., 1926. The Theory of the Gene. Yale University Press, New Haven, CT.

Morley, M., Molony, C.M., Weber, T.M., Devlin, J.L., Ewens, K.G., Spielman, R.S., et al., 2004. Genetic analysis of genome-wide variation in human gene expression. Nature 430, 743–747.

Morrison, R.N., Cooper, G.A., Koop, B.F., Rise, M.L., Bridle, A.R., Adams, M.B., et al., 2006. Transcription profiling of the gills of amoebic gill disease (AGD) affected Atlantic salmon (*Salmo salar*): a role for the tumor suppressor protein p52 in ACD – pathogenesis. Physiological Genomics 26.

Muller, H.J., 1926. The gene as the basis of life. Proceedings of the International Congress of Plant Science 1, 897–921.

Muller-Belecke, A., Horstegen-Schwark, G., 1995. Sex determination in tilapia (*Oreochromis niloticus*) sex ratios in homozygous gynogenetic progeny and their offspring. Aquaculture 137, 57–65.

Mushtaq, M.Y., Verpoorte, R., Kim, H.K., 2013. Zebrafish as a model for systems biology. Biotechnology and Genetic Engineering Reviews 29 (2).

Myers, J.M., Hershberger, W.K., Iwamoto, R.N., 1986. The induction of tetraploidy in salmonids. Journal of the World Aquaculture Society 17, 1–7.

National Human Genome Research Institute, 2010. A Brief Guide to Genomics. https://www.genome.gov/ Retrieved 2011-12-03.

Neira, R., Lhorente, J.P., Araneda, N.D., Bustos, E., Alert, A., 2005. Studies on carcass quality traits in two populations of Coho salmon (*Oncorhynchus kisutch*): phenotypic and genetic parameters. Aquaculture 241, 117–131.

Nirenberg, M., Leder, P., Bernfield, M., Brimacombe, R., Trupin, J., Rottman, F., et al., 1965. RNA codewords and protein synthesis, Vil. On the general nature of the RNA code. Proceedings of the National Academy of Sciences of the United States America 53 (5), 1161.

Norris, A.T., Bradley, D.G., Cunningham, E.P., 2000. Parentage and relatedness determination in farmed Atlantic salmon (*Salmo salar*) using microsatellite markers. Aquaculture 182, 73–83.

Ohyama, K., Fukuzawa, H., Kohchi, T., Shirai, H., Sano, T., Sano, S., 1986. Chloroplast gene organization deduced from complete sequence of liverwort *Marchantia polymorpha* chloroplast DNA. Nature 322 (6079), 572–574. Available from: https://doi.org/10.1038/322572a0. S2CID 4311952.

Oliver, S.G., van der Aart, Q.J., Agostoni-Carbone, M.L., Aigile, M., Alberghina, L., Alexandraki, D., et al., 1992. The complete DNA sequence of yeast chromosome III. Nature 357 (6373), 38–46. Available from: https://doi.org/10.1038/357038a0. PMID 1574125. S2CID 4271784.

Pante, M.J., Gjerde, B., Misztal, I.M., 2002. Estimation of the additive and dominance genetic variances for body weight at harvest in rainbow trout, *Oncorhynchus mykiss*. Aquaculture 204, 383–392.

Parker, R., 2000. Aquaculture Science. Delmar Thomson Learning, Independence, KY, p. 6.

Peatman, E., Liu, Z., 2007. Microarray fundamentals: basic principles and application in aquaculture. In: Liu, Z. (Ed.), Aquaculture Genome Technologies. Blackwell Publishing Ltd, Oxford. Available from: https://doi.org/10.1002/9780470277560.ch21.

Parkinson, J., Blaxter, M., 2009. Expressed Sequence Tags: An Overview. Methods in Molecular Biology 533, 1–12. Available from: https://doi.org/10.1007/978-1-60327-136-3-1.

Peatman, E., Baoprasertkul, P., Terhune, J., Xu, P., Nandi, S., Kucuktas, H., et al., 2007. Expression analysis of the acute phase response in channel catfish (*Ictalurus punctatus*) after infection with a Gram-negative bacterium. Developmental and Comparative Immunology 31, 1183–1196.

Penman, D.J., Beeching, A.J., Penn, S., Rahman, A., Sulaiman, Z., Maclean, N., 1991. Patterns of transgene inheritance in rainbow trout (*Oncorhynchus mykiss*). Molecular Reproduction and Development 30, 201–206.

Penman, D.J., Gupta, M.V., Dey, M.M. (Eds.), 2005. Carp Genetic Resources for Aquaculture in Asia. WorldFish Center Technical Report 65. WorldFish Center, Penang.

Pevsner, J., 2009. Bioinformatics and Functional Genomics, second ed. Wiley-Blackwell, Hoboken, NJ, ISBN 978-0-470-08585-1.

Pickrell, J.K., Marioni, J.C., Pai, A.A., Degner, J.F., Engelhardt, B.E., Nkadori, E., et al., 2010. Understanding mechanisms underlying human gene expression variation with RNA sequencing. Nature 464, 768–772.

Pongthana, N., Penman, D.J., Baopraserthkul, P., Hussain, M.G., Islam, M.S., Powell, S.F., et al., 1999. Monosex female production in the silver barb (*Puntius gonionotus Bleeker*). Aquaculture 173, 246–256.

Ponzoni, W.W., Hamhah, A., Tan, S., Kamaruzzaman, N., 2005. Genetic parameter and response to selection for live weight strain of Nile tilapia (*Oreochromis niloticus*). Aquaculture 247, 203–210.

Pullin, R.S., Hamzah, A., Tan, S., Kamaruzzaman, N., 1991. Genetic improvement of farmed tilapias (GIFT) project: the story so far. Naga, the ICLARM Quarterly 14 (2), 3–6.

Purcell, M.K., Marjara, I.S., Batts, W., Kurath, G., Hansen, J.D., 2011. Transcriptome analysis of rainbow trout infected with high and low virulence strains of infectious hematopoietic necrosis virus. Fish & Shellfish Immunology 30, 84–93.

Quillet, E., Dorson, M., Le Guillou, S., Benmansour, A., Boudinot, P., 2007. Wide range of susceptibility to rhabdoviruses in homozygous clones of rainbow trout. Fish and Shellfish Immunology 22, 510–519.

Refstie, T., 1990. Application of breeding schemes. Aquaculture 85, 163–169.

Rise, M.L., Jones, S.R., Brown, G.D., Vou Scholburg, K.R., Davison, W.S., Koop, B.F., 2004. Microarray analyses identify molecular biomarkers of Atlantic salmon macrophage and hematopoietic kidney response to *Piscirickettsia salmonis* infection. Physiological Genomics 20, 20–35.

Rise, M.L., von Scalburg, K.R., Cooper, B.F., 2007. Salmonid DNA microarrays and other tools for genomic research. In: Liu, Z. (Ed.), Aquaculture Genome Technologies. Blackwell Publishing, Ames, IA, p. 369.

Rodrigues, P., Campos, A., Kuruvilla, J., et al., 2017. Proteomics in aquaculture. Proteomics in Food Science. pp. 279–295.

Rutten, M.J.M., Komen, H., Bovenhus, H., 2005. Longitudinal genetic analysis of Nile tilapia (*Oreochromis niloticus* L.) body weight using a random regression model. Aquaculture 246, 101–113.

Sanger, F., 1980. Nobel Lecture: Determination of Nucleotide Sequences in DNA (PDF). https://www.nobelprize.org/ (accessed 18.10.10.).

Sanger, F., Nicklen, S., Coulson, A.R., 1977. DNA sequencing with chain-terminating inhibitors. Proceedings of the National Academy of Sciences of the United States of America 74 (12), 5463–5467. Available from: https://doi.org/10.1073/pnas.74.12.5463. PMC 431765. PMID 271968.

Sarder, M.R.I., Penman, D.J., Myers, J.M., Mcandrew, B.J., 1999.). Production and propagation of fully inbred clonal lines in the Nile tilapia (*Oreochromis niloticus* L.). Journal of Experimental Zoology 284, 675–685.

Schadt, E.E., Monks, S.A., Drake, T.A., Lusis, A.J., Che, N., Colinayo, V., et al., 2003. Genetics of gene expression surveyed in maize, mouse and man. Nature 422, 297–302.

Scheerer, P.D., Thorgaard, G.H., Allendorf, F.W., Knudsen, K.L., 1986. Androgenetic rainbow trout produced from inbred and outbred sperm sources show similar survival. Aquaculture 57, 289–298.

Schena, M., Shalon, D., Davis, R.W., Brown, P.O., 1995. Quantitative monitoring of gene expression patterns with a complementary DNA microarray. Science 270, 467–470.

Shinozaki, K., Ohme, M., Tanaka, M., Wakasugi, T., Hayashida, N., Matsubayashi, T., 1986. The complete nucleotide sequence of the tobacco chloroplasts genome: its gene organization and expression. The EMBO Journal 5 (9), 2043–2049. Available from: https://doi.org/10.1002/j.1460-2075.1986.tb04464.x. PMC 1167080. PMID 164536699.

Sinden, R.R., 1994. DNA Structure and Function. Academic Press, New York, ISBN 0-12-645750-6.

Skaala, Ø., Wennevik, V., Glover, K.A., 2006. Evidence of temporal genetic change in wild Atlantic salmon, *Salmo salar* L., populations affected by farm escapees. ICES Journal of Marine Science 63, 1224–1233.

Smith, T.I., 1988. Aquaculture of striped bass and its hybrids in North America. Aquaculture Magazine 14, 40–49.

Snake, M., Maluwa, A., Zidana, H., Chigwechokha, P., Simwaka, M., 2020. Production of predominantly male tilapia progeny using two *Malawian tilapias*, Oreochromis shiranus and Oreochromis karongae. Aquaculture Reports 16, 100274.

Stein, L., 2001. Genome annotation: from sequence to biology. Nature Reviews Genetics 2 (7), 493–503. Available from: https://doi.org/10.1038/35080529. PMID 11433356. S2CID 12044602.

Streisiner, G., Walker, C., Dower, N., Knauber, D., Singer, F., 1981. Production of clones of homozygous diploid zebra fish (*Brachydanio rerio*). Nature 291, 293–296.

Tautz, D., 1989. Hypervariability of simple sequence as a general source of polymorphic DNA markers. Nucleic Acids Research 17, 6463–6471.

Tebery, J., Piotrowska, M., Darden, L., 2020. Molecular biology. In: Zalta, E.N. (Ed.), Stanford Encyclopedia of Philosophy. (Summer 2020 edition). Available from: https://plato.stanford.edu/archives/sum2020/entries/molecular-biology/.

Thorgaard, G.H., Scheerer, P.D., Hershberger, W.K., Myers, J.M., 1990. Androgenetic rainbow trout produced using sperm from tetraploid males show improved survival. Aquaculture 85, 215–221.

Ton, C., Stamatiou, D., Dzau, V.J., Liew, C.C., 2002. Construction of a zebrafish cDNA microarray: gene expression profiling of the zebrafish during development. Biochemical and Biophysical Research Communications 296, 1134–1142.

Torresen, O.K., Star, B., Jentoft, S., Jakobsen, K.S., Nederbragt, A.J., 2016a. The new Era of genome sequencing using high throughput sequencing technology: generation of the first version of the Atlantic cod genome. Genomics in Aquaculture. Academic Press, New York. Available from: https://doi.org/10.1016/B978-0-12-801418-9.00001-9.

Torresen, O.K., Rise, M.L., Xingkun, J., Star, B., Mackenzie, S., Kjetill, S., et al., 2016b. An improved version of the Atlantic cod genome and advancements in functional genomics: implications for the future of cod farming. Genomics in Aquaculture 2016, 45–72.

Uraiwan, S., Doyle, R.W., 1986. Replicate variance and the choice of selection procedures for Tilapia (*Oreochromis niloticus*) stock improvement in Thailand. Aquaculture 57, 93–98.

Vandeputte, M., 2009. Genetic improvement of common carp (*Cyprinus carpio* L.). Cahiers Agricultures 18, 256–261.

Velculescu, V.E., Zhang, L., Vogelstein, B., Kinzler, K.W., 1995. Serial analysis of gene expression. Science 270, 484–487.

Volff, J.N., Schartl, M., 2001. Variability of genetic sex determination in poeciliid fishes. Genetica 111, 101–110.

von Schalburg, K.R., Rise, M.L., Cooper, G.A., Brown, G.D., Gibbs, A.R., Nelson, C.C., et al., 2005. Fish and chips: various methodologies demonstrate utility of a 16,006-gene salmonid microarray. BMC Genomics 6, 126.

Vornanen, M., Hassiven, M., Kookinem, H., Krasnov, A., 2005. The steady-state effects of temperature acclimation on the transcriptome of rainbow trout. American Journal of Integrative and Comparative Physiology 289 (4), R1177–R1184.

Waldbieser, G.C., Wolters, W.R., 1999. Application of polymorphic microsatellite loci in channel catfish (*Ictalurus punctatus*) breeding program. Journal of the World Aquaculture Society 30, 256–262.

Wang, S., Shang, J., Jiao, W., Li, J., Xun, X., Sun, Y., et al., 2017. Scallop genome provides insights with evolution of bilaterian karyotype and development. Nature Ecology & Evolution 1, 120.

Wassermann, G.J., Afonso, L.O.B., 2002. Validation of the aceto-carmine technique for evaluating phenotypic sex in Nile tilapia (*Oreochromis niloticus*). Ciência Rural 32 (1). Available from: https://doi.org/10.1590/S0103-84782002000100023. Santa Maria.

Wayne, M.L., McIntyre, L.M., 2002. Combining mapping and arraying: an approach to candidate gene identification. Proceedings of the National Academy of Sciences of the United States of America 99, 14903–14906.

WHO, 2002. Genomics and World Health: Report of the Advisory Committee on Health Research. WHO, Geneva.

Wittbrdt, J., Meyer, A., Schartl, M., 1998. More genes in fish? Bioassays, Problems and Paradigms. Wiley Online Library, New York. Available from: https://doi.org/10.1002/(SICI)1521-1878(199806)20:6<511::AID-BIES10>3.0.CO;2-3.

Wohlfarth, G.W., 1993. Heterosis for growth rate in common carp. Aquaculture 113, 31–46.

Wohlfarth, G.W., Hulata, G., 1983. Applied Genetics of Tilapias, second ed. ICLARM Studies and Reviews 6. ICLARM, Manila, 26 pp.

Wohlfarth, G.W., Wedekind, H., 1991. The heredity of sex determination in tilapias. Aquaculture 92, 143–156.

Wolters, W.R., Lilyestrom, C.G., Graig, R.J., 1991. Growth, yield and dress-out percentage of diploid and triploid channel catfish in earthen ponds. The Progressive Fish Culturist 53, 33–36.

World Population Barometer, 2020. Current World Population. https://www.worldmwters.info.

Wu, C.W., Zhang, D., Kan, M.Y., Lv, M.Y., Zhu, Z.M., Su, A.Y., et al., 2014. The draft genome of the large yellow croaker reveals well-developed innate immunity. Nature Communications 5 (1), 5227. Available from: https://doi.org/10.1038/ncomms6227.

Yadav, S.P., 2007. The wholeness in suffix *-omics*, *-omes*, and the word *om*. Journal of Biomolecular Techniques 18 (5), 277.

Zhu, Z., Li, G., He, L., Chen, S., 1985. Novel gene transfer into the fertilized eggs of goldfish (*Carassius auratus* 1758). Journal of Applied Ichthyology 1, 31–33.

Zou, Z.H., Zhang, Z.P., Wang, Y.L., Han, K.H., Fu, M.J., Lin, P., et al., 2011. EST analysis on the gonad development related organs and microarray screen for differentially expressed genes in mature ovary and testis of *Scylla paramamosain*. Comparative Biochemistry and Physiology D: Genomics & Proteomics 6, 150–157.

FURTHER READING

Baroiller, J.F., D'Cotta, H., Penman, D.J., Tomasino, E., Coutanceau, J.P., Pepey, E., et al., 2008. Genetics of sex determination in tilapiine species. Sexual Development 2008 (2), 43–54.

Brent, M.R., 2005. Genome annotation past, present and future. How to define ORF at each locus. Genome Research. https://doi.org/10.1101/3866105.

Liu, Z.J., 2009. Aquaculture Genome Technologies. Blackwell Publishing, Oxford.

Rahman, M.A., Maclean, N., 1999. Growth performance in transgenic tilapia containing an exogenous piscine growth hormone gene. Aquaculture 173, 333–346.

Schartl, M., 2004. Sex chromosome evolution in non-mammalian vertebrate. Current Opinion in Genetics and Development 14, 634–641.

Tan, B., Grattapaglia, D., Martins, G.S., Ferreira, K.Z., Sundberg, B., Ingvarsson, P.K., 2017. Evaluating the accuracy of genomic prediction of growth and wood traits in two Eucalyptus species and their F1 hybrids. BMC Plant Biology 17, 110. Available from: https://doi.org/10.1186/s12870-017-1059-6.

World Health Organization, National Human Genome Research Institute, 2010. A Brief Guide to Genomics. https://www.genome.gov/. Retrieved 2020-11-03.

Zhang, P.J., Hayat, M., Joyce, C., Gonzalez Villasenor, L.I., Lin, C.M., Dunham, R.A., et al., 1990. Gene-transfer, expression and inheritance of pRSV-rainbow trout-GH cDNA in the common carp, *Cyprinus carpio* (Linnaeus). Molecular Reproduction and Development 25, 3–13.

Chapter | two

Proteomics for Quality and Safety in Fishery Products

Ana Paula Farinha[1], Márcio Moreira[1,2,3], Cláudia Raposo de Magalhães[1,3], Denise Schrama[1,3], Marco Cerqueira[1], Raquel Carrilho[1,3] and Pedro M. Rodrigues[1,3]

[1]Centre of Marine Sciences (CCMAR), University of Algarve, Faro, Portugal, [2]National Institute for the Sea and Atmosphere (IPMA), EPPO – Aquaculture Research Station, Olhão, Portugal, [3]Faculty of Sciences and Technology, University of Algarve, Faro, Portugal

2.1 INTRODUCTION

Proteins are the key players in all life forms, as they catalyze and control virtually all cellular processes (Aebersold and Mann, 2016). They form collectively a highly structured entity known as the proteome, and the constituent proteins (from a cell, tissue, or organ) carry out their functions with precision in time and cellular location, in association (physically or functionally) with other proteins or biomolecules (Aebersold and Mann, 2016). Proteomics, the scientific methodological approach to study the "proteome," represents, therefore, a unique way to understand biological processes, regardless of the organism, by identifying, quantifying, and looking into the main intervenient in the regulation of cellular function, the proteins. It is not surprising that proteomics, more specifically MS-based proteomics prompted by recent advances in the sequencing of fish genomes, has gained increasing importance in biological questions related to fish quality and safety, as demonstrated in the works discussed along with this chapter. The main covered topics are highlighted in Fig. 2.1, where we aim to show that aquaculture sustainability, fish quality, and safety are highly interconnected. Aquaculture involves many stakeholders besides small farmers and the industrial sector, for example, consumers and governmental policymakers (and legislators),

Sustainable Fish Production and Processing. DOI: https://doi.org/10.1016/B978-0-12-824296-4.00007-4

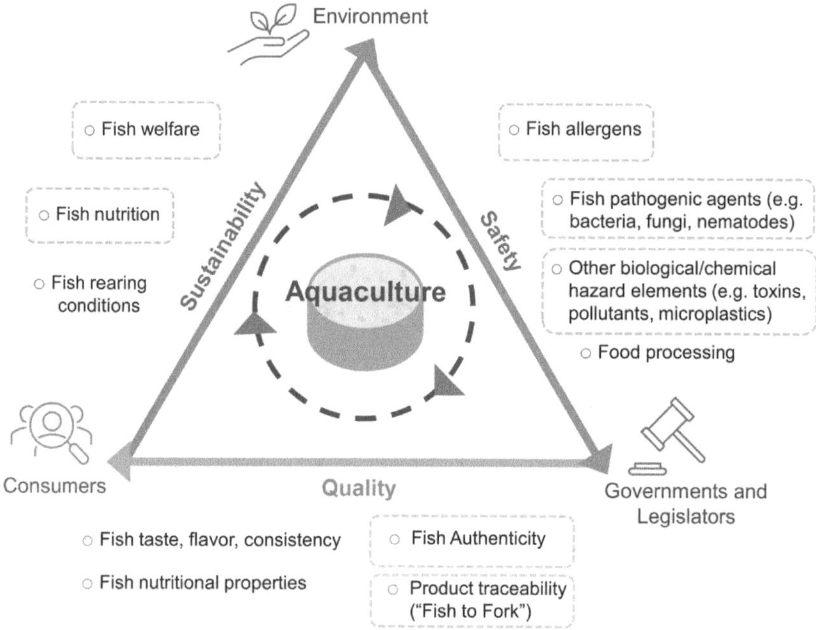

FIGURE 2.1 Proteomics applications toward sustainable aquaculture. Highlighted boxes indicate the main topics discussed in the present chapter.

that together are responsible for the implementation of quality and safety rules, in a continuous feedback loop, where environmental concerns are getting a central role (Fig. 2.1). The most common proteomics-based technologies employed in aquaculture have been recently reviewed (Rodrigues et al., 2012, 2016, 2017, 2018; Piras et al., 2016; Gomes et al., 2017; Carrera et al., 2020; Cerqueira et al., 2020a) and will be shortly presented before discussing its applications regarding fish quality and safety toward sustainable aquaculture.

2.2 PROTEOMIC TOOLS FOR A SUSTAINABLE AQUACULTURE

Currently, mass spectrometry (MS)-based approaches ("gel-free") have displaced "gel-based" workflows involving one- (1-DE) or two-dimensional gel electrophoresis (2-DE) for proteome analysis. 2-DE is a high-resolving power technique, where proteins first undergo isoelectric focusing based on their net charge at different pH values and then get separated in the second dimension, according to their molecular weight. This orthogonal approach allows detecting several hundreds of protein spots (if not thousands) in one single gel. 2-DE has been considered the "workhorse" of proteomics (Abdallah et al., 2012), becoming the method of choice for differential protein profiling in many distinct fields of biological research. It has progressed at different levels, for example, the quantitative two-dimensional difference in gel electrophoresis (2-D DIGE) and although it is no longer the preferential methodology across the proteomics community, still plays an important role, as recently discussed by Lee et al. (2020) and Marcus et al. (2020). 2-DE gel-based approaches are labor-intensive and have been steadily taken over by MS-based proteomics mostly due to its limited reproducibility, poor representation of low-abundant proteins, and eventual comigration of several proteins in the same gel position (spot) that could result in inaccurate protein identification (Abdallah et al., 2012). Additionally, proteins separated by 2-DE must be further identified by MS either through peptide mass fingerprinting (PMF) or tandem mass spectrometry (MS/MS).

In "gel-free"-based methodologies, protein samples remain "in solution" instead, and complex peptide mixtures are usually separated by reversed-phase liquid chromatography (LC) after protein proteolytic digestion. There is no doubt that liquid chromatography coupled to tandem mass spectrometry (LC-MS/MS) has become the technology of choice for the high-throughput characterization of proteomes (Aebersold and Mann, 2016). Despite the high-resolving power of 2-DE, MS-based approaches have taken a central role in proteomics, due to main advantages related to higher reproducibility, higher protein dynamic range, lower sample amount, and deeper proteome coverage. Furthermore, MS-based proteomics enables to identify and quantify several thousands of proteins (and their PTMs) across a significantly elevated number of samples in intricate experimental designs (Sinitcyn, et al., 2018).

2.2.1 "Top-Down" and "Bottom-Up" Proteomics

According to the need of analyzing proteins in their intact form or their derived peptides after proteolytic digestion with a protease (e.g., trypsin), there are two main options designated as "top-down" and "bottom-up" proteomics, correspondingly. In top-down proteomics, there is no protein breakdown/digestion before samples enter into the mass spectrometer, where both intact and protein fragment ions masses are measured. In bottom-up approaches, peptides must be generated (by enzymatic digestion of proteins) before introducing the mass spectrometer. The fragmentation-based identification of the peptides allows for the inferred identification of the original protein. There are several disadvantages associated with this approach, for example, peptide(s) may not be specific to an individual protein or protein form, some regions of the protein may not be identified. Therefore, information regarding PTMs or sequence variants may be missing, and modifications or sequence variations may occur on disparate peptides, causing their relationship to one another to be lost following digestion (Catherman et al., 2014). Despite its inherent disadvantages, peptide-based/bottom-up MS-based proteomics is still the most widely used approach since the high-throughput analysis of intact proteins raises many technical and computational difficulties (Catherman et al., 2014).

Nevertheless, these strategies are under continuous development. One of the leading applications involving the analysis of intact proteins is tissue imaging through matrix-assisted laser desorption/ionization time-of-flight (MALDI-TOF) based MS. However, top-down proteomic studies have primarily been implemented using electrospray ionization (ESI) methods coupled either to Fourier transform ion cyclotron resonance (FT-ICR) or Orbitrap mass analyzers (Catherman et al., 2014).

2.2.1.1 BOTTOM-UP PROTEOMICS WORKFLOWS

There are currently three main workflows used in bottom-up proteomics, all differing in the way that spectra data are acquired/selected by mass spectrometer instruments for further analysis: (1) untargeted (or discovery) data-dependent acquisition mode (DDA) based proteomics, (2) data-independent acquisition mode (DIA) based-proteomics, and (3) targeted proteomics (Aebersold and Mann, 2016). In DDA, only the most abundant ion precursors are selected for further MS/MS analysis. In contrast, in DIA tandem MS, data spectra acquisition is independent of the composition of precursor ions for their fragmentation. Therefore all the peptides within a defined *m/z* (mass to charge) frame are subjected to fragmentation (Zhang et al., 2020; Dupree et al., 2020). These two methods can be even combined and integrated into one single analytical run, called "data-dependent and independent acquisition" mode, as recently described by Sun et al. (2020).

The preferred strategy at the core technologies employed for proteome discovery, also known as "untargeted" proteomics (free from hypotheses), are bottom-up approaches using DDA workflows (Aebersold and Mann, 2016; Dupree et al., 2020). Shotgun DDA workflows have been most useful in the routine application of proteomics to cell biology and biochemical research, helping to disclose new and unexpected molecules and relationships, providing new biological insights regardless of the research area (Aebersold and Mann, 2016; Dupree et al., 2020).

More recently, DIA-based methods have emerged as a powerful driving technology in proteomics (Zhang et al., 2020; Dupree et al., 2020). This approach attempts to combine in-depth proteome coverage with quantitative consistency and accuracy in quantifying protein levels. DIA offers several advantages over DDA since it is not restricted to the profile of predefined peptides of interest, allowing to reproduce and quantify low abundant peptides. In DDA the mass spectra of all the ion species that coelute at a specific point in the gradient elution are sequentially detected. Regarding DIA-based methods the fragmentation of all detectable ions within an *m/z* window is systematically parallelized, regardless of ion intensity, thus providing a broader dynamic range of detected signals, better accuracy, and enhanced protein coverage (Aebersold and Mann, 2016; Dupree et al., 2020). The sequential window acquisition of all theoretical mass spectra (SWATH-MS) (Gillet et al., 2012) is a specific variant of DIA methods, most indicated for studies involving a large number of samples that require the highest accuracy and reproducibility quantitation, as the case of biomarker studies or clinical/drug perturbation studies (Ludwig et al., 2018). SWATH-MS is probably the DIA method most widely used so far, though other DIA methodologies are currently emerging as recently reviewed by Zhang et al. (2020). The major challenge in DIA-based methods is the computational deconvolution of the multiplexed fragmented spectra, only possible due to the use of highly sophisticated software tools for interpreting DIA data and the development of spectral libraries usually generated by DDA analysis. It is noteworthy that Monroe et al. (2020) have demonstrated very recently for the first time that SWATH-MS can be even applied to nonmodel organisms, using fish as a model.

Another possible workflow in bottom-up proteomics is targeted proteomics. In this approach the proteins of interest are predetermined and known. Based on previous information, characteristic and proteotypic peptides are selectively and recursively isolated and then fragmented over their chromatographic elution time. To achieve this selectivity the process is multiplexed to several fragments per peptide, also known as multiple reaction monitoring (MRM) or selected reaction monitoring (SRM), usually performed on triple quadrupole mass spectrometers. In contrast, another variant known as "parallel reaction monitoring" (PRM) utilizes the entire MS^2 spectrum recurring to quadrupole orbitrap instruments (Aebersold and Mann, 2016) instead.

2.2.2 General Considerations about Proteomics-Based Strategies

Proteomics research has benefited enormously from advances in genomics during the last decade, reflected on major improvements in the curation and annotation of public protein databases. Notably a significant number of fish including both model and nonmodel species has already been fully sequenced at the genome level, with more than 50 species being reported (Yuan et al., 2018). High-quality (well-annotated and curated) public protein databases are crucial for successful peptide/protein matching after database searching, and therefore for protein identification either through PMF or MS/MS. Ideally, all protein-coding gene sequences from a determinate organism should be known and all the encoded proteins should exist in a protein database with some degree of evidence (and properly annotated), to allow the identification by MS of the highest number of proteins in the proteome under study. Unfortunately, this is not the real scenario, and there is still much to be done, especially concerning nonmodel organisms, as in the case of farmed fish species. Significant efforts are, however, being conducted to best annotate protein databases through proteogenomic approaches by combining proteomic, genomic, and transcriptomic data, as highlighted by Dupree et al. (2019). Another essential aspect in the application of MS-based approaches is the need for publicly available bioinformatics tools for processing and interpreting the massive amounts of data that are generated, also requiring some technical expertise in the management of such devices, either in the case of free or purchased software.

Back to the beginning, when contrasting "gel-based" and "gel-free" approaches, or to make it simpler, 2-DE based strategies, for example, DIGE by contrast with typical bottom-up approaches, for example, shotgun proteomics, the trend seems obvious. Indeed, gel-based processes are becoming less applied with the routine implementation of MS-based proteomics, a global movement observed in biological sciences, particularly in human proteomics, but extended to other model and nonmodel organisms, for example, farmed fish. However, there is still room for gel-based workflows, especially when thinking that it represents a top-down approach, allowing to analyze intact proteins, and inspect quickly, for instance, their posttranslational modified "proteoforms" (Marcus et al., 2020). Gel-based proteomics can be applied in particular biological questions as a first screening method before more in-depth proteome analysis by MS-based approaches. A significant application of 1-DE and 2-DE is the study of the immunome/allergome, by using antibodies in Western-blot analyses to decipher which proteins of a pathogen/allergen are targeted by the immune system (Lopez-Pedrouso et al., 2020). These applications can be highly relevant in the study of fish allergens, as will be later discussed in this chapter. When deciding about the proteomics strategy (ies) to adopt in our work, it must prevail, which is/are the method(s) that best answers our biological question with the means we dispose of and considering the costs involved.

2.3 SUSTAINABILITY IN AQUACULTURE

The world is continuously changing, and fish and shellfish production are as important as any other animal production system to maintain harmony between the planet and the people's needs. To accommodate further growth of the seafood industry, providing a continuous supply of fish and shellfish beneficial to human sustenance without exceeding the capacity of the planet is crucial. Nevertheless, global fisheries capture stagnated in the late 1980s, standing to the overexploitation of wild stocks (FAO, 2020). Since then, exploiting wild reserves to and beyond their limit, alongside the ever fastest-growing human population, has promoted a turning point in fish and shellfish production. Aquaculture became the fastest-growing food production industry, accounting for more fish production than capture fisheries (Edwards et al., 2019) and total biomass than terrestrial meat. Aquaculture plays more than ever an essential role in the global food supply and security functioning as a purveyor of food (availability), livelihoods, and income. Currently, there are an estimated 59.5 million people engaged in the industry (Dupree et al., 2019). Consequently the continuous development of sustainable production practices is vital to safeguard the ability of humans' future generations to meet their own needs. Different strategies can be pursued to foster sustainability transitions in aquaculture, such as sustainable intensification and production of safe and healthy food.

2.3.1 The Importance of Fish Welfare in the Context of Sustainable Aquaculture

The intensification of the industry causes issues that go beyond food security or environmental constraints. It, therefore, is critical to consider the roles of aquaculture proponents in the governance of sustainability. Within this plan, fish welfare is central for access to safe and nutritious food and the global fish and seafood value chains. In the EU, fish welfare has become one of the main priorities in progressing toward sustainable aquaculture production, with several initiatives launched by the European Union within the framework of the 2030 agenda and within the scope of several of the UN Sustainable Development Goals (SDGs) (Giménez-Candela et al., 2020). These SDGs cover ending poverty and hunger, promote environmental sustainability, economic growth, and "life below water." Considering the welfare of fish, it would help to eliminate, or at least to minimize, the environmental impact of the industry, and ultimately to preserve the ocean, buffer against disease risks and improve the life of billions of fish produced every year.

The importance of animal welfare in aquaculture is derived not only from pragmatic reasons such as its contribution to food sustainability (i.e., preventing poor fish performance, health, and quality) but also grounded on ethical concerns. Studies increasingly suggest that fish recollect aversive experiences, display exact preferences, are likely to experience their world qualitatively, and to feel pain or suffer similarly to mammals if stressed beyond their ability to cope. Fish welfare definition is therefore based on the quality of life experienced by fish about its environment. It is considered an individual characteristic that can vary from low (or impaired) to very good (or ensured) on a continuum (Broom, 1991). Arguably, fish welfare is only possible if their biological needs (nutritional, environmental, physical health, and behavioral freedom) are positively challenged and fulfilled. Recent research shows that many aquatic animals can both positive and negative physical and psychological experiences (Cerqueira et al., 2017, 2020b). Therefore marine animals must be free to seek sufficient environmental challenges and social interaction as required for their physical and psychosocial allostatic development.

As attested by many other terrestrial meat animals, good welfare means good production, and in the aquaculture industry, this is not different. The benefits of fish welfare for aquaculture are multisectoral owing to direct implications for their products and the sustainability of the sector. Fish kept under the right welfare conditions have apparent economic benefits. Fish remain healthy, are less stressed, less prone to diseases, show better growth performance and food conversion efficiency, need fewer antimicrobials and treatments, and ultimately benefit the farming industry (Giménez-Candela et al., 2020). Moreover, integrating welfare management into the industry will significantly reduce the amounts of pollution brought by aquaculture farms, contributing to safer, more sustainable, and ethical products (White, 2017). Given that aquaculture production is rapidly expanding through the years without showing any signs of peaking, alongside the overall world fish market globalizing fish consumption, fish welfare labeling will be crucial to satisfy consumers' concerns about seafood safety and quality dimensions.

2.3.1.1 ASSESSING WELFARE IN AQUACULTURE

The quality and the safety of the aquaculture industry outcomes rely on the best assessment of the welfare state for the given species under the prevailing farming system (Toni et al., 2019). Adequate knowledge of the biology of the species and the production system is crucial to protect fish and its quality, and ultimately the consumer and the environment. To assess fish welfare holistically, stakeholders used available environmental- and animal-based inputs relying on specific welfare indicators (Kolarevic et al., 2018). These indicators are observations or measurements that provide information about how the animals' welfare needs are ensured or impaired. Hands for ensured welfare are yet questionable and, most likely, far away from everyday use. The main reason is that homeostasis is not challenged. Therefore there are no allostatic adjustments regarding potential physiological changes, that is, there is no need for a biological response pattern to arise to cope with it, as there is nothing to cope with.

Regarding impaired welfare, no limited number of welfare indicators exist, or ever will, since its digits are developing alongside the understanding of fish environmental needs and welfare requirements. To name a few, physiological, behavioral, physical, and quality indicators can be used to assess impaired welfare (Martos-Sitcha et al., 2020). Those need to be specific for the farmed species and the farming system. Monitor and assess impaired interest will need to combine different parameters to enhance decision-making outcomes. Therefore, under on-farm conditions, a set of welfare indicators combining divergent aspects of the rearing system is usually recommended, for example, the farming system features, environmental complexity (including chemical–physical characteristics of water), the size of fish, technology, and staff skills for fish welfare assessment. The most widely used indicators of impaired welfare are related to stress response, distress, disease, and pain tied to practices in daily farm management such as providing adequate nutrition, maintaining good water quality, and a suitable environment, among others.

Due to the intensification of the aquaculture industry, acute or chronic stressful situations (stressors) are inevitable consequences of standard farming practices. Routine procedures and manipulation strategies to maximize fish production, for example, high rearing densities, handling, and transportation, may have harmful effects on fish. A wide range of aquaculture stressors and their impact on fish have been well revised (Conte, 2004). The cumulative and long-term impacts of these stressors can be reflected in fish behavior and physiology, including growth, reproduction, and welfare, and may jeopardize fish health and survival (Eslamloo et al., 2014). Stress is intimately related to fish welfare and should be carefully monitored to ensure the animal's state of well-being.

As in other vertebrates, fish experiencing stress exhibit a set of physiological changes that have been extensively described in the literature. In the face of a perceived threat to homeostasis, fish launch an adaptive mechanism known as physiological stress response, which in the short-term helps the individual to cope with the adverse situation (eustress), not necessarily equating compromised welfare (Iwama, 2007; Schreck, 2010). However, the tertiary effects of this stress response, such as impaired reproductive and immune functions, are primarily associated with diminished interest (distress). These are usually the result of a prolonged, repeated, and unavoidable stressor that surpassed the organism's natural regulatory capacity (Boonstra, 2013).

The physiological stress response is a phased mechanism. The primary responsibility involves the immediate activation of the sympathetic pathway with the consequent release of catecholamines into the bloodstream, followed by a delayed activation of the hypothalamic–pituitary–interrenal (HPI) axis. A corticotropin-releasing factor from the hypothalamus stimulates the pituitary to synthesize and release a corticotropic hormone, which induces the synthesis of corticosteroids (cortisol in teleosts) in the interrenal cells (Pottinger, 2008). These hormones lead to a cascade of downstream responses and are responsible for most of the maladaptive effects at the whole-animal level described above. Alterations in energy metabolism characterize secondary reactions. A rapid energy source mobilization initiates, and specific pathways are activated to produce glucose, the essential substrate to meet the high energy demand during and after stress. The activation of these pathways, namely, the glycogenolysis in the liver or muscle and the hepatic gluconeogenesis, are carried out by the catecholamines and cortisol, respectively (Fabbri and Moon, 2016). Additionally, increased muscle activity can also be caused by stressful stimuli, which leads to the activation of the anaerobic glycolysis in the muscle, generating lactate. Both glucose and lactate are released into the bloodstream, and therefore, the plasma levels of these metabolites are often used alongside cortisol to assess the stress status of the fish (Ellis et al., 2012).

In an experimental context, cortisol is the most common and widely used indicator to characterize the stress response in fish (Pottinger, 2008). However, high levels of plasma cortisol indicate arousal or activation of the HPI axis, rather than being specific to diminished welfare (Ashley, 2007). The cortisol response varies among different species and even among individuals (coping styles) (Koolhaas et al., 2010; Castanheira et al., 2017). It can also be affected by several other parameters (e.g., domestication level, age, nutritional state, stressor severity, among others) (Gesto et al., 2015; Madaro et al., 2016). Furthermore, it has been shown that cortisol levels may return to prestress levels as time elapses, usually within 24 h (Naderi et al., 2018). This may be due to adaption or exhaustion of the HPI axis due to allostatic overload (Tort et al., 2001). This means that in the case of chronic stress exposure the reliability of this hormone becomes even lower. Therefore cortisol levels represent only a snapshot of the physiological stress response and should be used with caution when evaluating the magnitude of the reaction mounted and the level of stress experienced.

2.3.1.2 THE PROTEOMICS' CONTRIBUTION TO FISH WELFARE ASSESSMENT

Aquaculture stressors affect fish well-being and the entire chain of production and sustainability, negatively impacting fish quality, thus reducing the merchantable value of the final product (Huntingford et al., 2006). It is not surprising that the development of methodologies addressing fish welfare became a common goal among producers. The scientific knowledge on the physiological effects of different stressors in fish is, therefore, crucial to help in the improvement of the production and sustainability in aquaculture, and also in the development and implementation of best practices toward animal welfare across farmers and the industry itself.

Measuring fish welfare objectively has been a prevalent issue (Carenzi and Verga, 2009). It is unanimous that no single measure of interest is reliable when assessed separately. Instead a multidisciplinary approach that considers all levels of cellular organization, from molecular to whole-animal responses, should be conducted for correct interpretation of the fish welfare state (Ashley, 2007). An advanced high-throughput technology-based- "omics" research strategy started to gain popularity prompted by the increasing interest in farmed fish welfare, as recently reviewed by Raposo de Magalhães et al. (2020a). Proteomics emerged in aquaculture welfare studies as a unique tool to address the complexity of stress physiology in fish, with a primary focus on finding reliable indicators of stress and compromised welfare and optimizing fish capacity to cope with unavoidable challenges (Almeida et al., 2014; Di Girolamo et al., 2014). Proteomics plays a significant role in discovering stress biomarkers since it allows the identification and quantitation of a vast number of proteins present in a determinate tissue/organ, and proteins are most likely to be ubiquitously affected by environmental factors (Marco-Ramell et al., 2016). The identification and standardization of reliable molecular and physiological indicators of stress in fish that could be used as an early warning system before whole-animal, irreversible, maladaptive changes occur, is a recurrent matter of study and a complex task. Nonetheless, it would greatly contribute, together with behavioral observations, to the development of adequate welfare assessment measures and improved husbandry protocols. Ideally a reliable stress indicator should be sensitive, consistent, measurable on a commercial farm within a realistic framework, and more importantly, the effect of the stressful stimulus on it should happen over a slower timescale and preferably be a permanent change, or at least last long enough to be reliably quantified.

Most proteomic studies on fish welfare focus on the liver, as the main organ responsible for regulating the metabolism and immune-related pathways involved in stress responses. Studies using fish blood plasma and skin mucus are more recent. Still, they offer many advantages, as samples are obtained in a noninvasive way, with no need for animal sacrifice, allowing at the same fish monitoring along with different time points (Sanahuja and Ibarz, 2015; Patel and Brinchmann, 2017; Raposo de Magalhães et al., 2020b). On the other hand the majority of stress-related proteomic works settles on comparative approaches attempting to evaluate the effects of standard aquaculture practices and environmental stressors in fish, such as handling procedures (Yang et al., 2015), high stocking densities (Naderi et al., 2018, 2017a; Cordero et al., 2016), hypoxia (Tiedke et al., 2015; Pédron et al., 2017; Zhang et al., 2017), anoxia (Smith et al., 2009), diseases and parasites (Buján et al., 2015), temperature (Richard et al., 2016; Schrama et al., 2017; Nuez-Ortín et al., 2018) and pre and postslaughter stress (Morzel et al., 2006). The literature available continues to increase, highlighting the importance of proteomics in identifying molecular signatures beyond physiological stress response and adaptation. From these studies, many proteins have already been proposed as putative biomarkers to assess fish welfare, namely acute-phase proteins, microglobulins, macroglobulins, apolipoproteins, alpha-1-antitrypsin, transferrins, hemopexin, heat shock proteins, plasminogen, and complement system proteins (Raposo de Magalhães et al., 2020b; Cordero et al., 2017; Mahanty et al., 2016; Yue et al., 2018; Swirplies et al., 2019). Nevertheless, all these candidate biomarkers still require further validation, and the link between these molecular changes and fish physiological performance and fitness still needs to be effectively proved.

2.3.2 Sustainable Ingredients and Supplementation in Fish Nutrition

Fish nutrition is crucial for the aquaculture sector, with studies gaining the highest interest as the demand for farmed fish increased (Jobling, 2015). Research on fish nutrition usually addresses fish growth and survival rate, diet performance (feed intake, nutrient requirements, interaction, utilization, and synthesis), reproduction, early development, and, more recently, welfare and sustainability (Jobling, 2015; Prabu et al., 2017). Proteomics can be used in the study of any tissue/organ related to nutrition, that is, the liver due to its involvement in the regulation of critical metabolic pathways; the muscle that controls energy metabolism and growth; and the gut/intestine, necessary for the immune system and nutrient intake through the hepatopancreas, controlling the digestion. As the analysis of these tissues implies the sacrifice of animals, other tissues like blood plasma and skin mucus collected through a noninvasive method can be an excellent alternative to study the effects of nutrition on fish welfare, growth, and health. Nutritional proteomics may also include the analysis of whole body, since in some cases it is impossible to isolate fish tissues, such as at the larval stage (de Vareilles et al., 2012a). Proteomics employing either untargeted or targeted approaches have been extensively used to study biological effects associated with nutrition and food safety (Carrera et al., 2020). The influence of different nutrients on fish health, growth, and welfare can be most effective when complementing proteomics with other technologies like genomics, transcriptomics, and metabolomics.

Fish and seafood species have different nutritional requirements and behaviors that should be considered when formulating more sustainable feeds. Fish feeding diets are generally produced with fishmeal and fish oil rich in high-quality proteins and omega-3 polyunsaturated fatty acids (Carrera et al., 2006). Production of these fish feed is expensive, and to keep up with sustainable, economic, and nutritious diets, several modifications/improvements were performed over the years (Prabu et al., 2017). Substitution of the costly and unsustainable fish meal and fish oil in feed by alternative ingredients such as processed animal proteins, sustainable plant and vegetable proteins, and fats has become an essential subject in aquaculture. Table 2.1 shows a few examples of affected proteins/metabolisms by the substitution of ingredients. Several sustainable substitutions resulted in affected proteins involved in stress and immune response and proteins involved in energy, lipid, carbohydrate, and amino acid metabolisms (Ghisaura et al., 2014; Ghaedi, et al., 2016; Nuez-Ortin et al., 2016; Boonanuntanasarn et al., 2019; Estruch et al., 2020; Nakharuthai et al., 2020). Instead of the substitution of fish oil a total absence was also studied in rainbow trout and showed that the amino acid and fatty acid metabolism, glycolysis, and energy metabolism were affected (Kolditz et al., 2008). As vegetable ingredients do not possess the same amounts of essential amino acids (EAA) as animal-based ingredients, supplementations with these EAA try to compensate for this low abundance. Table 2.1 shows examples for proteomic analysis of supplementations with lysine and tryptophan (Gomez-Requeni et al., 2011; De Vareilles et al., 2012b; Cerqueira et al., 2020c). More specific supplementations have been studied for the benefit of fish health and welfare, such as β-glucan and vitamins (Ghaedi et al., 2016; Richard et al., 2014). Additives such as nucleotides, glycerol, maslinic acid, and natural compounds were also studied in several fish species using proteomics (Keyvanshokooh and Tahmasebi-Kohyani, 2012; Silva et al., 2012; Rufino-Palomares et al., 2011; Matos et al., 2013; Jensen et al., 2015) (Table 2.1). Other interesting ingredients that might contribute to sustainable feeds include microalgae, as they contain important omega-3 fatty acids. Supplementation of feed with specific minerals and vitamins together with the basic nutritional requirements of fish species may contribute to an enhanced health of the intestinal and gut microbiota (Francis et al., 2001). It is noteworthy that the development of fish health might be achieved at earliest developmental stages (Dawood and Koshio, 2018).

As demonstrated by proteomic analysis summarized in Table 2.1, many proteins (and most probably their posttranslational modifications) are affected by fish nutrition. To keep pace with the production of more sustainable and adequate fish feeding diets, future research should be performed in a multidisciplinary way by combining molecular analyses with fish physiological and nutritional performance parameters. This could be settled on the basis of improved experimental designs (e.g., the right number of animals and replicates for statistical analyses), the use of high-throughput and complementary technologies (e.g., proteomics combined with other omics, namely metabolomics), and proper data analysis tools to integrate and translate the

TABLE 2.1 Summary of Some Substitutions, Absence, or Supplementations of Ingredients in Fish Feed

Substitution, Supplementation, or Absence of Ingredients	Fish Species	Affected Proteins/Metabolisms	Publication
Substitution by soy, wheat, corn, green pea, hemoglobulin	Gilthead seabream	Apolipoproteins, transferrin, Wap65, fibrinogen, F-type lectin, and α-1-antitrypsin	Ghisaura, et al. (2014)
Substitution by soy	Rainbow trout	Keratin II, β-tubulin, HSP70, HSP78, HSP108, aldolase B, protein phosphatase 2A, apolipoproteins, and anabolic metabolism	Martin, et al. (2003)
Substitution by poultry, tuna, and flaxseed oil	Atlantic salmon	Oxidative stress, detoxification pathways, protein biogenesis in the endoplasmic reticulum and secretory protein trafficking,	Nuez-Ortin, et al. (2016)
Substitution by maize, wheat, peas, rapeseed	Rainbow trout	Energy metabolism and proteolysis	Vilhelmsson, et al. (2004)
Substitution by palm, linseed, and soybean oil	Nile tilapia	Oxidative stress, immune, inflammatory process, lipid, carbohydrate, and amino acid metabolism	Boonanuntanasarn, et al. (2019), Nakharuthai et al. (2020)
Substitution by wheat, fava bean, soybean, pea, sunflower, squid, and krill meal	Gilthead seabream	Binding proteins, immune and inflammatory response	Estruch, et al. (2020)
Substitution by cod liver oil and supplementation with soybean lecithin	Pikeperch	Methionine metabolism, structural proteins, and stress response	Hamza, et al. (2010)
Absence of fish oil	Rainbow trout	Fatty acid metabolism, glycolysis, amino acid catabolism, and energy production	Kolditz, et al. (2008)
Supplementation with lysine	Zebrafish	Cytoskeletal network, contractile apparatus, energy, and lipid metabolism	Gomez-Requeni, et al. (2011), De Vareilles et al. (2012b)
Supplementation with tryptophan	Gilthead seabream	Amino acid metabolism, immune response, and energy metabolism	Cerqueira, et al. (2020c)
Supplementation with phosphorus	Pufferfish	Energy metabolism and lipid transportation	Ye et al. (2016)
Supplementation with nucleotides	Rainbow trout	Metabolic and structural proteins	Keyvanshokooh and Tahmasebi-Kohyani (2012)
Supplementation with vitamin K	Senegalese sole	Skin, muscle, and bone development	Richard, et al. (2014)

Continued

TABLE 2.1 continued

Substitution, Supplementation, or Absence of Ingredients	Fish Species	Affected Proteins/Metabolisms	Publication
Supplementation with vitamin E and selenium	Rainbow trout	Heat shock proteins, glycolysis, and gluconeogenesis	Naderi et al. (2017b)
Supplementation with glycerol	Gilthead seabream	Detoxification processes, oxidative stress, and signaling	Silva et al. (2012)
Supplementation with maslinic acid	Gilthead seabream	Glucose and detoxification metabolism, oxidative stress, amino acid and purine metabolism, protein synthesis, and heat shock proteins	Rufino-Palomares, et al. (2011)
Supplementation with phospholipids, vitamins C and E, and taurine	Gilthead seabream	Oxidative stress, endoplasmic reticulum stress, aerobic metabolism, gluconeogenesis, immune response, and lipid metabolism	Richard et al. (2016), Schrama, et al. (2017)
Supplementation with identical natural compounds	Atlantic salmon	Heat shock proteins and cell structural proteins	Jensen, et al. (2015)
Supplementation with β-glucan	Rainbow trout	Structural proteins	Ghaedi, et al. (2016)

results obtained at the molecular and physiological level (e.g., zootechnical fish performance) to benefits across all aquaculture stakeholders. It is most likely that shortly, plant and vegetable ingredients and the supplementation of some specific beneficial ingredients will play a significant role in the achievement of more sustainable, cost-effective, and high-quality/healthy feed formulations in aquaculture.

2.4 PROTEOMICS IN THE ASSESSMENT OF FISH QUALITY AND SAFETY

Changes in consumers' dietary habits have a significant impact on modern food science with new and more powerful technologies emerging in the field. Within these technologies and particularly in fishery products, proteomics as a high-throughput approach is one of the best options to conventional methodologies that are becoming inadequate to comply with the EU's strict new regulations. Assessment of seafood authenticity and origin and fish products detection in foods are two of the main areas related to the quality and safety of food products where the applications of proteomics methodologies can have a significant impact.

2.4.1 Fish Authenticity

Fish authenticity is a primary concern in the food industry, as fish species substitution is still a common practice in many countries to guarantee some monetary benefits by selling lower-priced fish species at the price of more valuable species. Not only is this practice illegal but it is also potentially harmful to consumers, as the substituted species might present a potential danger to fish allergic patients (Sicherer and Sampson, 2010). Several labeling regulations at national and European levels have been implemented to ensure correct labeling and identification of fish species. Complying with these regulations created a need to develop new molecular techniques capable of detecting and identifying a commercial fish species, where proteomics technologies presented a promising alternative to the more traditional DNA-based methods. These proteomics-based workflows comprised both discovery and target-driven phases (Carrera et al., 2006, 2010, 2011).

Two-DE based methods (Fig. 2.2) were the first to emerge, with several works published describing the methodology to differentiate closely related hake species belonging to the genus *Merluccius* (Carrera et al., 2006, 2007; Piñeiro et al., 2001). Nevertheless, these studies highlighted that the absence of fish protein sequences in databases and the time-consuming 2-DE techniques could pose a significant drawback to this methodological approach. This led to the integration of high-resolution MS and MALDI-TOF MS molecular profiling strategies. In 2008 Mazzeo et al. (2008) successfully used this approach to discriminate between 25 different fish species belonging to Perciformes, Gadiformes, and Pleuronectiforms. Further studies by Carrera et al. (2010) combining a classical bottom-up proteomic approach with the accurate molecular mass determination by FT-ICR-MS of intact proteins and the selected MS/MS ion monitoring (SMIM) of peptide mass gaps, led to the unambiguous identification of any species of the Merlucciidae family (Carrera et al., 2010, 2011). A slightly different approach was developed by Wulff et al. (2013), generating, as a first step, a set of reference spectral libraries using muscle tissues from the orders of Salmoniformes, Siluriformes, Perciformes, Zeiformes, Pleuronectiformes, and Gadiformes, covering a total of 22 species. The method, instead of selecting specific spectra acquired in the first step for *de novo* sequencing and comparison between sequences, uses all acquired spectra and is independent of the selection of species-specific peptide biomarkers and availability of genomic information (Cordeiro et al., 2012). In a later study a shotgun proteomics-based method to analyze gilthead seabream muscle proteome was developed, allowing to discriminate between wild and farmed gilthead seabream (Piovesana et al., 2016), showing the importance of proteomics methodologies in fish origin traceability besides authenticity.

2.4.2 Detection of Fish Allergens

Seafood products (cartilaginous fish and bony fish) and shellfish (mollusks and crustaceans) can cause severe type I IgE-mediated reactions, with fish being

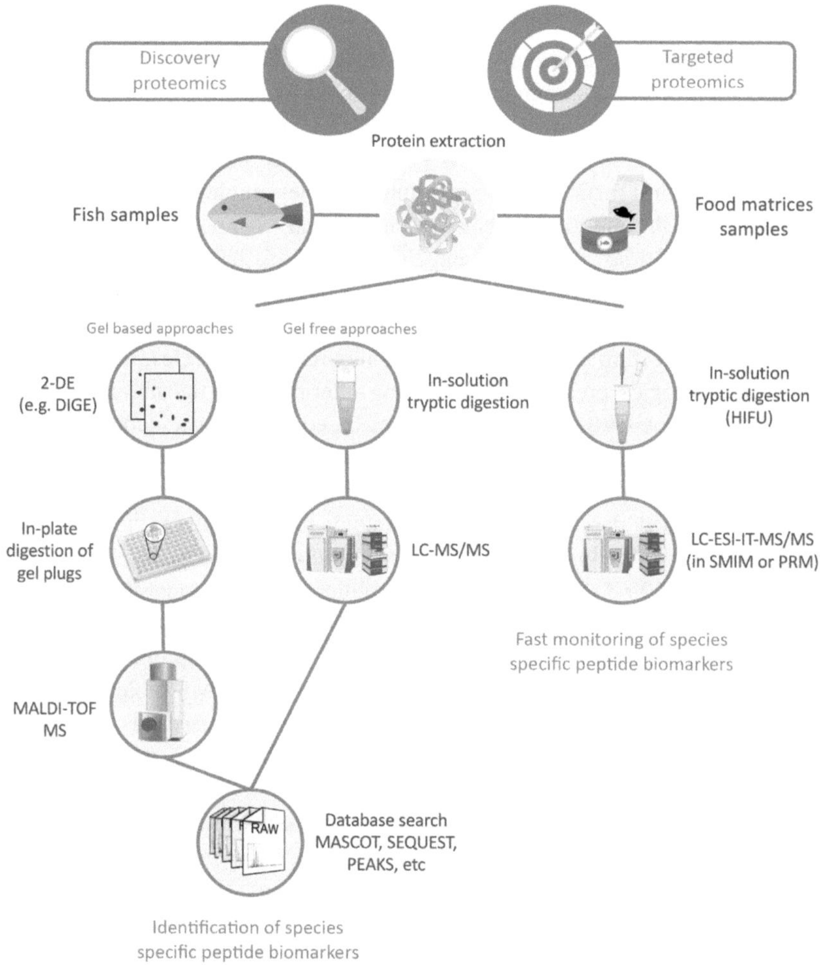

FIGURE 2.2 Proteomics workflow is used to identify and quantify fish allergens in complex food products.

one of the top eight allergenic foods (Sharp and Lopata, 2014). Seafood allergies are considered a public health problem, and to guarantee consumer's safety, regulations like "Directive 2007/68/EC" concerning the labeling of food allergens have been implemented (EFSA, 2014). Allergenic reactions to seafood products are caused by ingestion and handling and inhalation (Jeebhay et al., 2019). The only effective therapy to date for fish allergic patients is avoiding any food containing fish (Kuehn et al., 2014). Recent developments in fish allergy modulation reported that the enrichment of fish diets with molecules like ethylenediaminetetraacetic acid or creatine is promising methodologies able to reduce IgE-reactivity, are possible strategies to produce a potential hypoallergenic farmed fish (Schrama et al., 2018; De Magalhaes et al., 2020). In order to preserve the health of fish allergic patients, it is crucial to establish methodologies that can accurately, sensitively, and rapidly detect and identify seafood allergens as some products in the market may contain traces of seafood, originated most times during the manufacturing process.

Approximately 95% of fish allergies are triggered by a specific protein named β-parvalbumin (β-PRVB). In contrast, allergens such as aldolase or enolase, among others, are responsible for a minority of IgE-mediated reactions (Klueber et al., 2019). A list of the identified fish proteins with IgE reactivity is presented in Table 2.2. As observed in Table 2.2, β-parvalbumin is the primary allergen in fish, already identified in 18 different fish species. This protein is highly stable to proteolytic, chemical, and thermal denaturation, and it is expressed in fish muscle, presenting several isoforms differing in their allergenicity. The detection of these fish allergens can be made using a more classical approach, that is, through ELISA (enzyme-linked immune sorbent assay), implying their separation by 2-DE and further identification by MS. Isolated and identified allergens can then be used in the production of monoclonal and polyclonal antibodies, required for the elaboration of the ELISA assay. However, this methodology can compromise the immunoreactivity of the ELISA, as proteins can suffer structural alterations during food processing. Hence MS-based proteomics technologies are much better suited to detect fish allergens, also presenting the advantage of seeing multiple allergens at the same time.

The most commonly used proteomics workflow to identify and quantify fish allergens in complex food products comprises two phases: a discovery phase and a target-driven phase (Fig. 2.2) (Carrera et al., 2010, 2011). In the discovery phase the proteome of the chosen fish species is characterized to identify the allergenic proteins. A bottom-up approach is usually used with the protein separation step being done either by a gel-based or gel-free process, also named "shotgun proteomics." The more conventional gel-based method (2-DE) is usually used for nonsequenced organisms with protein identification based on peptide comparison from related species or *de novo* MS sequencing. Both approaches must be followed by a validation step in which allergy relevance of the identified proteins is evaluated, usually by IgE interaction studies. These identified peptides selected in the discovery phase, specific to each one of the protein allergens, are then used in the target-driven stage of the workflow, in which they are scanned with high sensitivity, accuracy, and reproducibility in a so-called monitoring mode (Borras and Sabido, 2017). This workflow was the basis for developing a rapid detection method of β-PRVBs in foodstuffs, including processed and precooked products (Carrera et al., 2012). The process monitors the several β-PRVBs peptide biomarkers identified in the discovery phase, using selected MS/MS ion monitoring (SMIM) in a linear ion trap mass spectrometer, allowing for its rapid detection in less than 2 h. More recently a new method based on LC-MS/MS and LC-MRM-MS/MS was developed to identify and quantify the β-PRB of flounder (*Paralichthys olivaceus*). This newly

TABLE 2.2 Fish Allergens Retrieved from the World Health Organization and International Union of Immunological Societies (http://www.allergen.org/) and Accessed on 11th October 2020

Fish Protein	Molecular Weight (kDa)	Function	Source
Aldolase A	40	Glycolysis	Atlantic cod Striped catfish Atlantic salmon Yellowfin tuna
Collagen alpha	13–140	Structural	Barramundi Atlantic salmon
Creatine Kinase	43	Metabolism	Striped catfish Atlantic salmon
β-Enolase	50	Glycolysis	Atlantic cod Striped catfish Atlantic salmon Yellowfin tuna
Glucose 6-phosphate isomerase	60	Glycolysis	Striped catfish
Glycealdehyde-3-phosphatedehydrogenase	36	Glycolysis	Striped catfish
L-lactate dehydrogenase	34	Metabolism	Striped catfish
Parvalbumin	11–12	Ca^{2+} binding	Indian mackerel Atlantic mackerel
β-Parvalbumin	9–13	Ca^{2+} binding	Atlantic herring Grass carp Common carp Baltic cod Atlantic cod Barramundi Megrim Whiff, Turbot Rainbow trout Striped catfish Atlantic salmon Pacific pilchard Ocean perch Redfish Snapper Yellowfin tuna Swordfish
Pyruvate kinase PKM-like	65	Metabolism	Striped catfish
Triosephosphate isomerase	28	Glycolysis	Striped catfish Atlantic salmon
Tropomyosin	33–37	Structural	Mozambique tilapia Striped catfish Atlantic salmon
β-prime-component of vitellogenin	18	Yolk protein	Chum salmon

validated method was then successfully applied to determine the presence of β-PRBS in different food matrices (Sun et al., 2019).

Overall these studies reflect the success of the applied proteomics-based methods to fish authenticity, traceability, and safety, offering industry and consumers a reliable, sensitive, and fast approach in alternative to more traditional molecular-based methods.

2.4.3 Health, a Cornerstone on Seafood Quality, Safety, and Sustainability

Aquaculture provides solutions to feed a rapidly growing human population and alleviate poverty in many countries (FAO, 2020; World Bank, 2013). In aquaculture, quality and health rules and conditions can be linked to sustainability, environmental concerns, and product consumption safety (Freitas et al., 2020). The risk of propagation of infectious or toxic agents and occurrence of disease outbreaks assessed by better disease diagnosis methodologies has increased in the last years mainly related to high-density production systems, alterations in the ecological systems caused by pollution and climatic changes, increase in the number of international transactions of aquaculture products and their derivatives, and consumption of raw or minimally processed fish (Brugere et al., 2017; Chintagari et al., 2018). This contributed to alterations in the dynamics of the relationship between organisms, infectious agents, and people that influences pathogen rates of replication and proliferation, leading to a broader geographic distribution of pathogenic agents and an increase of affected species (Brugere et al., 2017). Therefore increasing the chance of introducing possible hazards to the consumer may lead to market access exclusion and significant economic loss or elevated costs to the producer (Chintagari, et al., 2018; Aung and Chang, 2014; Trienekens and Zuurbier, 2008). To address health issues in aquaculture, several aspects related to pathologies and disease control, toxins, and other contaminants, besides zoonotic diseases, should be taken into consideration and will be further discussed in the following sections.

2.4.3.1 PATHOLOGIES AND DISEASE CONTROL

Disease outbreaks in aquaculture have become a critical factor for aquaculture production, being a severe constraint to production cycles, sustainability, and economic viability of aquaculture facilities worldwide (Shinn et al., 2015; Rigos and Katharios, 2010). Moreover the complex network of interactions on aquatic systems and their exchange with several zootechnical aspects plus possible pathogenic agents present a series of unique challenges in the health of marine organisms (Peeler and Taylor, 2011; Hedrick, 1996; Cameron, 2002; Oidtmann et al., 2014; Sitjà-Bobadilla and Oidtmann, 2017).

Regarding infectious pathologies in farmed aquatic organisms, there are the ones caused by agents present in the aquaculture habitat (or microbiota), for example *Vibrio* spp. and *Aeromonas* spp., and those derived from environmental contamination (e.g., Enterobacteriaceae, *Escherichia coli*, and *Salmonella* spp.) that can pose severe threats to fishery quality and consumer's health (Jahncke and Schwarz, 2002). To address this problem, epidemiological studies on central areas of aquatic animal health like transboundary, disease monitoring, surveillance, and surveys of the spread and impact of aquatic animal pathogens are of crucial importance (Peeler and Taylor, 2011). Also animal health surveillance should be performed with several objectives, such as disease prevalence monitorization, early detection of emerging exotic and new diseases, and infection and disease-free organisms demonstration (Oidtmann et al., 2014). Moreover, to counteract infectious diseases, biosecurity programs are crucial to prevent, control, and eradicate diseases by improving the quality management of the aquaculture operations and subsequentially of the farmed organisms (Sitjà-Bobadilla and Oidtmann, 2017; Scarfe and Palić, 2020). However, to obtain good epidemiological models, adequate animal health surveillance, and good biosecurity programs, it is necessary first to gather information regarding disease diagnosis, physiological and immunological response to pathogens, pathogen characterization, disease-related biomarkers, organism response to disease treatment, and vaccine development (Rodrigues et al., 2018; Cash, 2009). Classical, immunological, and molecular pathogen detection methods have been routinely and extensively used to address these areas (Burge et al., 2016). Proteomics is one of the most promising approaches to understand fish diseases and epidemiology better for further health management with significant benefits or the aquaculture industry and consumers (Rodrigues et al., 2012, 2016, 2018; Cash, 2009; Alves et al., 2010).

2.4.3.2 PROTEOMICS-BASED APPROACHES TO STUDY FISH RESPONSES TO PATHOGENIC AGENTS AND TREATMENTS

Proteomics is the most powerful tool that can be used for pathogen identification. In the case of several pathogenic bacteria isolates (Piamsomboon et al., 2020) or proteins associated with a bacterial infection, for example, *Aeromonas veronii* (Li et al., 2020a). It can be further used to characterize the host responses to pathogens (Ahmed et al., 2019), as the case of Pacific white leg shrimp (*Litopenaeus vannamei*) to White spot syndrome virus (Robalino et al., 2009), the immune response of marine bivalves to bacteria from the genera *Vibrio* (Huan et al., 2011), or the response of gilthead seabream (*Sparus aurata*) to the parasitic dinoflagellate *Amyloodinium ocellatum* (Moreira et al., 2017). Meanwhile, Tavares et al. (2018) addressed aspects related to the influence of abiotic factors in the pathogen virulence, as temperature effect on *Streptococcus agalactiae* virulence.

On the other hand the treatments associated with infectious pathologies may also constitute a possible hazard to farmed aquatic organisms, besides to the environment and consumers, if not properly handled (Lieke et al., 2020). One of the most studied aspects is antibiotic resistance from excessive use of antibiotics in aquaculture, imposing environmental and health risks, compromising sustainability, as recently reviewed by (Lulijwa et al., 2020). Proteomics constitutes a cutting-edge tool in this matter (Carrera et al., 2020) as indicated in the study of antibiotic resistance mechanisms (Peng et al., 2019; Su et al., 2020), in work reporting the effects of antibiotic treatments on pathogen protein regulation, the case of the use of oxytetracycline in *Aeromonas hydrophila* (Yao et al., 2016) or in the study on the impact of new therapies (Li et al., 2016). Other methods used in the combat of infectious diseases in aquaculture, such as vaccination, have been recently reported in grass carp (*Ctenopharyngodon idella*) against *Vibrio mimicus* (Li et al., 2020b).

2.4.3.3 PROTEOMICS IN THE EVALUATION OF ABIOTIC RISKS: POLLUTANTS, TOXINS, AND MICROPLASTICS

Aquatic organism's health and quality might also be seriously threatened by abiotic hazards, for example, pollutants (aromatic hydrocarbons, dioxins, and heavy metals), toxins produced during harmful algal blooms, or new environmental contaminants such as microplastics (Carrera et al., 2020; Huss, 1994). Pollutants are a well-known risk to fishery products, with well-documented effects on aquatic organisms' metabolism and a consumer (Kroon et al., 2017). Amongst other techniques, proteomics has been widely used to assess the impact of several pollutant expositions to the host organism, as the effect of copper in gilthead seabream (Isani et al., 2011) or the effect of 17β-estradiol and the estrogen mimic 4-nonylphenol in zebrafish (*Danio rerio*) (Shrader et al., 2003).

Another problem arising in fishery products is the mortalities associated with toxin contamination and toxin effects on the consumer (Tomanek, 2011). There are several examples of studies on toxins that affect fish and the consumers, as cyanotoxins like microcystin-LR (Mezhoud et al., 2008), toxins produced by the dinoflagellate *Karlodinium micrum* (Deeds et al., 2006), or studies aiming at distinguishing between the toxic and nontoxic dinoflagellate *Alexandrium tamarense* strains (Chan et al., 2006). Proteomics has been successfully employed in the course of the effects of microcystin-LR toxin in aquatic organisms (Karim et al., 2011) and the results of neurotoxin brevetoxin-1 in medaka (*Oryzias latipes*) (Tian et al., 2011).

More recently, microplastics emerged as a new and unknown pollutant that constitutes a risk in fishery products with many implications on fish safety since they can represent vectors for both chemical and biological contamination for the host and the consumer (Ziccardi et al., 2016; Naik et al., 2019). Proteomics proved to be also useful in this field (Liu et al., 2020), helping to elucidate the response of organisms after exposure to microplastics (Green et al., 2019) and in the study of the response of organisms to pollutants adsorbed n microplastics (Mu et al., 2018; Dong et al., 2018).

2.4.3.4 PROTEOMICS TO ADDRESS ZOONOTIC DISEASES

The globalization of aquaculture markets implies some risks for food safety. The associated viruses, bacteria, and parasites that infest aquatic organisms are a vital aspect of consumer's food safety and quality (Matos et al., 2017) as aforementioned. The aquaculture sector should not underestimate the ease of spread of pathogenic agents with globalization and the zoonotic capabilities of some of these microorganisms that may trigger unintended allergic reactions, especially concerning the consumers (Piras et al., 2016; Freitas et al., 2020).

In this aspect, there is a particular focus on the parasitic nematodes from the Anisakidae family, mostly belonging to the genus *Anisakis*, *Pseudoterranova, Hysterothylacium*, and *Contracaecum* since they cause a fish-borne zoonosis known as anisakidosis (anisakiasis when caused by *Anisakis* spp.) (Aibinu et al., 2019). This disease is responsible for severe economic losses in the fishing industry due to additional costs associated with inspection procedures to control and remove visible parasites (Llarena-Reino et al., 2015; Buchmann and Mehrdana, 2016; Debenedetti et al., 2019; Goffredo et al., 2019; Mercken et al., 2020; Smaldone et al., 2020; Tejada et al., 2014). In the case of *Anisakis* spp. the zoonotic capabilities of this parasite allow the accidental infestation through consumption of fish parasitized with live larvae in the third stage of development (L3) (Audicana and Kennedy, 2008; Gómez-Mateos et al., 2016), mostly when the fish is consumed raw or when the processing, storage or cooking processes are insufficient to kill the larvae (Mattiucci et al., 2018). Exposure to the parasite is mainly characterized by gastrointestinal manifestations of epigastralgia, nausea, abdominal pain, and diarrhea (Audicana and Kennedy, 2008; Shimamura et al., 2016), and also by a characteristic IgE-mediated allergic reaction causing hypersensitivity, angioedema, urticaria, or anaphylaxis (Daschner et al., 2013; Ventura et al., 2013; Nieuwenhuizen and Lopata, 2014; Mattiucci et al., 2017; Moneo et al., 2017). Other allergenic fish parasites, such as *Gymnorhynchus gigas*, can also be implicated in zoonosis (Armentia et al., 2017).

The European Union (EU), the Food and Agriculture Organization of the United Nations (FAO), and the World Health Organization (WHO) have established several legislations to overcome this problem (Freitas et al., 2020). However, the visual inspection methods specified by EU legislation and FAO guidelines, as the central procedure to detect these parasites, are neither efficient nor practical for the detection or quantitative determination of parasites in fish (Bao et al., 2019). There are already available genetic and genomic technical tools that are being used in its detection and quantitation (Smaldone et al., 2020; Arafa et al., 2019; Mostafa et al., 2020), as well as in the characterization of allergenic reactions (Palomba et al., 2019). Proteomics as the most potent tool in the evaluation of safety in fishery products (Rodrigues et al., 2012, 2017; D'Amelio et al., 2020) has also already been applied in the identification and detection of anisakids (Carrera et al., 2016; Fæste et al., 2016), proteome profiling of its life stages (Stryiński et al., 2019), and in the detection of new allergens associated with *Anisakis* sp. Fæste et al. (2014). All these molecular-based technologies should therefore be included in new legislations toward the assessment of zoonosis, as more objective, precise, and reliable when compared with those based on visual inspection only.

2.5 FINAL REMARKS

The technological advances in MS instrumentation, together with innovations in analytical and computational methods, supported by better annotated protein databases, have prompted MS-based approaches as the main driving force in proteomics research. Although aquaculture research is still far from major hallmarks in human proteomics where 90.4% of the human proteome has been cataloged (including PTMS, splice variants, interactions, and functions) (Adhikari et al., 2020), many signs of progress have already been undertaken in recent years. The use of MS-based approaches undoubtedly had a critical role in these advances in aquaculture research, as we have shown in this chapter. It is not surprising if the number of works based on proteomic tools will continue to increase in the next few years, using interdisciplinary approaches by combining proteomics with other technologies, for example, omics-based (transcriptomics and metabolomics) and advanced phenotyping techniques. The major challenge, as in human proteomics, is to translate all this information/data into real benefits to major participating partners in the aquaculture sector, not forgetting consumers and the environment itself. Herein scientists will have a primordial role as main interlocutors with farmers and the industry sector, policymakers, and consumers, toward "translational" aquaculture.

ACKNOWLEDGMENTS

This work received national funding through the Foundation for Science and Technology (FCT) through project UID/Multi/04326/2020 and projects WELFISH (Refª 16–02–05-FMP-12, "Establishment of Welfare Biomarkers in farmed fish using a proteomics approach") and ALLYFISH (Refª 16–02–01-FMP-0014, "Development of a farmed fish with reduced allergenic potential") both financed by Mar2020, in the framework of the program Portugal 2020. Márcio Moreira, Cláudia Raposo de Magalhães, and D Schrama acknowledge FCT Ph.D. fellowships SFRH/BD/118601/2016, SFRH/BD/138884/2018, and SFRH/BD/136319/2018, respectively.

REFERENCES

Abdallah, C., Dumas-Gaudot, E., Renaut, J., Sergeant, K., 2012. Gel-based and gel-free quantitative proteomics approaches at a glance. International Journal of Plant Genomics 2012, 494572.

Adhikari, S., Nice, E.C., Deutsch, E.W., Lane, L., Omenn, G.S., Pennington, S.R., et al., 2020. A high-stringency blueprint of the human proteome. Nature Communications 11 (1), 5301.

Aebersold, R., Mann, M., 2016. Mass-spectrometric exploration of proteome structure and function. Nature 537 (7620), 347–355.

Ahmed, F., Kumar, G., Soliman, F.M., Adly, M.A., Soliman, H.A.M., El-Matbouli, M., et al., 2019. Proteomics for understanding pathogenesis, immune modulation and host pathogen interactions in aquaculture. Comparative Biochemistry and Physiology Part D: Genomics and Proteomics 32, 100625.

Aibinu, I.E., Smooker, P.M., Lopata, A.L., 2019. *Anisakis* nematodes in fish and shellfish-from infection to allergies. International Journal for Parasitology: Parasites and Wildlife 9, 384–393.

Almeida, A.M., Bassols, A., Bendixen, E., Bhide, M., Ceciliani, F., Cristobal, S., et al., 2014. Animal board invited review: advances in proteomics for animal and food sciences. Animal. 9 (1), 1–17.

Alves, R.N., Cordeiro, O., Silva, T.S., Richard, N., de Vareilles, M., Marino, G., et al., 2010. Metabolic molecular indicators of chronic stress in gilthead seabream (*Sparus aurata*) using comparative proteomics. Aquaculture 299 (1–4), 57–66.

Arafa, W.M., Hassan, A.H.A., Mahrous, L.N., Abdel-Ghany, A.E., Aboelhadid, S.M., 2019. Occurrence and molecular characterization of zoonotic *Anisakis simplex sensu stricto* and *Anisakis pegreffii* larvae in retail-marketed fish. Journal of Food Safety 39 (5), e12682.

Armentia, A., Santos, J., Serrano, Z., Martín, B., Martín, S., Barrio, J., et al., 2017. Molecular diagnosis of allergy to *Anisakis simplex* and *Gymnorhynchus gigas* fish parasites. Allergologia et Immunopathologia 45 (5), 463–472.

Ashley, P.J., 2007. Fish welfare: current issues in aquaculture. Applied Animal Behaviour Science 104 (3–4), 199–235.

Audicana, M.T., Kennedy, M.W., 2008. *Anisakis simplex*: from obscure infectious worm to inducer of immune hypersensitivity. Clinical Microbiology Reviews 21 (2), 360–379.

Aung, M.M., Chang, Y.S., 2014. Traceability in a food supply chain: safety and quality perspectives. Food Control 39, 172–184.

Bao, M., Pierce, G.J., Strachan, N.J.C., Pascual, S., González-Muñoz, M., Levsen, A., 2019. Human health, legislative and socioeconomic issues caused by the fish-borne zoonotic parasite *Anisakis*: challenges in risk assessment, Trends in Food Science & Technology, 86. pp. 298–310.

Boonanuntanasarn, S., Nakharuthai, C., Schrama, D., Duangkaew, R., Rodrigues, P.M., 2019. Effects of dietary lipid sources on hepatic nutritive contents, fatty acid composition and proteome of Nile tilapia (*Oreochromis niloticus*). Journal of Proteomics 192, 208–222.

Boonstra, R., 2013. Reality as the leading cause of stress: rethinking the impact of chronic stress in nature. Functional Ecology 27 (1), 11–23.

Borras, E., Sabido, E., 2017. What is targeted proteomics? A concise revision of targeted acquisition and targeted data analysis in mass spectrometry. Proteomics 17 (17–18). Available from: https://doi.org/10.1002/pmic.201700180.

Broom, D.M., 1991. Animal welfare: concepts and measurement. Journal of Animal Science 64 (10), 4167–4175.

Brugere, C., Onuigbo, D.M., Morgan, K.L., 2017. People matter in animal disease surveillance: challenges and opportunities for the aquaculture sector. Aquaculture 467, 158–169.

Buchmann, K., Mehrdana, F., 2016. Effects of *anisakid* nematodes *Anisakis simplex* (s.l.), *Pseudoterranova decipiens* (s.l.) and *Contracaecum osculatum* (s.l.) on fish and consumer health. Food and Waterborne Parasitology. 4, 13–22.

Buján, N., Hernández-Haro, C., Monteoliva, L., Gil, C., Magariños, B., 2015. Comparative proteomic study of *Edwardsiella tarda* strains with different degrees of virulence. Journal of Proteomics 127, 310–320.

Burge, C.A., Friedman, C.S., Getchell, R., House, M., Lafferty, K.D., Mydlarz, L.D., et al., 2016. Complementary approaches to diagnosing marine diseases: a union of the modern and the classic. Philosophical Transactions of the Royal Society B: Biological Sciences 371 (1689), 20150207.

Cameron, A., 2002. Survey Toolbox for Aquatic Animal Diseases: A Practical Manual and Software Package. The Australian Centre for International Agricultural Research, Canberra.

Carenzi, C., Verga, M., 2009. Animal welfare: review of the scientific concept and definition. Italian Journal of Animal Science 8, 21–30.

Carrera, M., Canas, B., Pineiro, C., Vazquez, J., Gallardo, J.M., 2006. Identification of commercial hake and grenadier species by proteomic analysis of the parvalbumin fraction. Proteomics 6 (19), 5278–5287.

Carrera, M., Cañas, B., Piñeiro, C., Vázquez, J., Gallardo, J.M., 2007. *De novo* mass spectrometry sequencing and characterization of species-specific peptides from nucleoside diphosphate kinase B for the classification of commercial fish species belonging to the family merlucciidae. Journal of Proteome Research 6 (8), 3070–3080.

Carrera, M., Cañas, B., Vázquez, J., Gallardo, J.M., 2010. Extensive de novo sequencing of new parvalbumin isoforms using a novel combination of bottom-up proteomics, accurate molecular mass measurement by FTICR-MS and selected MS/MS ion monitoring. Journal of Proteome Research 9 (9), 4393–4406.

Carrera, M., Canas, B., Lopez-Ferrer, D., Pineiro, C., Vazquez, J., Gallardo, J.M., 2011. Fast monitoring of species-specific peptide biomarkers using high-intensity-focused-ultrasound-assisted tryptic digestion and selected MS/MS ion monitoring. Analytical Chemistry 83 (14), 5688–5695.

Carrera, M., Canas, B., Gallardo, J.M., 2012. Rapid direct detection of the major fish allergen, parvalbumin, by selected MS/MS ion monitoring mass spectrometry. Journal of Proteomics 75 (11), 3211–3220.

Carrera, M., Gallardo, J.M., Pascual, S., González, Á.F., Medina, I., 2016. Protein biomarker discovery and fast monitoring for the identification and detection of *Anisakids* by parallel reaction monitoring (PRM) mass spectrometry. Journal of Proteomics 142, 130–137.

Carrera, M., Piñeiro, C., Martinez, I., 2020. Proteomic strategies to evaluate the impact of farming conditions on food quality and safety in aquaculture products. Foods. 9 (8), 1050.

Cash, P., 2009. Proteomics in the study of the molecular taxonomy and epidemiology of bacterial pathogens. Electrophoresis 30 (S1), S133–S141.

Castanheira, M.F., Conceição, L.E.C., Millot, S., Rey, S., Bégout, M.-L., Damsgård, B., et al., 2017. Coping styles in farmed fish: consequences for aquaculture. Reviews in Aquaculture 9 (1), 23–41.

Catherman, A.D., Skinner, O.S., Kelleher, N.L., 2014. Top down proteomics: facts and perspectives. Biochemical and Biophysical Research Communications 445 (4), 683–693.

Cerqueira, M., Millot, S., Castanheira, M.F., Felix, A.S., Silva, T., Oliveira, G.A., et al., 2017. Cognitive appraisal of environmental stimuli induces emotion-like states in fish. Scientific Reports 7 (1), 13181.

Cerqueira, M., Farinha, A., Schrama, D., Moreira, M., Raposo de Magalhâes, C., Rodrigues, P.F., 2020a. Proteomics for food authentication. In: Nollet, L.M.L., Otles, S. (Eds.), Food Analysis & Properties. CRC Press. Taylor & Francis Group, pp. 177–196.

Cerqueira, M., Millot, S., Felix, A., Silva, T., Oliveira, G.A., Oliveira, C.C.V., et al., 2020b. Cognitive appraisal in fish: stressor predictability modulates the physiological and neurobehavioural stress response in sea bass. Proceedings of the Royal Society of London. Series B 287, 20192922.

Cerqueira, M., Schrama, D., Silva, T.S., Colen, R., Engrola, S.A.D., Conceição, L.E.C., et al., 2020c. How tryptophan levels in plant-based aquafeeds affect fish physiology, metabolism and proteome. Journal of Proteomics 221, 103782.

Chan, L.L., Sit, W.-H., Lam, P.K.-S., Hsieh, D.P.H., Hodgkiss, I.J., Wan, J.M.-F., et al., 2006. Identification and characterization of a "biomarker of toxicity" from the proteome of the paralytic shellfish toxin-producing dinoflagellate *Alexandrium tamarense* (Dinophyceae). Proteomics 6 (2), 654–666.

Chintagari, S., Hazard, N., Edwards, G., Jadeja, R., Janes, M., 2018. Risks associated with fish and seafood. Preharvest Food Safety 123–142.

Conte, F.S., 2004. Stress and the welfare of cultured fish. Applied Animal Behaviour Science 86 (3–4), 205–223.

Cordeiro, O.D., Silva, T.S., Alves, R.N., Costas, B., Wulff, T., Richard, N., et al., 2012. Changes in liver proteome expression of Senegalese sole (*Solea senegalensis*) in response to repeated handling stress. Marine Biotechnology 14 (6), 714–729.

Cordero, H., Morcillo, P., Cuesta, A., Brinchmann, M.F., Esteban, M.A., 2016. Differential proteome profile of skin mucus of gilthead seabream (*Sparus aurata*) after probiotic intake and/or overcrowding stress. Journal of Proteomics 132, 41–50.

Cordero, H., Brinchmann, M.F., Cuesta, A., Esteban, M.A., 2017. Chronic wounds alter the proteome profile in skin mucus of farmed gilthead seabream. BMC Genomics 18 (1), 939.

D'Amelio, S., Lombardo, F., Pizzarelli, A., Bellini, I., Cavallero, S., 2020. Advances in omic studies drive discoveries in the biology of *anisakid* nematodes. Genes. 11 (7), 801.

Daschner, A., Fernández-Fígares, V., Valls, A., de Frutos, C., Rodero, M., Ubeira, F.M., et al., 2013. Different fish-eating habits and cytokine production in chronic urticaria with and without sensitization against the fish-parasite *Anisakis simplex*. Allergology International 62 (2), 191–201.

Dawood, M.A.O., Koshio, S., 2018. Vitamin C supplementation to optimize growth, health and stress resistance in aquatic animals. Reviews in Aquaculture 10 (2), 334–350.

De Magalhaes, C.R., Schrama, D., Fonseca, F., Kuehn, A., Morisset, M., Ferreira, S.R., et al., 2020. Effect of EDTA enriched diets on farmed fish allergenicity and muscle quality; a proteomics approach. Food Chemistry 305, 125508.

de Vareilles, M., Richard, N., Gavaia, P.J., Silva, T.S., Cordeiro, O., Guerreiro, I., et al., 2012a. Impact of dietary protein hydrolysates on skeleton quality and proteome in *Diplodus sargus* larvae. Journal of Applied Ichthyology 28 (3), 477–487.

De Vareilles, M., Conceicao, L.E., Gomez-Requeni, P., Kousoulaki, K., Richard, N., Rodrigues, P.M., et al., 2012b. Dietary lysine imbalance affects muscle proteome in zebrafish (*Danio rerio*): a comparative 2D-DIGE study. Marine Biotechnology 14 (5), 643–654.

Debenedetti, Á.L., Madrid, E., Trelis, M., Codes, F.J., Gil-Gómez, F., Sáez-Durán, S., et al., 2019. Prevalence and risk of anisakid larvae in fresh fish frequently consumed in Spain: an overview. Fishes 4 (1), 13.

Deeds, J.R., Reimschuessel, R., Place, A.R., 2006. Histopathological effects in fish exposed to the toxins from *Karlodinium micrum*. Journal of Aquatic Animal Health 18 (2), 136–148.

Di Girolamo, F., D'Amato, A., Lante, I., Signore, F., Muraca, M., Putignani, L., 2014. Farm animal serum proteomics and impact on human health. International Journal of Molecular Sciences 15, 15396–15411.

Dong, X., Qiu, X., Meng, S., Xu, H., Wu, X., Yang, M., 2018. Proteomic profile and toxicity pathway analysis in zebrafish embryos exposed to bisphenol A and di-n-butyl phthalate at environmentally relevant levels. Chemosphere 193, 313–320.

Dupree, E.J., Crimmins, B.S., Holsen, T.M., Darie, C.C., 2019. Developing well-annotated species-specific protein databases using comparative proteogenomics. In: Woods, A., Darie, C. (Eds.), Advancements of Mass Spectrometry in Biomedical Research Advances in Experimental Medicine and Biology. Springer, Cham, pp. 389–400.

Dupree, E.J., Jayathirtha, M., Yorkey, H., Mihasan, M., Petre, B.A., Darie, C.C., 2020. A critical review of bottom-up proteomics: the good, the bad, and the future of this field. Proteomes. 8 (3), 14.

Edwards, P., Zhang, W., Belton, B., Little, D.C., 2019. Misunderstandings, myths and mantras in aquaculture: its contribution to world food supplies has been systematically over reported. Marine Policy 106, 103547.

Ellis, T., Yildiz, H.Y., López-Olmeda, J., Spedicato, M.T., Tort, L., Øverli, Ø., et al., 2012. Cortisol and finfish welfare. Fish Physiology and Biochemistry 38 (1), 163–188.

EFSA, 2014. Scientific opinion on the evaluation of allergenic foods and food ingredients for labelling purposes. EFSA Journal. 12 (11), 3894.

Eslamloo, K., Akhavan, S.R., Fallah, F.J., Henry, M.A., 2014. Variations of physiological and innate immunological responses in goldfish (*Carassius auratus*) subjected to recurrent acute stress. Fish & Shellfish Immunology 37 (1), 147–153.

Estruch, G., Martinez-Llorens, S., Tomas-Vidal, A., Monge-Ortiz, R., Jover-Cerda, M., Brown, P.B., et al., 2020. Impact of high dietary plant protein with or without marine ingredients in gut mucosa proteome of gilthead seabream (*Sparus aurata*, L.). Journal of Proteomics 216, 103672.

Fabbri, E., Moon, T.W., 2016. Adrenergic signaling in teleost fish liver, a challenging path. Comparative Biochemistry and Physiology. B, Comparative Biochemistry 199 (199), 74–86.

Fæste, C.K., Jonscher, K.R., Dooper, M.M.W.B., Egge-Jacobsen, W., Moen, A., Daschner, A., et al., 2014. Characterisation of potential novel allergens in the fish parasite *Anisakis simplex*. EuPA Open Proteomics 4, 140–155.

Fæste, C.K., Moen, A., Schniedewind, B., Haug Anonsen, J., Klawitter, J., Christians, U., 2016. Development of liquid chromatography-tandem mass spectrometry methods for the quantitation of *Anisakis simplex* proteins in fish. Journal of Chromatography. A 1432, 58–72.

FAO, 2020. The State of World Fisheries and Aquaculture 2020. Sustainability in Action. FAO, Rome.

Francis, G., Makkar, H.P.S., Becker, K., 2001. Antinutritional factors present in plant-derived alternate fish feed ingredients and their effects in fish. Aquaculture 199, 197–227.

Freitas, J., Vaz-Pires, P., Câmara, J.S., 2020. From aquaculture production to consumption: freshness, safety, traceability and authentication, the four pillars of quality. Aquaculture 518, 734857.

Gesto, M., López-Patiño, M.A., Hernández, J., Soengas, J.L., Míguez, J.M., 2015. Gradation of the stress response in rainbow trout exposed to stressors of different severity: the role of brain serotonergic and dopaminergic systems. Journal of Neuroendocrinology 27 (2), 131–141.

Ghaedi, G., Keyvanshokooh, S., Mohammadi Azarm, H., Akhlaghi, M., 2016. Proteomic analysis of muscle tissue from rainbow trout (*Oncorhynchus mykiss*) fed dietary β-glucan. Iranian Journal of Veterinary Research 17 (3), 184–189.

Ghisaura, S., Anedda, R., Pagnozzi, D., Biosa, G., Spada, S., Bonaglini, E., 2014. Impact of three commercial feed formulations on farmed gilthead sea bream (*Sparus aurata*, L.) metabolism as inferred from liver and blood serum proteomics. Proteome Science 12 (1), 44.

Gillet, L.C., Navarro, P., Tate, S., Röst, H., Selevsek, N., Reiter, L., et al., 2012. Targeted data extraction of the MS/MS spectra generated by data-independent acquisition: a new concept for consistent and accurate proteome analysis. Molecular & Cellular Proteomics 11 (6).

Giménez-Candela, M., Saraiva, J.L., Bauer, H., 2020. The legal protection of farmed fish in Europe – analysing the range of EU legislation and the impact of international animal welfare standards for the fishes in European aquaculture. Derecho Animal Forum of Animal Law Studies 11 (1), 65.

Goffredo, E., Azzarito, L., Di Taranto, P., Mancini, M.E., Normanno, G., Didonna, A., et al., 2019. Prevalence of anisakid parasites in fish collected from Apulia region (Italy) and quantification of nematode larvae in flesh. International Journal of Food Microbiology 292, 159–170.

Gomes, T., Albergamo, A., Costa, R., Mondello, L., Dugo, G., 2017. Potential use of proteomics in shellfish aquaculture: from assessment of environmental toxicity to evaluation of seafood quality and safety. Current Organic Chemistry 21 (5), 402–425.

Gómez-Mateos, M., Valero, A., Morales-Yuste, M., Martín-Sánchez, J., 2016. Molecular epidemiology and risk factors for *Anisakis simplex* s.l. infection in blue whiting (*Micromesistius poutassou*) in a confluence zone of the Atlantic and Mediterranean: differences between A. simplex s.s. and *A. pegreffii*. International Journal of Food Microbiology 232, 111–116.

Gomez-Requeni, P., de Vareilles, M., Kousoulaki, K., Jordal, A.E., Conceicao, L.E., Ronnestad, I., 2011. Whole body proteome response to a dietary lysine imbalance in zebrafish *Danio rerio*. Comparative Biochemistry and Physiology Part D: Genomics and Proteomics. 6 (2), 178–186.

Green, D.S., Colgan, T.J., Thompson, R.C., Carolan, J.C., 2019. Exposure to microplastics reduces attachment strength and alters the haemolymph proteome of blue mussels (*Mytilus edulis*). Environmental Pollution 246, 423–434.

Hamza, N., Silvestre, F., Mhetli, M., Khemis, I.B., Dieu, M., Raes, M., et al., 2010. Differential protein expression profile in the liver of pikeperch (*Sander lucioperca*) larvae fed with increasing levels of phospholipids. Comparative Biochemistry and Physiology Part D: Genomics and Proteomics. 5 (2), 130–137.

Hedrick, R.P., 1996. Movement of pathogens with the international trade of live fish: problems and solutions. Revue Scientifique et Technique (International Office of Epizootics) 15 (2), 523–531.

Huan, P., Wang, H., Liu, B., 2011. Comparative proteomic analysis of challenged Zhikong scallop (*Chlamys farreri*): a new insight into the anti-Vibrio immune response of marine bivalves. Fish & Shellfish Immunology 31 (6), 1186–1192.

Huntingford, F.A., Adams, C., Braithwaite, V.A., Kadri, S., Pottinger, T.G., Sandøe, P., et al., 2006. Current issues in fish welfare. Journal of Fish Biology 68 (2), 332–372.

Huss, H.H., 1994. Assurance of Seafood Quality. Food and Agriculture Organization of the United Nations, Rome.

Isani, G., Andreani, G., Carpenè, E., Di Molfetta, S., Eletto, D., Spisni, E., 2011. Effects of waterborne Cu exposure in gilthead sea bream (*Sparus aurata*): a proteomic approach. Fish & Shellfish Immunology 31 (6), 1051–1058.

Iwama, G.K., 2007. The welfare of fish. Diseases of Aquatic Organisms 75 (2), 155–158.

Jahncke, M.L., Schwarz, M.H., 2002. Public, animal and environmental aquaculture health issues in industrialized countries. In: Jahncke, M., Garrett, E.S., Reilly, A., et al.,Public, Animal and Environmental Aquaculture Health Issues. John Wiley & Sons, New York, pp. 67–102.

Jeebhay, M.F., Moscato, G., Bang, B.E., Folletti, I., Lipinska-Ojrzanowska, A., Lopata, A. L., et al., 2019. Food processing and occupational respiratory allergy – an EAACI position paper. Allergy 74 (10), 1852–1871.

Jensen, L.B., Provan, F., Larssen, E., Bron, J.E., Obach, A., 2015. Reducing sea lice (*Lepeophtheirus salmonis*) infestation of farmed Atlantic salmon (*Salmo salar* L.) through functional feeds. Aquaculture Nutrition 21 (6), 983–993.

Jobling, M., 2015. Fish nutrition research: past, present and future. Aquaculture International 24 (3), 767–786.

Karim, M., Puiseux-Dao, S., Edery, M., 2011. Toxins and stress in fish: proteomic analyses and response network. Toxicon 57 (7–8), 959–969.

Keyvanshokooh, S., Tahmasebi-Kohyani, A., 2012. Proteome modifications of fingerling rainbow trout (*Oncorhynchus mykiss*) muscle as an effect of dietary nucleotides. Aquaculture 324–325, 79–84.

Klueber, J., Schrama, D., Rodrigues, P., Dickel, H., Kuehn, A., 2019. Fish allergy management: from component-resolved diagnosis to unmet diagnostic needs. Current Treatment Options in Allergy. 6 (4), 322–337.

Kolarevic, J., Stien, L.H., Espmark, A.M., Izquierdo-Gomez, D., Sæther, B.S., Nilsson, J., et al., 2018. Welfare Indicators for farmed Atlantic salmon: Part B–Fit for purpose OWIs for different production systems. In: Noble, C., Gismervik, K., Iversen, M.H., Kolarevic, J., Nilsson, J., Stien, L.H., et al.,Welfare Indicators for Farmed Atlantic Salmon: Tools for Assessing Fish Welfare. pp. 146–237.

Kolditz, C.I., Paboeuf, G., Borthaire, M., Esquerré, D., SanCristobal, M., Lefèvre, F., et al., 2008. Changes induced by dietary energy intake and divergent selection for muscle fat content in rainbow trout (*Oncorhynchus mykiss*), assessed by transcriptome and proteome analysis of the liver. BMC Genomics 9, 506.

Koolhaas, J.M., de Boer, S.F., Coppens, C.M., Buwalda, B., 2010. Neuroendocrinology of coping styles: towards understanding the biology of individual variation. Frontiers in Neuroendocrinology 31, 307–321.

Kroon, F., Streten, C., Harries, S., 2017. A protocol for identifying suitable biomarkers to assess fish health: a systematic review. PLoS One 12 (4), e0174762.

Kuehn, A., Swoboda, I., Arumugam, K., Hilger, C., Hentges, F., 2014. Fish allergens at a glance: variable allergenicity of parvalbumins, the major fish allergens. Frontiers in Immunology 5, 179.

Lee, P.Y., Saraygord-Afshari, N., Low, T.Y., 2020. The evolution of two-dimensional gel electrophoresis – from proteomics to emerging alternative applications. Journal of Chromatography. A 1615, 460763.

Li, W., Yao, Z., Sun, L., Hu, W., Cao, J., Lin, W., et al., 2016. Proteomics analysis reveals a potential antibiotic cocktail therapy strategy for *Aeromonas hydrophila* infection in biofilm. Journal of Proteome Research 15 (6), 1810–1820.

Li, Y., Yang, B., Tian, J., Sun, W., Wang, G., Qian, A., et al., 2020a. An iTRAQ-based comparative proteomics analysis of the biofilm and planktonic states of *Aeromonas veronii* TH0426. International Journal of Molecular Sciences 21 (4), 1450.

Li, J.-N., Zhao, Y.-T., Cao, S.-L., Wang, H., Zhang, J.-J., 2020b. Integrated transcriptomic and proteomic analyses of grass carp intestines after vaccination with a double-targeted DNA vaccine of vibrio mimicus. Fish & Shellfish Immunology 98, 641–652.

Lieke, T., Meinelt, T., Hoseinifar, S.H., Pan, B., Straus, D.L., Steinberg, C.E.W., 2020. Sustainable aquaculture requires environmental-friendly treatment strategies for fish diseases. Reviews in Aquaculture 12 (2), 943–965.

Liu, W., Zhao, Y., Shi, Z., Li, Z., Liang, X., 2020. Ecotoxicoproteomic assessment of microplastics and plastic additives in aquatic organisms: a review. Comparative Biochemistry and Physiology Part D: Genomics and Proteomics 36, 100713.

Llarena-Reino, M., Abollo, E., Regueira, M., Rodríguez, H., Pascual, S., 2015. Horizon scanning for management of emerging parasitic infections in fishery products. Food Control 49, 49–58.

Lopez-Pedrouso, M., Lorenzo, J.M., Gagaoua, M., Franco, D., 2020. Current trends in proteomic advances for food allergen analysis. Biology. 9 (9), 247.

Ludwig, C., Gillet, L., Rosenberger, G., Amon, S., Collins, B.C., Aebersold, R., 2018. Data-independent acquisition-based SWATH-MS for quantitative proteomics: a tutorial. Molecular Systems Biology 14 (8), e8126.

Lulijwa, R., Rupia, E.J., Alfaro, A.C., 2020. Antibiotic use in aquaculture, policies and regulation, health and environmental risks: a review of the top 15 major producers. Reviews in Aquaculture 12 (2), 640–663.

Madaro, A., Fernö, A., Kristiansen, T.S., Olsen, R.E., Gorissen, M., Flik, G., et al., 2016. Effect of predictability on the stress response to chasing in Atlantic salmon (*Salmo salar* L.) parr. Physiology & Behavior 153, 1–6.

Mahanty, A., Purohit, G.K., Banerjee, S., Karunakaran, D., Mohanty, S., Mohanty, B.P., 2016. Proteomic changes in the liver of *Channa striatus* in response to high temperature stress. Electrophoresis 37 (12), 1704–1717.

Marco-Ramell, A., De Almeida, A.M., Cristobal, S., Rodrigues, P., Roncada, P., Bassols, A., 2016. Proteomics and the search for welfare and stress biomarkers in animal production in the one-health context. Molecular Biosystems 12, 2024–2035.

Marcus, K., Lelong, C., Rabilloud, T., 2020. What room for two-dimensional gel-based proteomics in a shotgun proteomics world? Proteomes. 8 (3), 17.

Martin, S.A.M., Vilhelmsson, O., Médale, F., Watt, P., Kaushik, S., Houlihan, D.F., 2003. Proteomic sensitivity to dietary manipulations in rainbow trout. Biochimica et Biophysica Acta 1651 (1–2), 17–29.

Martos-Sitcha, J.A., Mancera, J.M., Prunet, P., Magnoni, L.J., 2020. Editorial: welfare and stressors in fish: challenges facing aquaculture. Frontiers in Physiology 11, 162.

Matos, E., Silva, T.S., Wulff, T., Valente, L.M.P., Sousa, V., Sampaio, E., et al., 2013. Influence of supplemental maslinic acid (olive-derived triterpene) on the post-mortem muscle properties and quality traits of gilthead seabream. Aquaculture 396–399, 146–155.

Matos, E., Dias, J., Dinis, M.T., Silva, T.S., 2017. Sustainability versus quality in gilthead seabream (*Sparus aurata* L.) farming: are trade-offs inevitable? Reviews in Aquaculture 9 (4), 388–409.

Mattiucci, S., Colantoni, A., Crisafi, B., Mori-Ubaldini, F., Caponi, L., Fazii, P., et al., 2017. IgE sensitization to *Anisakis pegreffii* in Italy: Comparison of two methods for the diagnosis of allergic anisakiasis. Parasite Immunology 39 (7), e12440.

Mattiucci, S., Cipriani, P., Levsen, A., Paoletti, M., Nascetti, G., 2018. Molecular epidemiology of *Anisakis* and anisakiasis: an ecological and evolutionary road map. In: Rollinson, D., Stothard, J.R. (Eds.), Advances in Parasitology, vol. 99. Academic Press, Cambridge, MA, pp. 93–263.

Mazzeo, M.F., De Giulio, B., Guerriero, G., Ciarcia, G., Malorni, A., Russo, G.L., et al., 2008. Fish authentication by MALDI-TOF mass spectrometry. Journal of Agricultural and Food Chemistry 56 (23), 11071–11076.

Mercken, E., Van Damme, I., Serradell, A., Gabriël, S., 2020. Presence of Anisakidae in commercial fish species imported into the Belgian food markets: a systematic review and *meta*-analyses. International Journal of Food Microbiology 318, 108456.

Mezhoud, K., Bauchet, A.L., Château-Joubert, S., Praseuth, D., Marie, A., François, J.C., et al., 2008. Proteomic and phosphoproteomic analysis of cellular responses in medaka fish (*Oryzias latipes*) following oral gavage with microcystin-LR. Toxicon 51 (8), 1431–1439.

Moneo, I., Carballeda-Sangiao, N., González-Muñoz, M., 2017. New perspectives on the diagnosis of allergy to *Anisakis* spp. Current Allergy and Asthma Reports 17 (5), 27.

Monroe, A.A., Zhang, H., Schunter, C., Ravasi, T., 2020. Probing SWATH-MS as a tool for proteome level quantification in a nonmodel fish. Molecular Ecology Resources 20, 1647–1657.

Moreira, M., Schrama, D., Soares, F., Wulff, T., Pousão-Ferreira, P., Rodrigues, P., 2017. Physiological responses of reared sea bream (*Sparus aurata* Linnaeus, 1758) to an *Amyloodinium ocellatum* outbreak. Journal of Fish Diseases 40, 1545–1560.

Morzel, M., Chambon, C., Lefèvre, F., Paboeuf, G., Laville, E., 2006. Modifications of trout (*Oncorhynchus mykiss*) muscle proteins by preslaughter activity. Journal of Agricultural and Food Chemistry 54 (8), 2997–3001.

Mostafa, E., Omar, M., Hassan, S.S., Samir, M., 2020. Occurrence and molecular identification of *Anisakis* larval type 1 (Nematoda: Anisakidae) in marketed fish in Egypt. Journal of Parasitic Diseases 44 (3), 536–545.

Mu, X., Huang, Y., Li, J., Yang, K., Yang, W., Shen, G., et al., 2018. New insights into the mechanism of phthalate-induced developmental effects. Environmental Pollution 241, 674–683.

Naderi, M., Keyvanshokooh, S., Salati, A.P., Ghaedi, A., 2017a. Effects of chronic high stocking density on liver proteome of rainbow trout (*Oncorhynchus mykiss*). Fish Physiology and Biochemistry 43 (5), 1373–1385.

Naderi, M., Keyvanshokooh, S., Salati, A.P., Ghaedi, A., 2017b. Proteomic analysis of liver tissue from rainbow trout (*Oncorhynchus mykiss*) under high rearing density after administration of dietary vitamin E and selenium nanoparticles. Comparative Biochemistry and Physiology Part D: Genomics and Proteomics. 22, 10–19.

Naderi, M., Keyvanshokooh, S., Ghaedi, A., Salati, A.P., 2018. Effect of acute crowding stress on rainbow trout (*Oncorhynchus mykiss*): a proteomics study. Aquaculture 495, 106–114.

Naik, R.K., Naik, M.M., D'Costa, P.M., Shaikh, F., 2019. Microplastics in ballast water as an emerging source and vector for harmful chemicals, antibiotics, metals, bacterial pathogens and HAB species: a potential risk to the marine environment and human health. Marine Pollution Bulletin 149, 110525.

Nakharuthai, C., Rodrigues, P.M., Schrama, D., Kumkhong, S., Boonanuntanasarn, S., 2020. Effects of different dietary vegetable lipid sources on health status in Nile tilapia (*Oreochromis niloticus*): haematological Indices, immune response parameters and plasma proteome. Animals. 10 (8), 1377.

Nieuwenhuizen, N.E., Lopata, A.L., 2014. Allergic reactions to *Anisakis* found in fish. Current Allergy and Asthma Reports 14 (8), 455.

Nuez-Ortin, W.G., Carter, C.G., Wilson, R., Cooke, I., Nichols, P.D., 2016. Preliminary validation of a high docosahexaenoic acid (DHA) and alpha-linolenic acid (ALA) dietary oil blend: tissue fatty acid composition and liver proteome response in Atlantic salmon (*Salmo salar*) Smolts. PLoS One 11 (8), e0161513.

Nuez-Ortín, W.G., Carter, C.G., Nichols, P.D., Cooke, I.R., Wilson, R., 2018. Liver proteome response of pre-harvest Atlantic salmon following exposure to elevated temperature. BMC Genomics 19 (1), 133.

Oidtmann, B.C., Peeler, E.J., Thrush, M.A., Cameron, A.R., Reese, R.A., Pearce, F.M., et al., 2014. Expert consultation on risk factors for introduction of infectious pathogens into fish farms. Preventive Veterinary Medicine 115 (3–4), 238–254.

Palomba, M., Paoletti, M., Colantoni, A., Rughetti, A., Nascetti, G., Mattiucci, S., 2019. Gene expression profiles of antigenic proteins of third stage larvae of the zoonotic nematode *Anisakis* pegreffii in response to temperature conditions. Parasite 26, 52.

Patel, D.M., Brinchmann, M.F., 2017. Skin mucus proteins of lumpsucker (*Cyclopterus lumpus*). Biochemistry and Biophysics Reports 9, 217–225.

Pédron, N., Artigaud, S., Infante, J.L.Z., Le Bayon, N., Charrier, G., Pichereau, V., et al., 2017. Proteomic responses of European flounder to temperature and hypoxia as interacting stressors: differential sensitivities of populations. The Science of the Total Environment 586, 890–899.

Peeler, E.J., Taylor, N.G.H., 2011. The application of epidemiology in aquatic animal health –opportunities and challenges. Veterinary Research 42 (1), 94.

Peng, B., Li, H., Peng, X., 2019. Proteomics approach to understand bacterial antibiotic resistance strategies. Expert Review of Proteomics 16 (10), 829–839.

Piamsomboon, P., Jaresitthikunchai, J., Hung, T.Q., Roytrakul, S., Wongtavatchai, J., 2020. Identification of bacterial pathogens in cultured fish with a custom peptide database constructed by matrix-assisted laser desorption/ionization time-of-flight mass spectrometry (MALDI-TOF MS). BMC Veterinary Research 16 (1), 52.

Piñeiro, C., Vázquez, J., Marina, A.I., Barros-Velázquez, J., Gallardo, J.M., 2001. Characterization and partial sequencing of species-specific sarcoplasmic polypeptides from commercial hake species by mass spectrometry following two-dimensional electrophoresis. Electrophoresis 22, 1545–1552.

Piovesana, S., Capriotti, A.L., Caruso, G., Cavaliere, C., La Barbera, G., Chiozzi, R.Z., et al., 2016. Labeling and label free shotgun proteomics approaches to characterize muscle tissue from farmed and wild gilthead sea bream (*Sparus aurata*). Journal of Chromatography. A 1428, 193–201.

Piras, C., Roncada, P., Rodrigues, P.M., Bonizzi, L., Soggiu, A., 2016. Proteomics in food: quality, safety, microbes, and allergens. Proteomics 16 (5), 799–815.

Pottinger, T.G., 2008. The stress response in fish-mechanisms, effects and measurement. In: Branson, E.J. (Ed.), Fish Welfare. Blackwell Publishing Ltd, Oxford, pp. 32–48.

Prabu, E., Felix, S., Felix, N., Ahilan, B., Ruby, P., 2017. An overview on significance of fish nutrition in aquaculture industry. International Journal of Fisheries and Aquatic Studies 5 (6), 349–355.

Raposo de Magalhães, C.S.F., Cerqueira, M.A.C., Schrama, D., Moreira, M.J.V., Boonanuntanasarn, S., Rodrigues, P.M.L., 2020a. A Proteomics and other Omics approach in the context of farmed fish welfare and biomarker discovery. Reviews in Aquaculture 12 (1), 122–144.

Raposo de Magalhães, C., Schrama, D., Farinha, A.P., Revets, D., Kuehn, A., Planchon, S., et al., 2020b. Protein changes as robust signatures of fish chronic stress: a proteomics approach to fish welfare research. BMC Genomics 21 (1), 309.

Richard, N., Fernandez, I., Wulff, T., Hamre, K., Cancela, L., Conceicao, L.E., et al., 2014. Dietary supplementation with vitamin K affects transcriptome and proteome of Senegalese sole, improving larval performance and quality. Marine Biotechnology 16 (5), 522–537.

Richard, N., Silva, T.S., Wulff, T., Schrama, D., Dias, J.P., Rodrigues, P.M., et al., 2016. Nutritional mitigation of winter thermal stress in gilthead seabream: associated metabolic pathways and potential indicators of nutritional state. Journal of Proteomics 142, 1–14.

Rigos, G., Katharios, P., 2010. Pathological obstacles of newly-introduced fish species in Mediterranean mariculture: a review. Reviews in Fish Biology and Fisheries 20 (1), 47–70.

Robalino, J., Carnegie, R.B., O'Leary, N., Ouvry-Patat, S.A., de la Vega, E., Prior, S., et al., 2009. Contributions of functional genomics and proteomics to the study of immune responses in the Pacific white leg shrimp *Litopenaeus vannamei*. Veterinary Immunology and Immunopathology 128 (1), 110–118.

Rodrigues, P.M., Silva, T.S., Dias, J., Jessen, F., 2012. Proteomics in aquaculture: applications and trends. Journal of Proteomics 75 (14), 4325–4345.

Rodrigues, P.M., Schrama, D., Campos, A., Osório, H., Freitas, M., 2016. Applications of proteomics in aquaculture. In: Salekdeh, G.H. (Ed.), Agricultural Proteomics Volume 1: Crops, Horticulture, Farm Animals, Food, Insect and Microorganisms. Springer International Publishing, Cham, pp. 165–199.

Rodrigues, P.M., Campos, A., Kuruvilla, J., Schrama, D., Cristobal, S., 2017. Proteomics in aquaculture: quality and safety. Proteomics in Food Science. Elsevier, Amsterdam, pp. 279–295.

Rodrigues, P.M., Martin, S.A.M., Silva, T.S., Boonanuntanasarn, S., Schrama, D., Moreira, M., et al., 2018. Proteomics in fish and aquaculture research. In: de Almeida, A.M.,

Eckersall, D., Miller, I. (Eds.), Proteomics in Domestic Animals: from Farm to Systems Biology. Springer International Publishing, Cham, pp. 311–338.

Rufino-Palomares, E., Reyes-Zurita, F.J., Fuentes-Almagro, C.A., de la Higuera, M., Lupianez, J.A., Peragon, J., 2011. Proteomics in the liver of gilthead sea bream (*Sparus aurata*) to elucidate the cellular response induced by the intake of maslinic acid. Proteomics 11 (16), 3312–3325.

Sanahuja, I., Ibarz, A., 2015. Skin mucus proteome of gilthead sea bream: a non-invasive method to screen for welfare indicators. Fish & Shellfish Immunology 46 (2), 426–435.

Scarfe, A.D., Palić, D., 2020. Aquaculture biosecurity: practical approach to prevent, control, and eradicate diseases. Aquaculture Health Management. Elsevier, Amsterdam, pp. 75–116.

Schrama, D., Richard, N., Silva, T.S., Figueiredo, F.A., Conceicao, L.E., Burchmore, R., et al., 2017. Enhanced dietary formulation to mitigate winter thermal stress in gilthead sea bream (*Sparus aurata*): a 2D-DIGE plasma proteome study. Fish Physiology and Biochemistry 43 (2), 603–617.

Schrama, D., Cerqueira, M., Raposo, C.S., Rosa da Costa, A.M., Wulff, T., Goncalves, A., et al., 2018. Dietary creatine supplementation in gilthead seabream (*Sparus aurata*): comparative proteomics analysis on fish allergens, muscle quality, and liver. Frontiers in Physiology 9, 1844.

Schreck, C.B., 2010. Stress and fish reproduction: the roles of allostasis and hormesis. General and Comparative Endocrinology 165 (3), 549–556.

Sharp, M.F., Lopata, A.L., 2014. Fish allergy: in review. Clinical Reviews in Allergy & Immunology 46 (3), 258–271.

Shimamura, Y., Muwanwella, N., Chandran, S., Kandel, G., Marcon, N., 2016. Common symptoms from an uncommon infection: gastrointestinal anisakiasis. Canadian Journal of Gastroenterology and Hepatology 2016, 5176502.

Shinn, A.P., Pratoomyot, J., Bron, J.E., Paladini, G., Brooker, E.E., Brooker, A.J., 2015. Economic costs of protistan and metazoan parasites to global mariculture. Parasitology 142 (1), 196–270.

Shrader, E.A., Henry, T.R., Greeley, M.S., Bradley, B.P., 2003. Proteomics in zebrafish exposed to endocrine disrupting chemicals. Ecotoxicology 12 (6), 485–488.

Sicherer, S.H., Sampson, H.A., 2010. Food allergy. The Journal of Allergy and Clinical Immunology 125 (2), S116–S125.

Silva, T.S., Matos, E., Cordeiro, O.D., Colen, R., Wulff, T., Sampaio, E., et al., 2012. Dietary tools to modulate glycogen storage in gilthead seabream muscle: glycerol supplementation. Journal of Agricultural and Food Chemistry 60 (42), 10613–10624.

Sinitcyn, P., Rudolph, J.D., Cox, J., 2018. Computational methods for understanding mass spectrometry–based shotgun proteomics data. Annual Review of Biomedical Data Science 1 (1), 207–234.

Sitjà-Bobadilla, A., Oidtmann, B., 2017. Integrated pathogen management strategies in fish farming. In: Jeney, G. (Ed.), Fish Diseases. Academic Press, Cambridge, MA, pp. 119–144.

Smaldone, G., Abollo, E., Marrone, R., Bernardi, C.E.M., Chirollo, C., Anastasio, A., et al., 2020. Risk-based scoring and genetic identification for anisakids in frozen fish products from Atlantic FAO areas. BMC Veterinary Research 16 (1), 65.

Smith, R.W., Cash, P., Ellefsen, S., Nilsson, G.E., 2009. Proteomic changes in the crucian carp brain during exposure to anoxia. Proteomics 9 (8), 2217–2229.

Stryiński, R., Mateos, J., Pascual, S., González, Á.F., Gallardo, J.M., Łopieńska-Biernat, E., et al., 2019. Proteome profiling of L3 and L4 *Anisakis simplex* development stages by TMT-based quantitative proteomics. Journal of Proteomics 201, 1–11.

Su, Y.-B., Kuang, S.-f, Peng, X.-x, Li, H., 2020. The depressed P cycle contributes to the acquisition of ampicillin resistance in *Edwardsiella piscicida*. Journal of Proteomics 212, 103562.

Sun, L., Lin, H., Li, Z., Sun, W., Wang, J., Wu, H., et al., 2019. Development of a method for the quantification of fish major allergen parvalbumin in food matrix via liquid chromatography-tandem mass spectrometry with multiple reaction monitoring. Food Chemistry 276, 358–365.

Sun, F., Tan, H., Li, Y., De Boevre, M., Zhang, H., Zhou, J., et al., 2020. An integrated data-dependent and data-independent acquisition method for hazardous compounds screening in foods using a single UHPLC-Q-Orbitrap run. Journal of Hazardous Materials 401, 123266.

Swirplies, F., Wuertz, S., Baßmann, B., Orban, A., Schäfer, N., Brunner, R.M., et al., 2019. Identification of molecular stress indicators in pikeperch *Sander lucioperca* correlating with rising water temperatures. Aquaculture 501, 260–271.

Tavares, G.C., Carvalho, A.F., Pereira, F.L., Rezende, C.P., Azevedo, V.A.C., Leal, C.A. G., et al., 2018. Transcriptome and proteome of fish-pathogenic *Streptococcus agalactiae* are modulated by temperature. Frontiers in Microbiology 9, 2639.

Tejada, M., Karl, H., de las Heras, C., Vidacek, S., Solas, M.T., Garcia, M.L., 2014. Does the intensity of anisakis infection affect the quality of hake muscle? Journal of Aquatic Food Product Technology 23 (3), 221–236.

Tian, L., Wang, M., Li, X., Lam, P.K.S., Wang, M., Wang, D., et al., 2011. Proteomic modification in gills and brains of medaka fish (*Oryzias melastigma*) after exposure to a sodium channel activator neurotoxin, brevetoxin-1. Aquatic Toxicology 104 (3), 211–217.

Tiedke, J., Borner, J., Beeck, H., Kwiatkowski, M., Schmidt, H., Thiel, R., et al., 2015. Evaluating the hypoxia response of ruffe and flounder gills by a combined proteome and transcriptome approach. PLoS One 10 (8), e0135911.

Tomanek, L., 2011. Environmental proteomics: changes in the proteome of marine organisms in response to environmental stress, pollutants, infection, symbiosis, and development. Annual Review of Marine Science 3 (1), 373–399.

Toni, M., Manciocco, A., Angiulli, E., Alleva, E., Cioni, C., Malavasi, S., 2019. Review: assessing fish welfare in research and aquaculture, with a focus on European directives. Animal 13 (1), 161–170.

Tort, L., Montero, D., Robaina, L., Fernandez-Palacios, H., Izquierdo, M.S., 2001. Consistency of stress response to repeated handling in the gilthead sea bream *Sparus aurata* Linnaeus, 1758. Aquaculture Research 32 (7), 593–598.

Trienekens, J., Zuurbier, P., 2008. Quality and safety standards in the food industry, developments and challenges. International Journal of Production Economics 113 (1), 107–122.

Ventura, M., Napolitano, S., Menga, R., Cecere, R., Asero, R., 2013. *Anisakis simplex* hypersensitivity is associated with chronic urticaria in endemic areas. International Archives of Allergy and Immunology 160 (3), 297–300.

Vilhelmsson, O.T., Martin, S.A., Medale, F., Kaushik, S.J., Houlihan, D.F., 2004. Dietary plant-protein substitution affects hepatic metabolism in rainbow trout (*Oncorhynchus mykiss*). British Journal of Nutrition 92 (1), 71–80.

White, P., 2017. Aquaculture Pollution: An Overview of Issues with a Focus on China, Vietnam and the Philippines. World Bank, Washington, DC.

World Bank, 2013. Fish to 2030—Prospects for Fisheries and Aquaculture. World Bank, Washington, DC.

Wulff, T., Nielsen, M.E., Deelder, A.M., Jessen, F., Palmblad, M., 2013. Authentication of fish products by large-scale comparison of tandem mass spectra. Journal of Proteome Research 12 (11), 5253–5259.

Yang, B., Wang, C., Tu, Y., Hu, H., Han, D., Zhu, X., et al., 2015. Effects of repeated handling and air exposure on the immune response and the disease resistance of gibel carp (*Carassius auratus* gibelio) over winter. Fish & Shellfish Immunology 47 (2), 933–941.

Yao, Z., Li, W., Lin, Y., Wu, Q., Yu, F., Lin, W., et al., 2016. Proteomic analysis reveals that metabolic flows affect the susceptibility of aeromonas hydrophila to antibiotics. Scientific Reports 6 (1), 39413.

Ye, C.-X., Wan, F., Sun, Z.-Z., Cheng, C.-H., Ling, R.-Z., Fan, L.-F., et al., 2016. Effect of phosphorus supplementation on cell viability, anti-oxidative capacity and comparative proteomic profiles of puffer fish (*Takifugu obscurus*) under low temperature stress. Aquaculture 452, 200–208.

Yuan, Z., Liu, S., Zhou, T., Tian, C., Bao, L., Dunham, R., et al., 2018. Comparative genome analysis of 52 fish species suggests differential associations of repetitive elements with their living aquatic environments. BMC Genomics 19, 141.

Yue, X., Huan, P., Hu, Y., Liu, B., 2018. Integrated transcriptomic and proteomic analyses reveal potential mechanisms linking thermal stress and depressed disease resistance in the turbot *Scophthalmus maximus*. Scientific Reports 8 (1), 1896.

Zhang, G., Zhang, J., Wen, X., Zhao, C., Zhang, H., Li, X., et al., 2017. Comparative iTRAQ-based quantitative proteomic analysis of *Pelteobagrus vachelli* liver under acute hypoxia: implications in metabolic responses. Proteomics 17 (17–18). Available from: https://doi.org/10.1002/pmic.201700140.

Zhang, F., Ge, W., Ruan, G., Cai, X., Guo, T., 2020. Data-independent acquisition mass spectrometry-based proteomics and software tools: a glimpse in 2020. Proteomics 20 (17–18), e1900276.

Ziccardi, L.M., Edgington, A., Hentz, K., Kulacki, K.J., Kane Driscoll, S., 2016. Microplastics as vectors for bioaccumulation of hydrophobic organic chemicals in the marine environment: a state-of-the-science review. Environmental Toxicology and Chemistry 35 (7), 1667–1676.

Chapter | three

Postharvest Practices in Small-Scale Fisheries

Shehu Latunji Akintola[1], Kafayat Adetoun Fakoya[1], Isa Olalekan Elegbede[2], Esther Odunsi (Odunayo)[3] and Toheeb Jolaosho (Lekan)[1]

[1]Fisheries Department, Lagos State University, Lagos, Nigeria, [2]Environmental Planning, Cottbus-Senftenberg Brandenburg University of Technology, Cottbus, Germany, [3]Department of Aquaculture and Fisheries Management, Federal University of Agriculture, Abeokuta, Nigeria

3.1 INTRODUCTION

Globally, increasing food production at any system to feed the increasing world population has to be done not to upset the planet further. According to the United Nations, the projection indicates that the global population could grow to around 8.5 billion in 2030, 9.7 billion in 2050, and 10.9 billion in 2100. The challenge for fisheries research and policy is feeding people at these different scales sustainably without increasing pressure on the already fragile aquatic systems. It means the need to give priority to postharvest practice with a special focus on the small-scale fisheries that carry the burden of food and nutrition security in different contexts.

Most of the fisherfolks activities in the small-scale fisheries are generally community-based. They can boast of cultural, traditional, and other social and ecological values in the developing and developed world. The Food and Agriculture Organization's definition of small-scale and artisanal fisheries is adopted: subsistence fishing consisting of fishing family units characterized by the low amount of capital and energy, small fishing vessels, and short fishing trips near the shore. Realistically, this definition differs among countries and contexts globally.

Small-scale fisheries may be predominantly identified as a growing division requiring laborious harvesting, postharvest, and dispersion mechanism to utilize

Sustainable Fish Production and Processing. DOI: https://doi.org/10.1016/B978-0-12-824296-4.00008-6

both oceanic and inshore aquatic stocks. This activity results in low output and low-quality products directed mainly for local use. Nevertheless, there is an opportunity to boost its performance with technological advances, particularly in postharvest practices being developed across the globe in terms of local and international contexts as modernized artisanal fisheries can be profitable with high-grade products for export markets.

Furthermore the significance of fisheries, particularly artisanal fisheries, to supply food, earnings, and livelihood for a high proportion of the world's coastal and marginalized people is often ignored in fisheries planning and financial records. It is essential as income generation by millions of folks living in coastal and village communities plays a significant role in eliminating hunger and poverty while increasing wealth, foreign monetary gains, and rural areas development. All its activities along the value chain are indispensable both at the local level and globally. They will play an essential role in reaching zero hunger by 2030 and beyond because of their food and nutrition security in different economic contexts. As opposed to industrial fisheries, where a substantial proportion of its catch is geared toward animal feed and other products, 50% of all fish eaten as food stems from artisanal fishing, and a high percentage of fish from the sector is used for direct human consumption even though this may be missing in many records. The Food and Agriculture Organization (FAO) estimates that 50 of the world's 51 million fishers are small-scale fishers and are mostly found in the developing countries.

Small-scale fisheries make a significant contribution of more than 60% to the home and territorial and market and export-based processing units in Africa. Globally, 84 million people hold down a job in fish processing, disbursement, and merchandise, and are predominantly female. FAO (2005a) estimated that 100 million people are involved in the postharvest harvest. Also, their supplementary hundreds of millions of people who are engaged in nonpermanent fishing ventures in marine regions and, more recurrently, in streams, estuaries, small inland seas and water bodies, periodic or nonpermanent ponds, swamps, and tidal plains.

The loss of food between production and consumption has attracted the attention of scholars. In sub Saharan Africa (SSA), about 5%–87% quality losses occur along entire fish value chains due to discoloration, bad weather, damage during handling and transportation, insect infestation, and spoilage depending on the fish species, leading to income losses that could exceed 32%–50%.

This chapter presents reasons for fish spoilage and, consequently, the various types of fish losses encountered from the point of view of small-scale fisheries practices globally. It also forays different technologies that have been deployed over the years in mitigating fish losses and which have defined the procedures of postharvest activities in small-scale fisheries. Recommendations in the light of adopting and implementing the SSF Guidelines as the meta governance upon which efforts to minimize fish loss through sustainable postharvest practice can be hinged upon. It emphasizes the importance of sustainable postharvest practices is germane and linked to achieving sustainable fish consumption and production and moving toward achieving the objectives of SDGs.

3.2 WHAT TO KNOW: POSTHARVEST PRACTICE

3.2.1 Fish Nutritional Composition

The nutritional composition of fresh fish and preserved amount varies from species, stages at the life cycle, feed intake, migratory swimming, and morphology as in lean or fatty fish, and sexual changes connected with spawning. Regardless, every bite of fish provides a lot of nutritional content and benefits. Fish has proved to be one of the most nutritious foodstuffs, supplying one-fifth of the protein intake of two-third of the world's population and is also the source of health-friendly oils, omega-3 (ω-3), polyunsaturated fatty acids (PUFAs), amino acids eaten fresh or preserved. The fatty acids composition of fresh, sun-dried, and smoked prawn is presented in Tables 3.1 and 3.2. Fishes and shellfishes are a rich source of many micronutrients like vitamin D and different minerals.

The composition of fish means so many to different users, and the knowledge becomes instrumental, especially with regard to postharvest practices. The processor needs to be aware of the fish's nature before he can apply the right technique. These vary from the period of Holocene through Industrial Revolution to the modern era from salting, sun drying, chilling, freezing, smoking, and canning. Interestingly, interest of taphonomic study is also attracted to knowing the composition of fish. The nutritionist wants to know the contribution of fish to the diet and health especially given incidences of noncommunicable diseases such as overweight and obesity, diabetes, cardiovascular diseases (coronary heart disease, acute myocardial infarction (AMI), and ischemic heart disease), blood pressure and hypertension. The consumer is significant and pertinent as the interest is not only in whether a particular fish tastes good, which is a matter of opinion but also in whether it is nutritious.

3.2.1.1 COMPONENT OF FISH MUSCLE

The muscle of fish is the primary concern to human and hunger eradication. Generally the nutritional composition of fish varies between species, individuals' fish, within the anatomical part of the fish, physiologically, sex differences exist, seasons, and location in this wise, scientists, researchers, and indigenous knowledge practice study what is called the proximate analysis of the fish. The three principal components of fish are water, protein, and lipid (fat or oil). Water in the extremes could make up 90% of the fish (http://www.fao.org/3/x5916e01.htm#The20principal20components20of20fish20muscle), unevenly distributed across the fish body, and usually is bounded into the protein and difficult to get rid of even under high pressure. Knowledge of the water composition of the fish is essential concerning postharvest practice. Protein in fish is said to be complete. It may range from 16% to 21%, and there are three main groups of proteins in fish, myofibrillar, sarcoplasmic, and stroma proteins, which constitute 70%–80%, 20%–30%, and 3% of total muscle proteins, respectively. Fish lipid consists of fat, cholesterol, waxes, etc.

Other components of fish include carbohydrates, vitamins, minerals, and extractives. Literature agreed that fish is low in carbohydrates and therefore

TABLE 3.1 Fatty Acid Profile of Raw, Smoked, and Sun-Dried Muscle of Penaeus Monodon (%).

Fatty Acids	Fresh	Smoked	Sun-Dried
Saturated Fatty Acids (SFA)			
C 10:0 Capric Decanoic	0.75 ± 0.02	0.33 ± 0.30	0.40 ± 0.22
C 12:0 Lauric Dodecanoic	0.62 ± 0.08	0.28 ± 0.25	0.38 ± 0.20
C 14:0 Myristic Tetradecanoic	7.69 ± 1.22	7.08 ± 0.54	8.32 ± 0.17
C 16:0 Palmitic Hexadecanoic	16.49 ± 0.06	15.11 ± 0.68	16.88 ± 1.35
C 18:0 Stearic Octadecanoic	3.09 ± 0.02	2.30 ± 0.20	3.25 ± 0.06
Σ SFA	28. 64	25.10	29.23
Monounsaturated Fatty Acids (MUFA)			
C16:1 n−7 Palmitoleic cis-9-hexadecenoic	11.66 ± 0.17	12.48 ± 1.07	11.54 ± 0.37
C18:1 n−9 Oleic cis-9-Octadecenoic	22.31 ± 0.62	27.11 ± 2.00	24.72 ± 0.95
C22:1 n−9 Erucic cis-13-Docosenoic	4.32 ± 0.03	5.38 ± 0.78	4.47 ± 0.08
Σ MUFA	38.29	44.97	40.73
Polyunsaturated Fatty Acids (PUFAs)			
C18:2 n−6 Linoleic 9−12,octadecadienoic	1.77 ± 0.11	0.86 ± 0.22	1.57 ± 0.24
C18:3 n−6 gamma Linolenic cis-6,9, 12-Octadecatrienoic	0.89 ± 0.08	0.61 ± 0.19	0.76 ± 0.06
C20:4 n−6 Arachidonic cis-5,8,11, 14-Eicosatetraenoic	1.10 ± 0.04	0.67 ± 0.20	0.95 ± 0.10
C20:5 n−3 Eicosapentaenoic cis-5,8,11,14, 17-Eicosapentanoic	17.41 ± 0.71	18.43 ± 1.01	16.95 ± 0.76
C22:6 n−3 Docosahexaenoic cis-4,7,10,13,16, 19-Docosahexaenoic	11.90 ± 0.27	9.36 ± 0.54	9.81 ± 0.53
Σ PUFA	33.07	29.93	30.04
DHA/EPA	0.68	0.51	0.58
Σ_{n-3} PUFA	29.11	27.84	26.76
Σ_{n-6} PUFA	3.75	2.14	3.28
ω−3/ ω−6	7.76	13.01	8.16
IA	0.67	0.58	0.71
IT	0.23	0.20	0.26

Values are given as means ± SD from triplicate determinations of three independent experiments.
Source: *Based on Akintola S.L., Brown A., Abdullahi B., Osowo O.D., Bello B.O., Effects of hot smoking and sun drying processes on nutritional composition of giant tiger shrimp (Penaeus monodon, Fabricius, 1798). Polish Journal of Food and Nutrition Sciences 63 (4), 227–237.*

TABLE 3.2 Amino Acids Composition in Raw, Smoked and Sun-Dried *Penaeus Monodon*.

Amino Acids	Fresh g/100 g	mg/g Protein	Smoked g/100 g	mg/g Protein	Sun-Dried g/100 g	mg/g Protein
Glycine	4.34 ± 0.01	66	3.93 ± 0.02	62	3.98 ± 0.00	58
Alanine	3.53 ± 0.01	54	3.71 ± 0.01	58	3.37 ± 0.02	49
Serine	2.77 ± 0.01a	42	2.30 ± 0.03	36	2.69 ± 0.21	39
Proline	5.17 ± 0.00	79	4.29 ± 0.01	68	4.74 ± 0.00	69
Valine	2.36 ± 0.01	36	2.54 ± 0.00	40	2.75 ± 0.00	40
Threonine	3.48 ± 0.03	53	3.71 ± 0.01	58	3.62 ± 0.01	53
Isoleucine	2.90 ± 0.00	44	2.31 ± 0.01	36	2.24 ± 0.01	33
Leucine	4.50 ± 0.03	68	4.44 ± 0.01	70	5.26 ± 0.02	77
Aspartate	6.71 ± 0.01	102	6.17 ± 0.06	97	6.03 ± 0.02	88
Lysine	4.03 ± 0.00	61	2.78 ± 0.00	44	3.63 ± 0.00	53
Glutamate	9.11 ± 0.00	138	9.40 ± 0.00	149	11.00 ± 0.00	158
Methionine	2.78 ± 0.00	42	1.85 ± 0.01	29	2.16 ± 0.00	32
Phenylalanine	2.54 ± 0.01	39	2.64 ± 0.02	42	3.14 ± 0.01	46
Histidine	1.90 ± 0.00	29	2.37 ± 0.01	37	2.65 ± 0.01	39
Arginine	4.11 ± 0.01	62	3.94 ± 0.01	62	4.86 ± 0.01	71
Tyrosine	2.21 ± 0.00	34	2.88 ± 0.00	45	2.82 ± 0.01	41
Cystine	3.37 ± 0.01	51	4.28 ± 0.02	67	3.68 ± 0.00	54
Total	65.81		63.54	1000	68.62	1000
Σ EAA	28.6		26.58		30.31	
Σ NEAA	37.21		36.96		38.31	
EAA/ NEAA	0.77		0.72		0.79	

Values are given as means ± SD from triplicate determinations of three independent experiments. N.B Except otherwise stated means were significantly different ($P < .05$) at 0.05 and 0.01 among treatments subjected to ANOVA and LSD. *PDCAAS*, Protein digestibility corrected-amino acid score at 95%.
Source: *Based on Akintola S.L., Brown A., Abdullahi B., Osowo O.D., Bello B.O., Effects of hot smoking and sun drying processes on nutritional composition of giant tiger shrimp (Penaeus monodon, Fabricius, 1798). Polish Journal of Food and Nutrition Sciences 63 (4), 227–237.*

often not accounted for in literature or reported as fiber or/and carbohydrate. Fiber and carbohydrate are often used interchangeably, fiber is a form of carbohydrate that cannot be broken down into sugar molecules, and instead, it passes through the body undigested, unlike other forms of carbohydrate which are broken down into sugar molecules. Either as digestible or indigestible, negligible carbohydrate amounts of these nutrients have been reported in fin and shellfishes by other authors. Authors have reported that vitamins and minerals are the most relevant micronutrients from fish and shellfish and stated that fish meat is a good source of the vitamin B complex. Fatty fish and shellfish are an excellent source of fat-soluble vitamins A, D, and E. Vitamins are organic and are classified as fat-soluble, for example, Vitamins A, D, E, and K. In contrast, the most relevant water-soluble vitamins found are thiamine, riboflavin, pyridoxamine, niacin, folic acid, pantothenic acid, vitamins B12 and B6, and vitamin C. Vitamins are essential in terms of postharvest as they are easily destroyed by heat, light, extended storage, cooking, and other food preparation processes.

Minerals are inorganic examples, including the essential minerals calcium, phosphorus, sodium, potassium, magnesium, manganese, sulfur, chloride, iron, iodine, fluoride, zinc, copper, selenium, chromium, and cobalt. Essentially, literature emphasized that vitamin concentration and presence differed from body parts such as the liver, and dorsal muscle, within and across fish species. Just like vitamins, minerals are also affected by the methods and practice of postharvest.

Extractives are called because they easily are extracted from fish flesh by water or water-based or nitrogenous base solutions as they are not bounded in protein structure but chemically related to ammonia. They generally contribute to the flavor of fish and are present at low concentrations. Literature has indicated that considerable variation exists both between and within species. An additional complication is how the attention of these compounds changes during storage and spoilage, which hold a very significant effect on the postharvest practices in the industry. Another important extractive substance is the bioactive peptides with many users such as antihypertensive, antithrombotic, immune-modulatory, and antioxidative properties.

In a similar vein, another important extractive substance is the nonprotein nitrogen (NPN). Compound making up NPN may be grouped as follows: (1) volatile bases (ammonia, mono-, di-, and trimethylamines), of minor significance in muscles of living fish but most essential to fish handling and processing as they are found in the typical pattern of spoilage; (2) trimethylammonium bases (trimethylamine oxide and betaines); (3) guanidine derivatives (creatine and arginine); (4) imidazole or glyoxaline derivatives (histidine, carnosine, and anserine); (5) miscellaneous (urea, amino acids, and purine derivatives).

3.2.1.2 FISH WASTES

Fish enzymes are components of fish extracted from fish discards and waste from the fishery industry, such as the offal. Fish and shellfish processing by-products are potential sources of fish enzymes, viscera being the most important source for obtaining them. Fish enzymes have many potential applications, allowing an integrated utilization of the fish industry with fish processing wastes. Joshi et al. (2013) reported that the hydrolytic enzymes from fish sauce and fish silage play in amino acids, most of the essential amino acids. reported on the current utilization of spoiled fish and fish waste such as fish hay, meal, and sauce.

3.2.2 Fish Spoilage Mechanism

Fish spoilage occurred when the fish losses its attributes for consumption through clear and not too precise conditions. Science has established that spoilage of fish, like any other food product, results from chemical, enzymatic, or microbial activities. According to Baird-Parker (2000), 25% of gross primary agricultural and fishery product loss every year is due to chemical deterioration. Microbial spoilage and losses in fish without interventions could be as high as 27.3% ± 14.3% or more, while loss can be minimized as much as 14.7% ± 11.9% with interventions.

3.2.2.1 STAGES OF SPOILAGE AND MECHANISMS

Generally, once the fish is caught or dies naturally like any other animal, it began to experience changes and can be wholesome only for a few periods. Putrefaction or spoilage sets in, which means the undesirable change in the color, texture, flavor, odor, and loss of wholesomeness due to the fish's contamination. The literature recognizes four phases in fish spoilage though the changes are affected by internal factors, which are fish species-specific and external factors the prevailing temperature. According to and as shown in Fig. 3.1 (and adaption of the work of), spoilage proceed in four phases.

Rigor mortis occurs between stages one and two. Fish spoilage results from three primary mechanisms, namely enzymatic autolysis, oxidation, microbial growth. Although many authors describe enzymatic autolysis as the starting point in the mechanism of fish spoilage, reported that glycolysis and decrease in pH precede autolysis. Carbohydrate, though low in fish muscle, plays an essential role in rigor mortis and fish spoilage. An explanation to their views, according to, is that when a fish is caught after vigorous struggles, blood glucose and a substantial quantity of tissue and liver glycogen produce energy through the glycolytic pathway. The presence of phosphorylase, which is

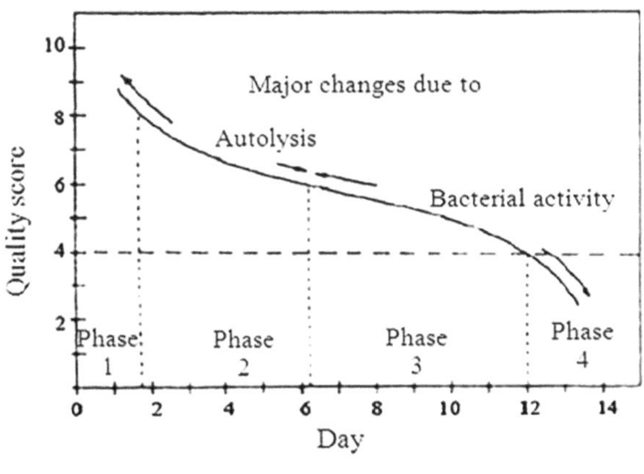

FIGURE 3.1 The four phases of fish spoilage. Based on Shawyer, M., Pizzali, M. (Eds.), 2003. The Use of Ice on Small Fishing Vessels. Food and Agriculture Organization of the United Nation, Rome, pp. 1–34.

distributed all around fish, and the almost neutral pH optimum condition, leads to lactic acid production, which is reconverted to glycogen. The dead fish lack this luxury.

3.2.2.1.1 Autolytic Enzymatic

Ghaly et al. (2010) stated that autolytic degradation could limit shelf-life and product quality even with relatively low levels of spoilage organisms. Digestive enzymes and protease in the gastrointestinal tract after that caused the softening of the stomach of the fish and shellfish, leading to rupturing stomach well and draining out of blood and water that consist of fat and protein. Some proteolytic enzymes such as cathepsins and calpains are present and influence degradation of protein and solubilization, sensorial or product alteration, in fish muscle and fish products during storage and processing.

3.2.2.1.2 Oxidative Spoilage

Oxidative spoilage is otherwise known as lipid oxidation or rancidity (chemical spoilage). It occurs in fish and shellfishes because they contain fats and oils. Its occurrence leads to deterioration of the flavors, odor, and colors making the product less acceptable. Researchers established that lipid oxidation involves a three-stage free radical mechanism: initiation, propagation, and termination. Initiation consists of forming lipid-free radicals through catalysts such as heat, metal ions, and irradiation. The termination phase occurs when a buildup of these free radicals interacts to form nonradical products. Oxidation typically involves the reaction of oxygen with the double bonds of fatty acids. Therefore fish lipids, which consist of PUFAs, are highly susceptible to oxidation.

3.2.2.1.3 Microbial Spoilage

Gram et al. (2002) stated that approximately a quarter of the global food supply is wasted or lost because of postharvest microbial spoilage. Microbial spoilage in fish is driven by two main microbes, two candidates, namely, bacteria and fungi in finfish and shellfish. Knowledge on bacteria spoilage quite has many studies since the work of introduced the concept of Specific Spoilage Organism (SSO) and subsequently the clarity on the types of bacteria involved in fish seafood spoilage. Main spoilage microorganisms in different seafood and seafood products include *Shewanella* spp., *Pseudomonas* spp., *Aeromonas* spp., *Photobacterium phosphoreum*, *Enterobacteriaceae*, and LAB. Bacterial growth also causes spoilage, leading to the formation of off-odors and off-flavors. Spoilage is typically detected when microbial counts are above 10^6–10^7 CFU/g. Mycoflora is another group of microbes that has received quite voluminous literature and attention because of their role in food spoilage. In the past, there have been confusion and debate on its uneven functions in food spoilage. They are resilient, survive, and thrives in what ordinary may prove difficult for other microbes because of their structural features and adaptive mechanism to leave on any substrate.

Fungi remain quite an enigma in the sense that they are unable to produce energy on their own that can regulate spatial and temporal flows of nutrients and energy yet they are ubiqutous and dominates virtually all their host. Smith et al. (1992) explained this rare attribute of fungi, thus that through their extensive, often, long-lived mycelial and rhizomorphic networks extending over tens or hundreds of meters.

In the context of spoilage and toxicity, understanding the role, impact, and influence mycology postharvest practices cannot be ignored. Quality assurance managers should continue to show strong interest and questions at handling, storage, transportation, and final release of fish and fishery products. Commonly found fungi in fresh or preserved fish include *Aspergillus flavus*, *Aspergillus fumigatus*, *Aspergillus glaucus*, *Aspergillus niger*, *Aspergillus restrictus*, *Aspergillus ochraceus*, *Aureobasidium* spp., *Basipetospora halophila*, *Cladosporium herbarum*, *Gleomastix* spp., *Penicillium chalybeum*, *Penicillium expansum*, *Penicillium atrovenetum*, *Penicillium glaucum. Saccharomyces cerevisiae*, *Mucor hiemalis*, *Rhizopus* sp., *Alternaria* sp., *Fusarium solani*, *Fusarium avenaceum*, *Fusarium equiseti*, *Fusarium sporotrichides*, *Mucor racemosus*, *M. hiemalis*, *Rhizopus stolonifer*, *Eurotium* sp., *Trichoderma harzianum and Rhizopus oryzae.* Detected aflatoxigenic molds isolated from fish and their products are presented in Table 3.3. There is a similar behavior in terms of specificity in bacteria and fungi regarding involvement in fish seafood spoilage identical to the concept SSO.

3.2.2.2 QUALITY ASSESSMENT OF FISH SPOILAGE

An essential quality of fish is freshness. This attribute is associated with newly caught, processed, and preserved fish and shellfish. It is apt to affirm that spoilage begins with the loss of freshness of the fish or shellfish. Adenosine triphosphate (ATP) is the primary source of energy for metabolic activity. In its absence, all biosynthetic activities come to a halt.

The role and importance of biochemical activity in spoilage have received considerable attention. Food spoilage is a complex event in which a combination of microbial and (bio)chemical activities may interact. This has become increasingly important that issue of shelf-life extention continuous to receive attention to keep the fish and shellfish wholesome from being landed, stored, preserved, transported, distributed, and finally consumed. Postharvest methods like processing and preservation, such as icing, smoking, and drying, frying, salting, freezing, and fermentation and host methods adopted to keep the fish and shellfish in good quality have proven to fall short of quality indices of healthy fish and fishery product. Preserved fish and shellfish are known to show signs of spoilage, and this counter to the critical objective of postharvest practice, which is protecting the freshness and the quality of seafood to keep the consumer and perhaps avoid legal implications.

Several chemicals, microbial, and sensory methods have been employed to assess the fish and seafood quality. Several ATP-related compounds are used as a freshness indicator. Among the chemical methods for measuring freshness the

TABLE 3.3 Incidence of Fungal Species in Examined Samples of Fishes.

Identified Mold Spp.	*Tilapia Nilotica* (40)		Salted Fish (30)		Smoked Fish (30)	
	No of +ve Samples	%	No of +ve Samples	%	No of +ve Samples	%
Aspergillus spp.	20	50	25	83.3	30	100
A. flavus	20	50	20	66.6	21	70
A. niger	16	40	22	73.3	11	36.6
A. terreus	0	0	4	13.3	-	0
Penicillium spp.	28	70	12	40.0	18	60
P. citrinum	8	20	10	33.3	10	33.3
P. expansum	6	15	10	33.3	8	26.6
Alternaria spp.	36	90	0	0.00	5	16.6
Cladosporium spp.	28	70	6	20.0	14	46.6
Rhizopus spp.	8	20	2	6.6	8	26.6
Fusarium spp.	2	5	4	13.3	3	13.3
Mucor spp.	8	20	4	13.3	6	20.0
Candida spp.	28	70	28	93.3	22	73.3
Rhodotorula spp.	28	70	24	80	20	66.6

All tables expressed as the percentage was calculated in relation to the number of examined sample.
Source: *Based on Hassan, A.A., Hassan, M.A., El Shafei, H.M., El Ahl, R.M.H.S., Abd El-Dayem, R.H. Detection of aflatoxigenic molds isolated from fish and their products and its public health significance. Nature and Science 9 (9), 106–114.*

concentrations of ATP and its breakdown products such as adenosine diphosphate (ADP), adenosine monophosphate (AMP), inosine 5′-monophosphate (IMP), inosine (Ino), and hypoxanthine (Hx) are used extensively to calculate a number of different specific indices of freshness in a wide variety of fish before bacterial spoilage commences. Also, other indices such as total volatile base nitrogen (TVB-N), tri-methyl amine (TMA), di-methyl amine (DMA), and several biogenic amines such as histamine

3.3 FOODBORNE PATHOGENS AND TOXINS

3.3.1 Biotoxins

World Health Organisation (WHO) 2019 estimated that 600 million people, almost 1 in 10 people in the world, fall ill annually from consuming contaminated food, with diarrheal diseases being the most common form of these illnesses. The WHO 2015 stated that 31 foodborne hazards had been identified as primary causative agents resulting in 33 million disability-adjusted life years

(DALYs) worldwide. DALYs is a measure of overall disease burden, expressed as the number of years lost due to ill-health, disability, or early death. Fish and fisheries products also carry foodborne pathogens from bacteria, fungi, marine biotoxins, and viruses.

The European Association for Food Safety, SAFE consortium Section 3 covers food safety as related to Emerging Threats. Tang and Mullins (2017) stated there is strong evidence that the prevalence of food allergy is increasing in western countries, and Asian and African children raised in the west have the same incidence of allergy as Caucasian children.

Pathogenic bacteria can cause illness in humans, either by infection or intoxication. In the case of foodborne diseases, they are caused by swallowing live pathogens that grow within the body. Most of the pathogenic bacteria are by contamination that occurs during the handling of the fish and fishery products. If the time and temperature conditions are favorable, these organisms get an opportunity to grow and multiply at a faster rate. Consumption of such fish is dangerous, and it will lead to food poisoning. Bacteria of public health significance in fish and fish products include *Escherichia coli* (*E. coli*), Fecal *Streptococci*, *Staphylococcus aureus*, *Salmonella* spp., *Shigella* spp., *Vibrio cholerae*, *Vibrio parahaemolyticus*, *Vibrio vulnificus*, *Listeria monocytogenes*, *Campylobacter jejuni*, *Bacillus cereus*, *Clostridium botulinum*, *Clostridium perfringens*, *Aeromonas* spp., *Plesiomonas shigelloides*, and *Yersinia enterocolitica*.

3.3.2 Mycotoxins

Mycotoxins and their derivatives are produced by toxigenic fungi/molds, which contaminate many foods. They are secondary metabolites or specialized metabolites since these compounds are not involved in the average growth, development, or reproduction of the fungi/molds. Secondary metabolites are often produced for defense, competition, or communication, but have different effects in other species. In humans the consequences of most mycotoxin exposures are slow, for example, cancer, liver, or kidney failure. Mycotoxins likely cause a lot of diseases, but these cannot be related to a single food or even to a single mycotoxin.

The difficulty of establishing cause and effect may result in a less obvious urgency to invest in mycotoxin prevention. Over 400 toxins from fungi have been mentioned. Aspergillus species are the predominant species in smoked and dried fish from tropical nations. Tropical systems and postharvest practices provide enabling conditions for the growth of fungi molds. Fungi and mycotoxin contamination have been reported in sun-dried fish products.

In wrapping up this section, it is essential to note the severe implications of fungi/molds and their mycotoxins they produced as secondary metabolites and the practice of the entire fish value chain mechanism from preharvest, harvest, and postharvest. The reasons for this warning arise from the fact that the results of most mycotoxin exposures are slow, for example, cancer, liver, or kidney failure. Mycotoxins perhaps cause a lot of diseases, but these cannot be related to a single food or even to a single mycotoxin. The difficulty of establishing cause and effect may result in a less obvious urgency to invest in mycotoxin prevention. There is a lot of evidence that mycotoxin risk is growing in developing countries such as most African countries with very ineffective food safety mechanisms and even in Europe, thus monitoring should be expanded. Aflatoxin, for example, is a mycotoxin that can be in every component of our diet—smoked, sun-dried, and perhaps fermented fishes, and shellfishes. Mycotoxins are neither destroyed by heat nor chemicals nor drying and not even time destroys them. The only practical management is to not let them into the food supply chain in the first place, meaning they must be controlled at the farm.

3.3.3 Marine Biotoxins (Aquatic Biotoxins)

Marine biotoxins are secondary metabolites produced by algae, and they accumulate in shellfish or fish. Harmful algal blooms (HAB) are natural phenomena carried out by the overgrowth of marine phytoplankton. stated that among the thousands of microalgal species known in nature, about 300 are involved in harmful events—intoxication or even death in humans and animals. It is well known that a proportion of the toxic phytoplankton has a red-brown pigmentation, giving rise to the naming of algal blooms as red tides, but not all are toxic. The increase in the occurrence and incident of HAB globally from literature are due to rising ocean temperatures and growing coastal eutrophication, with ballast waters transporting encysted algae to new environments or massive algae spreading caused by aquaculture practices.

The pathway aquatic toxins get to humans is through the shellfish or fish feeding on phytoplankton and consequently ingesting HAB, which accumulates and become concentrated in shellfish such as bivalve mollusk or fish such as puffer. Therefore intoxication by a human through any of these pathways causes an illness with symptoms ranging from mild diarrhea and vomiting to memory loss, damage to nervous or intestinal system paralysis, and death.

3.4 AN OVERVIEW OF POSTHARVEST TOWARD SUSTAINABLE FISH PRODUCTION

As mentioned in the previous sections of this chapter, fish by its composition is subject to rapid changes and consequently loses in quality as a result of spoilage. Fisheries practice through preharvest, harvest, and postharvest are critical globally given the fisheries are vital for reasons associated with social, cultural, economic, and food and nutritional security. Specifically, in this context, small-scale fishing and the practice of postharvest practice are critical in promoting sustainable development and meeting the many aspirations of nations in the sustainable development goals (SDGs). Small-scale fisheries practitioners encounter many challenges that result in their loss and hunger, their poverty and our food and nutrition insecurity, and many more.

Fisheries were seen in the past with a focus mainly on increasing technology to harvest, albeit unsustainably until the decline set in.

In order to increase the opportunity for food and nutritional security, countries must be willing to continuously develop a robust program that seeks to ensure proper monitoring and updating of the postharvest practice. As discussed in sessions on the pathogen and toxins, issues of postharvest aside from the loss of food also deals with a topic relating to food safety, wellbeing, and welfare. The sustainable postharvest loss also impacts in many ways: informed decisions on policy and interventions on postharvest practices; careful analyses of mitigations of challenges and deployment of a strategic upgrade to solve problem associate with postharvest practices; improve income and poverty alleviation that reduces livelihood vulnerabilities of practitioners; need to promote inclusiveness in terms of gender equity and youth participation; need to balance fish production in line with the implementation of the code of conduct for responsible fisheries; improved infrastructure especially road networks and electricity in rural areas especially fishing communities; improved food administration and control efforts by national bodies; and many more.

There is a need to emphasize the capacity building in the delivery of small-scale fisheries courses even at different facet knowledge production systems, especially in developing countries, especially in Africa. Fish as food should be thought of at all levels as an essential element of entire knowledge production but not an ancillary course. Increased capacity building in all areas of postharvest practices at the apex knowledge system allows a cascading impact on field workers associated with postharvest handling, processing, storage, distribution, marketing, and trade, most especially across borders' sales of fish and fishery products. In the paradigm of improving postharvest practices, improved capacity building will allow the teaching and initialization of the current practice and deep understanding of the need to pursue aggressively by all institution's ancillary to postharvest methods such as those involved quality control and administration in relation to hygiene practice at the open market. Sales of fish in the free market remain a practice in many African countries.

Institutional capacity should be promoted to drive the process of postharvest practices toward the objective of food safety in which there is a reasonable regulatory framework for general food safety and specific biotoxin obligations such as practiced in the EU Food regulations. Postharvest practice needs to reflect the reality of modern life where culture, knowledge, education, and technology have benefitted immensely from the impact of globalization. In the world of food, this chapter opined that there is really no domain to the classification of developed or developing countries. Food is an identity of the phrase a globalized world. No measure is excessive in protecting every citizen of the globalized world from a failed or ineffective postharvest practice as the consequences are usually fatal and vibrate across the geographical expression. The global food industry must be encouraged to use all forms of protection to ensure a food safety regime even when trade my redefined for this singular purpose of keeping the world safe from itself. Food safety, luckily, and hunger appear to be mutually inclusive relationships!

It appears that the form of assistance to SSF generally in a developing country is focused more on the harvest, attention to postharvest is not adequate or done within the framework of the general provision of strategies and policy to encourage investors or ancillary investment in the rural economic development of linking the rural to the urban system. There has not been a deliberate plan to ensure this economic development plan to render social or economic growth with the intentional attention for the postharvest in order to achieve linked SDGs to small-scale fisheries developments.

The presence of postharvest in the SSF Guidelines is crucial as it recognizes the rights of fishers and fish workers, acting both individually and collectively, to improve their livelihoods through enhanced value chains, postharvest operations, and trade (FAO, 2015). A comprehensive guide that helps inform national and international communities on the relevance and positions of the postharvest to sustainable development is presented in part 2 of the SSF Guidelines. This framer of the SSF Guidelines, in my view, is a deliberate attempt to explain the exact link between postharvest practice to responsible fisheries and sustainable development. Chapter 7 of the SSF Guidelines is dedicated to value chains, postharvest operations, and trade.

3.4.1 Types of Postharvest Fish Loss

Postharvest fish losses occur in value chains throughout the world. The authors are sharing the views of many stakeholders in governments, private enterprises, NGOs, development agencies, and civil society, and many further asserted that it leads to reducing the amount of fish available for the consumer. Accurate assessments of postharvest losses in small-scale fisheries are difficult to obtain, as much of the catch is unrecorded, and trade is often informal. Nonetheless, it has been estimated by FAO that 10% of the world fish catch (in live weight equivalent) is lost due to poor handling, processing, storage, and distribution. Food quality loss, because of poor handling, is the most pervasive form of loss in small-scale fisheries.

Postharvest fish loss (PHFL) attracted many definitions from literature. The underlining and commonality are that it could be in terms of losses, that is, quality, nutrient, and economic, which consequently leads to depriving consumer access to food and nutrition and loss of income to different nodes in the value chain which include the fishers, processors, traders, investors, and consumers. Regardless of how the process of spoilage is slowed down through processing and preservation steps, fish and fishery products are susceptible to spoilage.

The best way to preserve fish is by icing it and keeping it cool. Cold storage facilities and the system remained a critical challenge to the fishing communities in developing countries of Africa, South East Asia, South

America, etc. In these countries, cold storage facilities or systems remained both a technical and economic problem, especially given the broken electricity challenges.

1. *Physical loss*: This occurs when the fish does not move after capturing beyond the stage of processing and preservation. It is either thrown away accidentally, voluntarily, or as authorized. A form of physical loss can be traced to delay or inability to access the fish after being caught such that fisher may find out that the fish is spoilt when he is to harvest the fish. Small-scale fishers often result to curing or salting, sun-drying, and smoking fish resulting in a less perishable product for several weeks compared to wet fish, however, unless the processing is done very carefully, physical losses can occur, which reduce quality and quantity and a lower value to the fish processor. This is one of the limitations experienced by small-scale fishers when they depend mainly on the weather and may suffer from spoilage when the weather suddenly becomes inclement. Fish is not sufficiently dried, creating an opportunity for infestation. Physical loss can also be caused by theft, by insects eating the fish, or by bird or animal predations.
2. *Quality loss*: Principally, when deterioration or spoilage is set in, the value associated with the fish reduces as the quality remains critical and crucial. The nodes in the value chain from the fishers to the processors, aggregators, and the consumer. Small-scale fisheries are prone to the loss occasioned by low product quality. reported that the majority of the mongers in Northern Ethiopia in their study prefer to throw away fish of low quality, and very few use it for domestic use. The practice of mixing old and fresh stock of smoked and sun-dried fish also promotes loss of quality. The chances that the ancient fish in which deterioration has set in creates possibility of infesting the fresher ones.

   Small-scale fisheries also experience quality loss traceable to the methodology used to catch the fish and which has some inference on the postharvest life of the fish. The poor handling during harvest impact negatively on the quality of the fish subsequently manifesting during the smoking period and consequently during packaging and at the point of sale in that order. Generally the losses incurred at the processing stage result from burning and breaking during smoking. The consequence of this is that it cannot keep for too long before it begins to show signs of spoilage, discoloration, and a general deterioration in form and quality. In fish and shellfish, qualities attribute to check for are categorized as sensory, biochemical, chemical, physical, and microbial.

   a. *Sensory*: In this wise, stakeholders within the value chain pay attention and can easily discern changes in the fish and shellfish with attention to appearance, odor, flavor, and texture of the fish and fishery product. It involves using human sensory organs, but it is mostly subjective such as color blindness, ability to taste the rancid flavor, and low response low to cold-storage flavor. In terms of appearance, consumers will look for the brightness of the skin, bloodspot on the gill cover, stiffness, belly, and smell. They observe the color of the eyes, gills, etc. closely, in terms of shape and odor. There are many quality index methods to standardized and reduce subjectivity regarding applying scores to sensory measurements in fish and shellfish.

b. *Biochemical and chemical*: The establishment of tolerance levels of chemical spoilage indicators eliminate basing decisions on human judgments and replaces more time-consuming microbiological methods. Such objective methods should, however, correlate with sensory quality evaluations.

c. *Physical*: Postmortem changes or degree of spoilage have long been determined using electrical properties of skin and tissue change after fish death. Other physical methods used or with possibility include texture measurement, pH, and Eh.

d. *Microbiological*: Total counts, spoilage and pathogenic bacteria, and spoilage reactions are used as measurements to determine the degree of microbial spoilage, food safety, and shelf life.

3. *Market forces loss*: Market force loss occurred when fishers sell below the right price due to social exigencies (e.g., a creditor relationship, family demand, and socially constructed relationships), excess supply, and inability to keep the fish fresh if not sold at the material time.

Literature has also a description of losses aside from physical, quality, and market forces losses. Therefore, economic and nutritional losses have been mentioned. The choice made by the fishers in processing or preservation has also been indicated to promote most of these losses.

4. *Economic loss*: Postharvest practices that lead to financial loss along the stages of processing and preservation, storage and distribution, and transportation before consumption of the fish and can only be seen from the balance sheet rather than the physical, quality, and market forces losses. Although many authors have associated this form of less weight of the fish or value of fish due to incidence of insect infestation, economic loss often is not directly related to physical, quality, and market force losses and often precede these losses. Financial loss occurs in an inefficient distribution channel.

In many instances across the developing countries of Africa and Asia, mongers and traders troop to landing site awaiting the fishermen's arrival. Fresh fish, in such a situation, commands high quality as the fishes are sold without any form of preservation, even in the least, iced. Therefore uniced fish in this scenario often command a premium over iced fish, and selling iced fish may result in a lower value, in which case, there has not been a drop-in quality but the perception of the consumers.

Financial or economic loss may also be experienced in the case where processing facilities are inadequate to cope with a glut, which may be due to seasonal abundance that is experienced in small-scale fisheries. Processors may discriminate in terms of patronage of fishers to buy from since there are an abundance and facilities are inadequate, resulting in the anglers being forces to face economic loss.

5. *Losses due to insect, mite, and mold infestation*: A common fish loss that impacts the various types of fish losses during postharvest is one that usually occurred during storage and is mainly caused by insect, mite, and mold. Invariably, insect, mite, and mold infestation cause or promote physical, quality, market forces, and economic losses since the infested fish does come up with attendance loss in value, quality, and maybe entirely discarded. Among the insect which causes damages to the preserved fish and leaving its attendance damage to the efforts of fish availability and losses to the processors, mongers and traders are the flies (Order: Diptera). This order- Diptera is highly diversified with over a million species known.
6. *Nutritional loss*: Postharvest practices have an impact on fish and shellfish, which could promote loss of nutrients such as Vitamin C, thiamine (B1), and pantothenic acid are all sensitive or highly sensitive to damage by heat and are exposed to sunlight. In the case of small-sized fish, almost all vitamin A is destroyed after sun-drying.

Abraha et al. (2018) stated that fish smoking contributes to the loss of lipids and micronutrients due to the dripping of fats and more water from the fish. The loss of fat (fat phenomenon) is not peculiar to fish smoking as Oparaku and Nwaka (2013) mentioned that solar dried fish loses fat more than in smoked sample in their study on the effect of processing on the nutritional qualities of three fish species (*Synodontis clarias*, *Trachurus trecae*, and *Clarias gariepinus*). Smoking of fish at high temperatures, according to Kabahenda et al. (2009), degrades protein, reduces the functionality of essential amino acids, and may lead to loss of vital nutrients such as antioxidants. It is well known that overheating in some of the traditional ovens may reduce the availability of lysine during the initial stages of smoking with fire. Idah and Nwakwo (2013) confirmed and noted that smoking causes some decrease in available lysine and that the loss of lysine is proportional to the temperature and duration of smoking.

The impact of postharvest practices on the nutritional composition of fish, shellfish, and fisheries products has attracted a whole lot of literature. There are studies indicating that processing may even promote and enhance the nutritional value of fish instead of fostering healthy loss. An examination of some of these studies and the seemingly on ending debate on the impact of fish processing on fish and fishery products are presented in a subsection on postharvest practices aimed at improving and reversing the different types of fish loss.

3.4.2 Ways to Mitigating Postharvest Fish Loss in Small-Scale Fisheries

The objective of the SDG 12 is to ensure sustainable consumption and production pattern. The focus should have been able to avoid wastages along the values chain. Common ways to reduce postharvest fish losses include improved fish handling, development of sustainable assessment of postharvest losses at a different scale, capacity building to safeguard fish losses, improvement in the value chain of postharvest, investment in postharvest projects, adoption of the postharvest practices as outlined in the SSF Guidelines, and mainstreaming postharvest practice in achieving the SDGs.

3.4.2.1 IMPROVED FISH HANDLING PRACTICES

The word handling or phrase fish handling has been used in literature extensively in fisheries across the domains of preharvest, harvest, and postharvest, and intensively postharvest. In this chapter specifically to postharvest practices, fish handling means all the forms of contacts direct or indirect with individual or surfaces with which the fish or fishery products are in touch from harvest to consumption. This clarification is necessary because fish handling is the first process and step toward reducing postharvest fish loss. Connections are made from the point of harvesting of the fish. The fish are handled at the end of processing and preservation, storage, distribution, marketing, and trade; thus, each of the stages improper or unhygienic handling of the fish and fishery product will promote any types of fish losses.

The key phrase about handling is extra care toward hygienic practices at each node of the movement of the fish. During the unloading of fish or in hauling nets the delay must be avoided. Fish processing activities such as splitting, gutting, and cutting of fish which in most small-scale fisheries settings take place on the ground facilitating conditions that are ideal for bacterial growth and contamination, thus accelerating spoilage and loss of fish. The use of simple hand-washable surfaces for cutting the fish, regular washing down of the characters with clean water, and washing the blood stained fish with clean water that ensures that all the viscera and blood are removed are acceptable practices of fish handling at the point of processing which will promote reduced fish contamination, spoilage and losses. Pilling up waste at the processing center should be avoided.

Traditional preservation practices such as curing and smoking should be carried out to ensure the fish is thoroughly dried to reduce the susceptibility of the fish to mold growth, particularly during storage. Similarly, as presented under the section of losses due to insects, efforts should be made to cover the fish during curing and smoking toward preventing infestation driving by blowfly and other forms of insects.

It is essential to reduce waiting time in fish processing centers. A carefully balancing of harvesting and processing should be planned as waiting time before a batch could be processed will ensure that the fish does not lose its quality before processing or preservation commences. The government should find the devil in the little detail in this balancing of fish harvest and postharvest practice in the quest for promoting SDG 12, where the objective is to attain sustainable consumption and production.

In closing up this section, it is essential that good acceptable hygienic practices should be developed and a strategic plan in the processing and preservation center. The use of Hazard Analysis and Critical Control Point (HACCP) should be streamlined into part of the efforts to ensure food safety practices at the level of small-scale fisheries, more so since their importance in food and nutrition security is well emphasized. The HACCP protocol should be taking down to the domestic market and ensure that governments apply this within the context of their routine sanitation measures for markets and food production centers without necessarily pursuing it as a source of revenue.

3.4.2.2 DEVELOPMENT OF SUSTAINABLE ASSESSMENT OF POSTHARVEST LOSSES PRACTICE AT DIFFERENT SCALE LEVELS

In spite of growing literature on postharvest practices and losses, considerable quality assessment studies of losses across countries mainly in Africa, Asia, and America subcontinents, there is still the need to increase research at context levels dealing upscaling to country levels. Given that considerable achievements have been made with both quantitative and qualitative methodologies in the assessment of fish losses based on the description by: informal (qualitative and indicative quantitative data on fish losses – tools and techniques of Rapid and Participatory Appraisal) and formal questionnaire survey.

Developing countries will need to up their games by ensuring that in pursuance of SDG 12 and specifically to Target 12.3, there is the need to push and carry out an assessment of fish losses toward building a database of fish information losses from postharvest practices. This will enable their approach to solving or reducing fish postharvest losses in a much pragmatic and sustainable manner and getting a bigger picture to devolve interventions be it a technical, economical, or functional upgrade. Development of a program to actually carry out a national assessment of losses from the context and upscaling will help in firming out sustainable postharvest practices suitable to the country's need and which reflect solutions to the small-scale fisheries needs.

3.4.2.3 CAPACITY BUILDING IN PRACTICES TO SAFEGUARD FISH LOSSES

There have been considerable efforts that should be deepened to building the capacity of the small-scale fisheries operators in the use of the modern postharvest practice. There have been many of such program which has received funding and supports from development agencies which promote the training of different stakeholders in the business of fish postharvest practices. Considerable numbers of manuals are out there developed to address regional problems being faced by small-scale fisheries practitioners from harvest and postharvest traditions.

Supervisory authorities in charge of postharvest should ensure that many of these manuals are translated to local languages, and training should not be restricted to urban or semiurban centers but should be taken to the rural communities where such training, skills, and knowledge will help abate losses giving that small-scale fishers often operate far from markets. There is the need to ensure that Post-Harvest Fisheries Research Programmes are launched across many developing programs such as has been done in Asia. There are many programs that have been designed and reconceptualized to suit the context and need to be done on consistent levels. Postharvest practices require that different stakeholders play different roles and functions in the value chain. Manuals and other training materials and methodology which is explicitly deplored to enhance skills and training of players in these functions and roles need to be promoted. Training manuals and material should be developed for preharvest (fishers) in relation to their importance upstream of postharvest practice; preharvest practice manuals need to be designed to improve the skills of those involved in storage, distribution, and marketing.

The truth from the field is that output from small-scale fisheries activities is essential in regional trades, and market requirements are moving toward global standards. In the world of globalization, consumption is being defined with references and particular attention to food safety. Players in regional markets are now being required to meet the agreements. The international regulatory framework for fish safety and quality within the framework of the World Trade Organisation (WTO).

3.4.2.4 INVESTMENT IN ROAD INFRASTRUCTURE AND POSTHARVEST PROJECTS

Reducing fish and fishery product loss and waste is demanding in terms of logistics requirements. Developed economies feared better than developing countries because of a well-developed network of road infrastructure and transportation facilities. Fish storage and other activities such as transportation, distribution, marketing, and trade benefit from access to good road networks, which invariably promoted the effective and efficient movement of fish from landing sites through to the consumer. The USA and EU countries have benefitted from well-developed road infrastructure, and consequently, the use of cold technologies and mechanical refrigeration in the development of fish supply chains system across the postharvest value chain functions between wholesaling and retailing.

Developing economies generally are confronted with challenges of logistics in rushing fish and shellfish across the postharvest value chain. Limited logistics and, many times, brutal weather and environmental conditions make the seafood supply chain challenges. Nevertheless, it is clear that cold chain is felt in the megacities of the developing world where investment in road and electricity infrastructures allows rapid expansion of cold chains and storage of fish in the wholesale markets and the metropolis. Conversely, retail markets, and rural areas, with their attendance, poor logistics, have very little from the cold storage system and refrigeration.

Policymakers in fisheries must encourage the government to make an investment in infrastructural development. In some countries, a typical example of Nigeria, often the governance structure for agriculture and rural development are merged as the Ministry of Agriculture and Rural Development. The implication of such an arrangement presupposes that the whole idea is to promote the link between rural development and the movement of raw material from the rural communities to the urban center. However, the missing link is that a political arrangement such as this is not a sufficient condition for the development of mutually inclusive growth in infrastructure and reduction in fish loss and waste since there has not been a deliberate policy in place to promote and invest in postharvest especially cold chain and refrigeration systems.

The system that was fashioned out in India presents much compelling attention to what works. According to the FAO the government of India established a separate department, the National Centre for Cold-chain Development (NCCD), to focus on the cold chain development for food loss waste reduction. The sector received substantial government aid, including the encouragement of foreign direct investment, tax exemption, and duty-free imports of cold chain equipment to help strengthen cold chain infrastructure, introduce advanced technologies and robust distribution channels. The Marine Products Exports Development Authority, established to increase exports and specify standards, also subsidized the construction of extensive cold storage facilities

3.4.2.5 ADOPTION OF THE SSF GUIDELINES IN POSTHARVEST PRACTICE

The SSF Guidelines provides the umbrella instrument for the meta governance of the SSF globally and the two order of small-scale fisheries governance (domestic constituents) for the governors. As indicated by nations to walk their talk concerning the implementation of the Guidelines. As mentioned earlier, Chapter 7 of the SSF Guidelines is dedicated to value chains, postharvest operations, and trade. The typical experience with other similar international instruments, conventions, and declarations is that it takes a lot of time. Usually, a wide implementation gap exists postsigning and adoption of this instrument, SSF Guidelines holds no exception and stated that there is a need to ensure that policy aimed at equal opportunity and inclusiveness is enshrined. The SSF Guidelines which to reverse this by providing an approach which involves all stakeholder is thrown up that involves vulnerable players or promote gender inclusiveness men or women depending on the context. It is well known that women dominate postharvest activities in the SSF as they constitute not less than 90% of the workforce in fish processing and preservation, storage, and trade. However, according to certain gender constraints limit women's abilities to process fish with minimal losses exist, especially in the Barotse Floodplain fishery, of Zambia.

Section 7.5 is a call that emphasizes the multisectorial approach to dealing with postharvest issues. This is a lens that strengthens existing efforts at CSO, NGOs, Research Institutions, and Development agencies others in their role to ensure the sustainable and environmentally friendly postharvest operation is promoted and entrenched. There is the need to ensure that the actual traditional means of the processing of fish are consciously replaced with modern kiln and oven systems.

3.4.3 Postharvest Handling, Processing, and Preservation Methods

3.4.3.1 POSTHARVEST HANDLING

Small-scale fisheries operate different types of fishing craft from traditional to modern supplicated vessels. At one end of the spectrum, for example, in North America, it is common to find aluminium-hulled, twin-engine, electronics-laden salmon gill-netter operating in the small-scale fisheries and in the in West Africa, small-scale fisher's carryout fishing operations using flat-bottomed wooden pirogue with or without outboard engines.

1. *Boarding*: The fish are transferred from the gear/craft through picking or hauling into receptacles such as baskets or steel containers. This process must be handled in line with the overall objective of achieving wholesome fish as the physical loss may occur through unintentional dropping off the fish. The method may also throw up other forms of losses. Rough handling by throwing and dropping the fish should be avoided. Delays in transferring fish from the boat/collecting boat to chilled storage facilities if it exists. Typically, in developing countries, landing sites do not have the luxury of chilled storage facilities.
2. *Holding*: Depending on the context the fish may be held in the container. In small-scale artisanal fishing communities in India, Bangladesh, and Sri Lanka, a low-cost, sturdy, well-insulated box made of local materials is employed.
3. *Sorting*: The fish is often sorted by size and species. It is advisable at the landing site or jetty where sorting is being carried out; acceptable hygiene practices should be carried out. It is common to see in developing countries that this operation is carried out without recourse to reduce contamination risk. Placing fish, with ice, in clean plastic containers reduces spoilage and contamination, and these practices need to be encouraged and promoted as a pact of keeping the fish wholesome in preparation for the postharvest operation, which involves more fish handling and contact.
4. *Bleeding and gutting*: In some fisheries, bleeding is done to enhance the fish's appearance toward better market valuation. In the case of filleted fish, it is taken that bloodstains reduce the market value as the push is for white flesh. Gutting involves the removal of the fish intestines limiting access to most spoilage bacteria, and also done with gills removal. However, for small fish like pelagic species, the bleeding and gutting stage may be skipped.
5. *Washing*: After bleeding and gutting the fish is transferred to the washing stage and usually carried out in clean water to eliminate the remains of blood and viscera residues. It also keeps the temperature of fresh fish low and advisable to be carried out quickly to avoid losing the nutrition substances.

   It is always better to sell chilled fish as chilling does not necessarily prevent spoilage completely.
6. *Storing*: Cured and smoked fishes may also be stored before being transported. Smoked or dried fish should be wrapped up in wax paper or foil and stored in a refrigerator where it can remain for up to 25 days or more

depending on methods adopted in smoking the fish. To hold it longer, we have to freeze it like the case of fresh.

3.4.3.2 POSTHARVEST PROCESSING AND PRESERVATION METHODS

Fish is a highly perishable food item that has been highlighted all through this chapter. The emphasis here is to present the frequently used techniques in small-scale fisheries to ensure that the shelf life of fish and fishery products are elongated after being handling process. Efforts are made to present the craftmanship that had defined development in these technologies over the years as they stiff defined the means by which they are preserved.

1. *Drying* is the oldest method used to remove the moisture content of the fish to slow down the activities of enzymes and microorganisms in the seafood and fish. It involves the removal of water from the internal structure of the fish to the surface and from the body. Wind and sun-drying have been used and the most convenient, depending on the natural endowment of the fisheries. Open sun-drying remains common in developing countries across the world, but it comes with its disadvantage of insect infestation, rodents, and vulnerability to being devoured by domestic animals such as goats, chickens, cats, and other physical losses. It is often carried out in large areas and subject to vagaries of weather and may be carried out with just covering on the ground or floor and may also be carried out on elevated platforms for better exposure to sunlight. Sun drying is laborious as the constant attention of the processors is required and involves their coming into the sun to turn around the products for better and continuous exposure to sunlight. The nutrient adequacies of sun-dried products have been reported in the literature.

   Solar dying represents a more recent technology but remains mostly less common and adopted the larger small-scale fisheries communities, although it minimizes or obviates some of the limitations of open sun drying mentioned earlier. indicated the various forms of the solar dryer, which have been designed in Nigeria, while other formats available in tropical and subtropical countries were reported in Bala and Debnath (2012).

   Freeze-drying, also known as lyophilization, is a method which relatively recent too. It is not generally used on fish for practical and economic reasons, but there are some references in the literature to its use on tuna, Atlantic and horse mackerel, and rainbow trout (Corapci and Guneri, 2020; Sarkardei and Howell, 2007).

   Osmotic dehydration is used as a supplementary treatment in processes such as salting, smoking, and marinating. It involves placing the fish or meat in contact with a further concentrated solution in a solute such as salt, sugar, and seasonings. It reduces color loss, promotes volatiles expel, enhances taste, inhibits enzymatic browning, reduces energy cost, and increases shelf life due to the salts and sugars used in osmotic solutions. Vacuum pulses are applied to the osmotic system to reduce processing time and thus improve the mass transfer rate.

2. *Salting*: It is by far the most effective form of fish preservation in many countries of the world. It involves the use of solute, salt, and spreading to have contact with the fish, thereby removing the water and consequently lowering the water activity. Salting found an association with the discovery of salt and its subsequent elevation historically used as a method of trade and currency. Common salt originating from the land (rock salt) is used in salting.

   Methods of fish salting in small-scale fisheries around the world include dry, brine, mixing of dry and brine, salting and chilling, salting by pressing, salting by a vacuum, pickle salting, and salting by the injector. Potassium nitrate (KNO_3) has been used as additives in fish salting. Fish curing underscores the importance of salting as most preservation methods such as fish drying, smoking, and chilling do not proceed without salting, especially in the big fishes and shellfishes such as prawns, shrimp, and oyster.
3. *Fish smoking*: It is the use of thermal destruction of wood and the resulting penetrating volatiles diffusing into the fish muscle, thereby preserving and increasing the flavor, odor, color, and taste. This is one of the oldest technologies used in extending the shelf-life of fish and shellfish globally and has also witnessed globalization as it has also been influenced by changing technologies. It may be carried out or supplemented with salting and drying, either as means of preserving fresh products or extending the shelf-life of previously smoked or dried fish once fish mongers notice the tendency of spoilage. Smoking has survived because aside from being used for preservation, it has gained eminence for adding specific sensory characteristics to the fish and fisheries products, especially the odor and flavor. The thermal process of burning the wood is referred to as pyrolysis.

   Cold smoking is usually carried out between 28°C and 32°C. The technologies for cold smoking involving the smoke generation and application of smoke components have witnessed tremendous changes, and stating that the preferences and use to be smouldering 65%, thermostatic plates 30%, and friction and smoke condensates 5%. Cold smoking has received a lot of scrutiny from consumers; thus, according to European Food Safety Authority, EFSA (2008) the efforts for reducing polycyclic aromatic hydrocarbons (PAH) concentration in foods by replacing traditional smoking processes with liquid smoke flavoring is considered an excellent strategy to minimize the PAH levels in commercially smoked food commodities were of the view that the trend of lowering the salt intake in diet often, associated with healthy nutrition and reduced incidence of coronary diseases, encouraged producers to develop more lightly processed products with a relatively limited shelf life. They averred further that in order to promote safety it became essential to control the possibilities of contamination by combining smoking and salting with preservation methods such as chilling, and specific packaging methods, of which the most common is vacuum packaging.

Hot smoking is the most common method of fish preservation in developing economies of Africa, and Asia usually carried out at temperatures above 80°C. In the process of cooking the flesh of the fish is also dried by the heat generated. In the process of hot smoking, bactericidal and bacteriostatic effects at different stages of processing and storage reduce the chances for the growth of microorganisms in the product categorized hot smoking of fish with mild heat treatments as a typical example where hurdle technology is in active use. Temperature and water attributes of hot smoking are well known for the smoked fish product. Hot smoking proceeds with the help of different kiln or ovens with the different types in Nigeria reported in Fig. 3.2 shows the traditional smoking system.

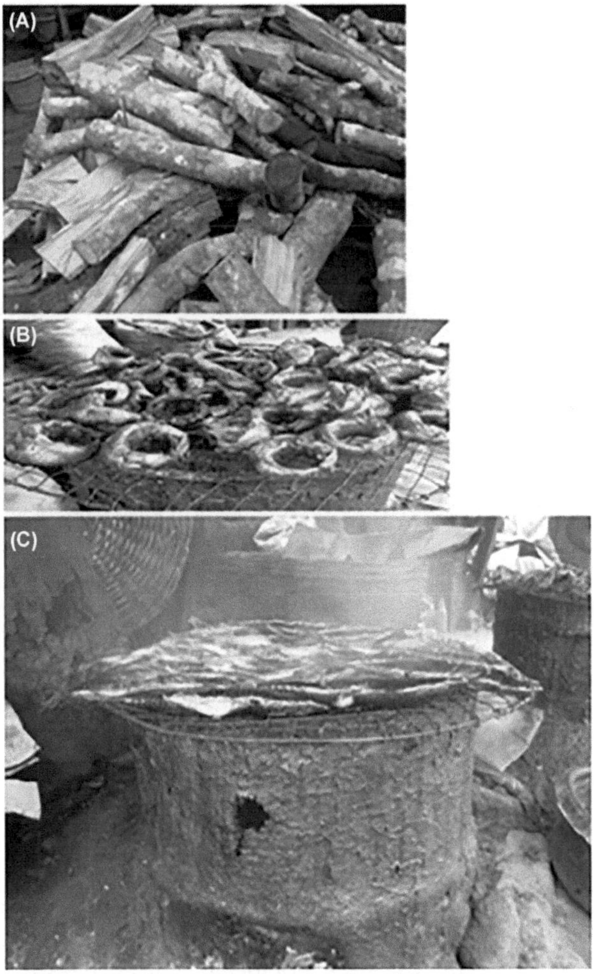

FIGURE 3.2 Typical traditional smoking systems are shown in (A) wood fuel (B) smoked fish, and (C) drum kiln.

Liquid smoking can be classified as one of the recent advancements in fish smoking and rarely used by small-scale fisheries processors. It is a new method invented in the late 1980s, which is used to reduce the presence of PAH in smoked fish. It involves the production of smoke condensate, which consists of the heating of wood that has been rinsed in water, natural hickory smoke flavor, vinegar, molasses, and Carmel color in the absence of air and collection of the smoked generated. Muscles of the fish are then dipped into the liquid condensates. Liquid smoke does have some of the same bactericidal properties as real smoke, flavor, and relatively safe as no or trace of trace amount of Benzo(a)pyrene is present. PAH is highly monitored in the European Union countries and the United States. The European Union applies the strictest restriction as a concentration within only 0.002 ppm allowance.

In the case of electrostatic smoking the fish is treated with infrared radiation. It involves the fish muscle being smoked by the creation of a positively charged electrical field while the fish is negatively charged. The electrical field works on the ionized smoke particles to accelerate the smoking process. Electrostatic smoking is fully mechanized, thus entailing savings on labor and production costs. Arason et al. (2014) stated that the electrostatic smoking process renders a higher quality in the final product compared with traditional smoking processes.

The combination of smoking methods is not commonly used, but there are studies in which these methods are combined with the objectives of providing information on the many attributes of fish that have been smoked, such as sensory attributes, bacteriostatic, and bactericidal properties.

There is also the FAO-Thiaroye processing technique (FTT-Thiaroye), which has been in African and is now being promoted in Asia, Sri Lanka. It builds on extant technologies prevailing and removing the drawback on these smoking kilns or ovens.

4. *Chilling/icing*: It is a process of cooling fish, shellfish temperature near freezing point but not below it. On the other hand, icing is one kind of chilling method, but according to Nowsad (2007), it is not adequately practiced in traditional fish handling and preservation in Bangladesh. The impacts of ice storage systems as reported by on the biochemical composition of the prawn, *Macrobrachium vollenhovenii* is shown in Table 3.4. Ice is the most widespread method of keeping fish fresh, but there are other traditional methods, for instance, placing the fish in containers with chilled water, general practice for small pelagic fish in European Nordic countries. They also reported that in Southern Europe, for some years now, small polystyrene containers with iced water have been used to transport small pelagic fish from the coast to significant city markets. Compared to other preservation methods so far discussed, it is the most recent and is commonly in small-scale fisheries of developed economies but less used in developing countries due to weak or broken systems in electricity. Different types of ice are available, including a block (which is usually contained in the nylon in Nigeria), flake, tube, and plate.

TABLE 3.4 Changes in Moisture, Ash, Crude Protein, and Total Volatile Based Nitrogen Contents of *M. Vollenhovenii* Affected by Ice Storage Conditions.

Parameter	Treatment	Storage (Day) 0	2	4	6	8	10
Moisture	DCI	77.86 ± 16.2^{Ee}	79.38 ± 20.8^{Rr}	81.76 ± 16.2^{Rr}	82.14 ± 27.2^{Tr}	83.94 ± 12.0^{Tr}	84.62 ± 29.2^{Tr}
	WCI	77.86 ± 16.2^{Ee}	77.74 ± 10.7^{Ee}	77.74 ± 16.3^{Ee}	77.16 ± 20.8^{Ee}	77.02 ± 21.0^{Ee}	77.60 ± 07.7^{Ee}
Ash(%)	DCI	1.64 ± 0.07^{Ee}	1.60 ± 0.12^{Ee}	1.62 ± 0.06^{Ee}	1.56 ± 0.08^{Ee}	1.52 ± 0.04^{Ee}	1.48 ± 0.08^{Ee}
	WCI	1.64 ± 0.07^{Ee}	1.58 ± 0.10^{Ee}	1.54 ± 0.05^{Ee}	1.60 ± 0.05^{Ee}	1.48 ± 0.15^{Ee}	1.36 ± 0.12^{Ee}
Crude protein (%)	DCI	19.12 ± 0.19^{Ee}	17.80 ± 0.50^{Rr}	17.10 ± 0.21^{Rr}	16.16 ± 0.39^{Tr}	15.22 ± 0.40^{Tr}	14.34 ± 0.42^{Tr}
	WCI	19.12 ± 0.19^{Ee}	10.30 ± 0.53^{Ee}	19.58 ± 0.37^{Ee}	19.72 ± 0.14^{Ee}	19.96 ± 0.54^{Ee}	19.49 ± 0.38^{Ee}
TVB-N (mg N/100 g)	DCI	17.35 ± 1.16^{Ee}	14.87 ± 0.94^{Rr}	11.33 ± 0.63^{Tr}	08.67 ± 0.33^{Tr}	06.88 ± 0.49^{Zr}	04.85 ± 1.06^{Er}
	WCI	17.35 ± 1.16^{Ee}	18.51 ± 2.06^{Re}	18.71 ± 0.95^{Ee}	20.17 ± 1.46^{ERe}	21.32 ± 0.49^{ERe}	21.89 ± 0.86^{Re}

Source: *Based on Akintola, S.L., Bakare, S.B. Effects of Ice Storage on the Biochemical Composition of Macrobrachium vollenhovenii (Herklots, 1857). Journal of Fisheries and Aquatic Science 8, 213–217.*

5. *Fermentation*: Fish are preserved and flavored by means of fermentation, especially in Southeast Asia. It involved the hydrolysis of fish in the presence of high salt concentration as the fish or seafood may retain their shape or the product comes out in the form of liquid or paste. Different types of fermented fish products exist with another method of preparation and shelf-life achieved. The fish or seafood may be fermented with cooked rice and mixture allowed to ferment for 7 days. In the process the product due to the action of LAB becomes acidic, thus ensuring a preservation shelf life of 2 weeks at room temperature. The fermented product prepared this way is standard in the Philippines and is called *burros*. Another development in the Philippines is the *Bagoong na isda*. In Korea the fermented fish product is called the *sik-hae* is widely consumed. The protocol for making the *sik-hae* involved, according to the eviscerating flatfish. described the fermented fish product of Indonesian called *trassi*-is, a paste made from small shrimp.

3.4.4 Recommendations

1. Countries should adopt and implement the SSF Guidelines. This is important to set the principles for the meta governance for sustainable postharvest practices toward overall arching objectives of ensuring food security and poverty are eradicated in small-scale fisheries.
2. Promote comprehensive scale studies on assessment of postharvest losses so as to generate a database of fish losses, which should be deplored as the basis for the overall strategies to sustaining food and nutritional security.
3. Any effective action to reduce losses will depend upon persuading fishers, fish processors, food regulatory agencies, extension agencies, and a whole lot of individuals involved in the management of postharvest losses and others to adopt improved practices along the postharvest value chain.
4. Sustainable postharvest practices should be mainstreamed to national the schemes of SDGs, especially SDGs 1, 2, 12.
5. Investment into infrastructural development as national economic strategies that will ensure linkage of fishing communities to towards sustainable postharvest logistics framework should be promoted.
6. Autonomous agencies should be established that primarily drives the overall architecture for postharvest practices separate from the governance challenges of managing fisheries within the national food production in developing countries.
7. Extant practices by nonstate actors such as development agencies, CSOs, NGOs involved in the promotion of sustainable postharvest practices through training and publication of postharvest manual should be sustained and spread across more countries.
8. Good hygiene practices at landing sites need to be enhanced in order to reduce fish losses and promote food safety at all levels, including domestic and regional trade. Often, large-scale fish producers are giving preference relative to small-scale fisheries actors in choice to criteria that enhance food safety.

9. Developing countries need to prioritize the adoption of cold chains, especially at the retail end markets of fish and seafood, through the promotion of solar cold chain technology within their energy infrastructure architecture as part of their rural development projects and activities.

3.5 Conclusion

Postharvest practices in small-scale fisheries are an essential element in sustainable fish and seafood production and processing and should be so accorded. Its importance can glance through many lenses, especially in the quest toward ensuring attainment and its direct and indirect linkages with many of the SDGs. The role of small-scale fisheries in terms of food and nutritional security cannot be overemphasized in spite of the many challenges and its overlooked contributions. Much attentions need to be accorded to postharvest practices since they served as a linkage between preharvest and harvest. Postharvest practices are critical, ultimately toward consumption of food to stave humans away from hunger and attainment of welfare in terms of health and other wellbeing. Fish and seafood, by nature of their chemical and biological composition, undergo spoilage or reduced quality and losses of all forms, which can be exacerbated by actions or inactions of all actors in the fish as food value chain when sustainable postharvest practices are not in place. Stakeholders in small-scale fisheries in particular and the food industry in general need to understand that sustainable postharvest practices germane to the overall desire for sustainable fish and seafood production. It provides efficient utilization of the harvested fish and seafood were canvased in this chapter. A starter to sustainable fish postharvest practices is recommended to be the adoption and implementation of the SSF Guidelines, among others.

References

Abbas, K.A., Saleh, A.M., Mohamed, A., Lasekan, O., 2009. The relationship between water activity and fish spoilage during cold storage: a review. Journal of Food, Agriculture and Environment 7, 86–90.

Affognon, H., Mutungi, C., Sanginga, P., Borgemeister, C., 2015. Unpacking postharvest losses in sub-Saharan Africa: a *meta*-analysis. World Development 66, 49–68.

Akande, G., Diei-Ouadi, Y., 2010. Post-Harvest Losses in Small-Scale Fisheries: Case Studies in Five Sub-Saharan African Countries. FAO Fisheries and Aquaculture Technical Paper No. 550. FAO, Rome.

Akintola, S.L., 2015. Effects of smoking and sun-drying on proximate, fatty and amino acids compositions of southern pink shrimp (*Penaeus notialis*). Journal of Food Science and Technology 52 (5), 2646–2656.

Akintola, S.L., Bakare, S.B., 2013. Effects of Ice Storage on the Biochemical Composition of *Macrobrachium vollenhovenii* (Herklots, 1857). Journal of Fisheries and Aquatic Science 8, 213–217.

Akintola, S.L., Fakoya, K.A., 2017. Small-scale fisheries in the context of traditional post-harvest practice and the quest for food and nutritional security in Nigeria. Agriculture and Food Security 6 (34), 1–17. Available from: https://doi.org/10.1186/s40066-017-0110-z.

Akintola, S.L., Brown, A., Abdullahi, B., Osowo, O.D., Bello, B.O., 2013. Effects of hot smoking and sun drying processes on nutritional composition of giant tiger shrimp (Penaeus monodon, Fabricius, 1798). Polish Journal of Food and Nutrition Sciences 63 (4), 227–237.

Alakali, J.S.M., Ariahu, C.C., Nkpa, N.N., 2006. Kinetics of osmotic dehydration of mango. Journal of Food Processing and Preservation 30, 597–607.

Arthur, T., 2017. Poverty Alleviation Strategies Employed by Fisher Folks in Elmina in the Komenda-Edina-Eguafo-Abrem (KEEA) Municipality (Doctoral dissertation). University of Cape Coast.

Atapattu, R., Samarajeewa, U., 1990. Fungi associated with dried fish in Sri Lanka. Mycopathologia 111, 55–59.

Borderías, J., Moreno, H.M., 2017. Recent advances in seafood technology an overview. In: Borda, D., Nicolau, A.I., Raspor, P. (Eds.), Trends in Fish Processing Technologies. CRC Press Taylor & Francis Group, United Kingdom, p. 356.

Carbonetti, B., Pomeroy, R., Richards, D.L., 2014. Overcoming the lack of political will in small scale fisheries. Marine Policy 44, 295–301.

Chiralt, A., Fito, P., Andrés, A., et al., 1999. Vacuum impregnation: a tool in minimally processing of foods. In: Oliveira, F.A.R., Oliveira, J.C. (Eds.), Processing of Foods: Quality Optimization and Process assessment. CRC Press, Boca Raton, FL, pp. 314–356.

Cole, S.M., McDougall, C., Kaminski, A.M., Kefi, A.S., Chilala, A., Chisule, G., 2018. Post-harvest fish losses and unequal gender relations: drivers of the social-ecological trap in the Barotse Floodplain fishery, Zambia. Ecology and Society 23 (2), 18. Available from: https://doi.org/10.5751/ES-09950-230218.

Corapci, B., Guneri, N., 2020. Comparative assessment of nutritional composition and physicochemical properties of fresh, freeze-dried and rehydrated rainbow trout (*Oncorhynchus mykiss* Walbaum, 1792) mince. Food Science and Technology 40 (Suppl. 1), 163–169.

Dalgaard, P., 1995. Qualitative and quantitative characterization of spoilage bacteria from packed fish. International Journal of Food Microbiology 26, 319–333.

Fafioye, O.O., Fagbohun, T.R., Olubanjo, O.O., 2008. Fungal infestation and nutrient quality of traditionally smoke-dried freshwater fish. Turkish Journal of Fisheries and Aquatic Sciences 8, 7–13.

FAO, 2005a. Increasing the contribution of small-scale fisheries to poverty alleviation and food security, FAO Technical Guidelines for Responsible Fisheries, vol. 10. Food and Agriculture Organization of the United Nations, Rome.

FAO, 2005b. Post-Harvest Changes in Fish. FAO Fisheries and Aquaculture Department, Food and Agriculture Organization, Rome.

FAO, 2005c. World inventory of fisheries. Reducing Post-Harvest Losses. Issues Fact Sheets, Fisheries and Aquaculture Department, Rome.

FAO, 2009. The State of World Fisheries and Aquaculture 2008. Rome.

FAO, 2011. Post-Harvest Fish Loss Assessment in Small-Scale Fisheries: A Guide for the Extension Officer. Rome.

FAO, 2015. Voluntary Guidelines for Securing Sustainable Small-Scale Fisheries in the Context of Food Security and Poverty Eradication. Food and Agriculture Organization of the United Nations, Rome.

Flynn, K., Villarreal, B.P., Barranco, A., Belc, N., Bjornsdottir, B., Fusco, V., et al., 2019. An introduction to current food safety needs. Trends in Food Science and Technology 84, 1–3. Available from: https://doi.org/10.1016/j.tifs.2018.09.012.

Fox, M., Service, M., Moore, H., Dean, M., Campbell, K., 2020. Barriers and facilitators to shellfish cultivation. Reviews in Aquaculture 12 (1), 406–437.

Ghaly, A.E., Dave, D., Budge, S., Brooks, M., 2010. Fish spoilage mechanisms and preservation techniques: review. American Journal of Applied Sciences 7 (7), 859–877. Available from: https://doi.org/10.3844/ajassp.2010.859.877.

Ghaly, A.E., Ramakrishnan, V.V., Brooks, M.S., Budge, S.M., Dave, D., 2013. Fish processing wastes as a potential source of proteins, amino acids and oils: a critical review. Journal of Microbial and Biochemical Technology 5, 107–129. Available from: https://doi.org/10.4172/1948-5948.1000110.

Gram, L., Ravin, L., Rasch, M., Bruhn, J.B., Christensen, A.B., Givskov, M., 2002. Food spoilage—interactions between food spoilage bacteria. International Journal of Food Microbiology 78, 79–97.

Greco, M., Pardo, A., Pose, G., 2015. Mycotoxigenic fungi and natural co-occurrence of mycotoxins in rainbow trout (*Oncorhynchus mykiss*) Feeds. Toxins 7, 4595–4609.

Hassan, A.A., Hassan, M.A., El Shafei, H.M., El Ahl, R.M.H.S., Abd El-Dayem, R.H., 2011. Detection of aflatoxigenic molds isolated from fish and their products and its public health significance. Nature and Science 9 (9), 106–114.

Jentoft, S., 2014. Walking the talk: implementing the international voluntary guidelines for securing sustainable small-scale fisheries. Maritime Studies 13, 16. Available from: https://doi.org/10.1186/s40152-014-0016-3.

Kimani, P., Wamukota, A., Manyala, J.O., Mlewa, C.M., 2020. Analysis of constraints and opportunities in marine small-scale fisheries value chain: a multi-criteria decision approach. Ocean and Coastal Management 189, 105151.

Kolding, J., Béné, C., Bavinck, M., 2014. Small-scale fisheries: Importance, vulnerability and deficient knowledge. In: Garcia, S.M., Rice, C., Charles, A. (Eds.), Governance of Marine Fisheries and Biodiversity Conservation. pp. 317–331.

Kumar, G.P., Martin Xavier, K.A., Nayak, B.B., Sanath Kumar, H., Venkateshwarlu, G., Balange, A.K., 2017. Effect of different drying methods on the quality characteristics of pangasius hypophthalmus. International Journal of Current Microbiology and Applied Sciences. 6 (10), 184–195. Available from: https://doi.org/10.20546/ijcmas.2017.610.024.

Lougovois, V., Kyrana, V., 2005. Freshness Quality and Spoilage of Chill-Stored Fish. Nova Science Publishers, Inc, New York.

McCarthy, M., Bane, V., García-Altares, M., vanPelt, F.N.A.M., Furey, A., O'Halloran, J., 2015. Assessment of emerging biotoxins (pinnatoxin G and spirolides) at Europe's first marine reserve: Lough Hyne. Toxicon 108, 202–209. Available from: https://doi.org/10.1016/j.toxicon.2015.10.007.

Mukundan, M.K., Antony, P.D., Nair, M.R., 1986. A review on autolysis in fish. Fisheries Research 4, 259–269.

National Research Council, 1988. Fisheries Technologies for Developing Countries. The National Academies Press, Washington, DC. Available from: https://doi.org/10.17226/1024.

Nowsad, A.K.M.A., 2007. Participatory Training of Trainers —A New Approach Applied in Fish Processing. Bangladesh Fisheries Research Forum, Dhaka. 328 p.

Oparaku, N.F., Nwaka, F.C., 2013. Effect of processing on the nutritional qualities of three fish species (Synodontis clarias, Trachurus trecae and Clarias gariepinus). International Journal of Biology and Biological Sciences 2 (10), 143–149.

Parlapani, F.F., Haroutounian, S.A., Nychas, G.-J.E., Boziaris, I.S., 2015. Microbiological spoilage and volatiles production of gutted European sea bass stored under air and commercial modified atmosphere package at 2°C. Food microbiology 50, 44–53.

Pushparajan, N., Varadharajan, D., Soundarapandian, P., 2013. Microbial studies of prepared curries stored in tin free steel cans. Journal of Food Processing & Technology 4 (6), 1000238.

Rahman, M.S., Al-Amri, O.S., Al-Bulushi, I.M., 2002. Pores and physico-chemical characteristics of dried tuna produced by different methods of drying. Journal of Food Engineering 53 (4), 301–313.

Ratner, B.D., Cohen, P., Barman, B., Mam, K., Nagoli, J., Allison, E.H., 2013. Governance of aquatic agricultural systems: analyzing representation, power, and accountability. Ecology and Society 18 (4), 59. Available from: http://doi.org/10.5751/ES-06043-180459.

Rossano, R., Mastrangelo, L., Ungaro, N., Riccio, P., 2006. Influence of storage temperature and freezing time on histamine level in the European anchovy *Engraulis encrasicholus* (L., 1758): a study by capillary electrophoresis. Journal of Chromatography B 830, 161–164.

Sarkardei, A., Howell, H.K., 2007. The effects of freeze-drying and storage on the FT-Raman spectra of Atlantic mackerel (*Scomber scombrus*) and horse mackerel (*Trachurustrachurus*). Food Chemistry 103, 62–70.

Shawyer, M., Pizzali, M. (Eds.), 2003. The Use of Ice on Small Fishing Vessels. Food and Agriculture Organization of the United Nation, Rome, pp. 1–34.

Singh, S., Shalini, R., 2016. Effect of hurdle technology in food preservation: a review. Critical Reviews in Food Science and Nutrition 56 (4), 641–649.

Smith, M.L., Bruhn, J.N., Anderson, J.B., 1992. The fungus *Armillaria bulbosa* is among the largest and oldest living organisms. Nature. 356, 428–431.

Tarr, H.L.A., 1965. Enzymatic degradation of glycogen and adenosine 5' triphosphate in fish muscle. In: Kruezer, R. (Ed.), Technology of Fish Utilisation. Fishing News, London, pp. 26–31.

Tesfay, S., Teferi, M., 2017. Assessment of fish post-harvest losses in Tekeze dam and Lake Hashenge fishery associations: northern Ethiopia. Agriculture & Food Security 6, 4. Available from: https://doi.org/10.1186/s40066-016-0081-5.

Varlet, V., Prost, C., Serot, T., 2007. Volatile aldehydes in smoked fish: analysis methods, occurrence and mechanisms of formation. Food Chemistry 105, 1536–1556.

Visciano, P., Schirone, M., Berti, M., Milandri, A., Tofalo, R., Suzzi, G., 2016. Marine biotoxins: occurrence, toxicity, regulatory limits and reference methods. Frontiers in Microbiology. 7, 1051. Available from: https://doi.org/10.3389/fmicb.2016.01051.

Wang, F., Fu, L., Bao, X., Wang, Y., 2017. The spoilage microorganisms in seafood with the existed quorum sensing phenomenon. Journal of Food Microbiology 1 (1), 14–19.

Ward, A., 1997. A Manual for Assessing Post-harvest Fisheries Losses. Natural Resources Institute, Chatham.

Zelasney, J., Ford, A., Westlund, L., Ward, A., Riego Peñarubia, O. (Eds.), 2020. Securing Sustainable Small-Scale Fisheries: Showcasing Applied Practices in Value Chains, Post-Harvest Operations and Trade. FAO Fisheries and Aquaculture Technical Paper No. 652. FAO, Rome, https://doi.org/10.406.

Chapter | four

Egyptian Marine Fisheries and Its Sustainability

Sahar Fahmy Mehanna
Fish Population Dynamics Lab, Fisheries Division, National Institute of Oceanography and Fisheries, Egypt

4.1 INTRODUCTION

Ocean, seas, and coastal areas offer various goods and ecosystem services vital to people's well-being, worldwide food security, and nutrition. They make an integrated and crucial component of the Globe's ecosystem and are critical to sustainable development. Fisheries including aquaculture are an essential source of food and animal protein for billions globally. Nearly 3 billion people obtain 20% of their daily animal protein intake from fish (FAO, 2020). Wild fisheries and aquaculture offer excellent opportunities to decrease hunger and improve nutrition, relieve poverty, create economic growth, and guarantee healthier use of natural resources. Employment in fisheries and aquaculture has grown continuously quicker than in the agriculture sector providing upto 55 million jobs worldwide. Fishery-related activities such as netting, fish processing, and packing have supported the livelihoods of around 10%–12% of the world's population. Stress caused by human activities on the marine ecosystems is highly acknowledged. Nowadays, 61% of commercially important considered aquatic fish stocks are fully exploited worldwide, 29% are overexploited and around 90% of large predatory fish stocks are now depleted (FAO, 2012, 2020).

Global fish production (Fig. 4.1) is estimated to have reached about 179 million tons in 2018, with a total trade value determined at the United States $401 billion, of which the United States $250 billion, the trade value of 82 million tons, derived from aquaculture production. Of the total global fish production, 156 million tons were consumed directly by a human, corresponding to an estimated supply of 20.5 kg per capita per annum. The outstanding 22 million tons were used for

Sustainable Fish Production and Processing. DOI: https://doi.org/10.1016/B978-0-12-824296-4.00010-4

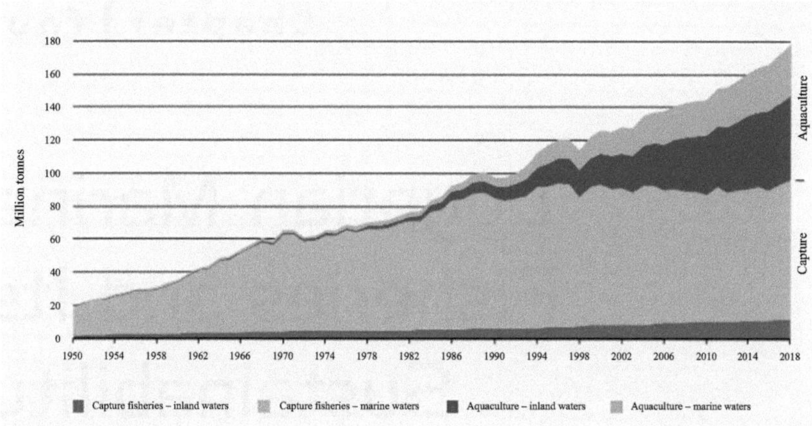

FIGURE 4.1 Global fish production excluding aquatic mammals, crocodiles, seaweeds and aquatic plants, alligators and caimans (FAO, 2020).

nonfood purposes, mostly for fishmeal and fish oil production (FAO, 2020). The major fish producer globally is China, accounting for 35% of worldwide fish production in 2018. Eliminating China, Asia was the primary fish producer in 2018, constituting about 34% of the global output, followed by the Americas which providing 14%, then Europe, which forms about 10%, Africa (7%) and Oceania (1%) (Fig. 4.2).

In 2018 about 60 million persons were involved in the full-time, part-time, or occasional manner in the primary sector of wild fisheries (39.0 million persons) and aquaculture (21 million persons). The sharing percentage of women in this sector was about 14% of the total, with 19% sharing in the aquaculture sector and 12% share in the wild fisheries sector. The number of people involved in primary fish production is much higher in developing countries rather than any other country, and mostly in small-scale and artisanal fisheries and aquaculture sectors.

Concerning the total number of fishing vessels in 2018, Asia had the biggest fishing fleet, estimated at 3.1 million fishing vessels, which constitutes 68% of the entire global fleet. While Africa's fishing vessels represented 20% of the worldwide fleet followed by that of the Americas (10%), Europe has a fleet size of slightly over 2% followed by Oceania, where its fleet size is less than 1% of the global fishing fleet. Asia had almost 75% (2.1 million fishing vessels) of the recorded motorized fishing fleet in 2018, while Africa has 280 thousand motorized fishing vessels and comes in the second degree after Asia. Also Asia still has the highest number of nonmotorized vessels, that estimated at 947 thousand vessels, then Africa which has just over 643 thousand nonmotorized vessels. Latin America, the Caribbean, Oceania, North America, and Europe have smaller numbers of nonmotorized fishing vessels. This type of nonmotorized and undecked vessels frequently has a length overall (LOA) class of smaller than 12 m (artisanal or small-scale fisheries). The LOA of at least 24 m and more (large scale or industrial fleet) accounts for about 67,800 fishing vessels in 2018 worldwide.

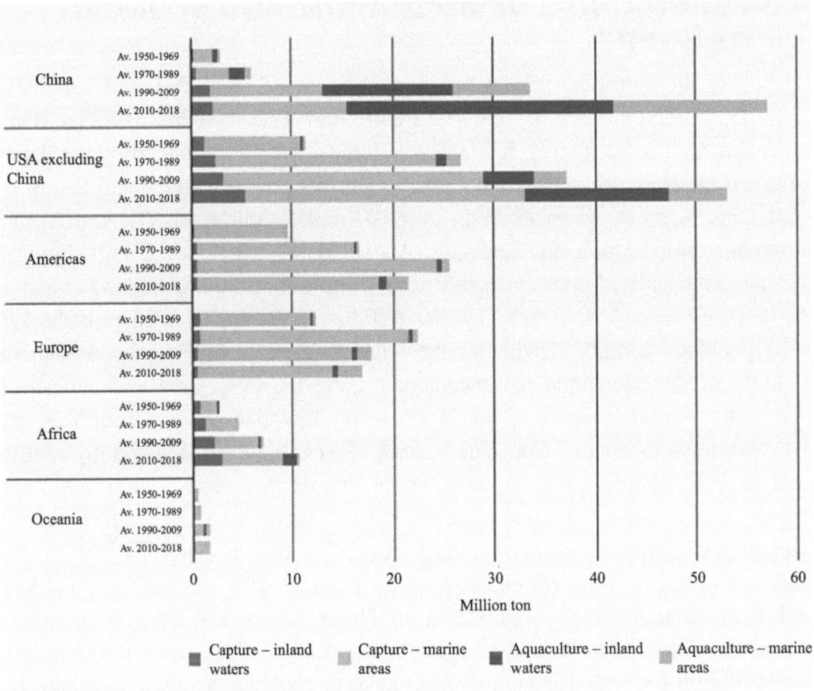

FIGURE 4.2 World fisheries and aquaculture production by region (FAO, 2020).

Fisheries and aquaculture supplied Egypt with 1.9 million tons of fish in 2018, mostly used as food for people. Fish production in Egypt has grown dramatically in the last three decades, where fish production increased from about 346 thousand tons in 1990 to upto 1.9 million tons in 2018 (459% increasing). Egyptian per capita food fish supply amplified from about 9 kg in the 1990s to upto 19 kg in 2018. In contrast to natural resources, which increased by only 30.5% in the period from 1990 to 2018, aquaculture production increased by 2502% during the same period. The majority of wild fisheries production in Egypt arises from coastal areas, where pollution is one of the main challenges that severely affected both the productivity and quality of fish stocks. Our marine resources are at the risk of irreversible loss to habitats, ecological functions, and biodiversity because of overfishing, pollution, destructive fishing methods, climatic changes, unsustainable coastal area development, and the competing demands from different industrial uses and human activities. Illegal, Unreported, and Unregulated (IUU) Fishing continues to be a big challenge to achieve sustainable fisheries. Furthermore, poor governance, management, and practices are a further challenge. Reducing overfishing, stimulating responsible and sustainable fisheries management, applying aquaculture new and friendly practices, and conserving the marine environment health are among the government's best opportunities to provide highly nutritious food to the increasing population in Egypt.

4.2 GEOGRAPHIC, DEMOGRAPHIC AND ECONOMY OF EGYPT

Egypt lies in northeast Africa between latitudes 22°N–32°N and longitudes 25°E–35°E with an area of upto one million km^2 (the land area is 995,450 km^2 and water area is 6000 km^2). It is a Mediterranean country bounded by the Aqaba Gulf and the Red Sea to the east, Libya to the west, the Mediterranean Sea to the north, and Sudan to the south. With over 100 million inhabitants (Central Agency for Public Mobilization and Statistics, Annual Statistical Report 2019), Egypt is the highest populated country in the Arab world and North Africa and comes in the third degree in Africa after Nigeria and Ethiopia. Worldwide, Egypt is the 13th most populous country. Owing to the extreme dryness of Egypt's climate the majority of the population is concentrated along the Delta area and the narrow Nile Valley. In other words, about 98.9% of the Egyptian population utilizes only 5.47% of the total area of land. The sources of wealth, which significantly contributed to Egypt's economy, are agriculture, natural gas, Suez Canal, petroleum imports, and tourism. Agriculture including the fisheries sector is a central component of the Egyptian economy, constituting 15% of the country's GDP (gross domestic product). The agricultural industry provides 28.4% of all Egyptian jobs, and more than 59% of employment in Upper Egypt and the countryside is agriculture-related. The longest river in the world, Nile River, flows in Egypt from south to north for upto 1532 km. North to Cairo, the Nile splits into two branches that feed the Mediterranean: the Rosetta (Rasheed Branch to the west and the Damietta Branch to the east, forming the Nile Delta).

4.3 FISHERY ADMINISTRATIVE

In Egypt, The General Authority for Fish Resources Development (GAFRD) is the only responsible governmental authority for the fisheries sector. GAFRD was established by presidential decree number 190 for the year 1983. It is responsible for fisheries data collection involving total landings in tons, landing by fishing gear in tons, landing by species in tons, catch composition, number of landings, number of fishing vessels, and types of fishing gear. It is responsible for establishing the decisions and laws of fisheries like the time and period of the closed season, the number of the fishing licenses for both fishing boats and fishers, restrictions on the fishing gears characterization, and the closed areas. GAFRD is the governmental body that represents the country on the national and international levels dealing with wild fisheries and aquaculture. It is also responsible for monitoring the governmental hatcheries and aquaculture farms, issues and renews the licenses of the private fish farms, and launching the annual statistical books which contain the fisheries statistical data (annual imports and exports, catch from different resources, annual catch by species, monthly catch by species by fishing area, number of fishing vessels for each gear, types of fishing gears, prices per species in the local market, hatcheries production, and aquaculture production according to the different ways used for fish farming). It is also authorized to collect broodstock or seed from natural populations and stocking the closed lakes with fingerlings and fish fries collected from wild marine resources. GAFRD is affiliated with the Ministry of agriculture and land reclamation. Its headquarter is located in Cairo with seven main sites and hundreds of offices covering Egypt; each site covers two or more governorates. These offices have qualified data collectors to introduce a precise and useful fishery data collection system that reflects the current situation of the fisheries on a consistent and regular basis. But on the land the data collection system lacks accuracy and biological, ecological, and socioeconomic changes are not fully documented or not documented at all. As the fishery statistics are vital for evaluating the current status of that fishery, an urgent need to improve the recording system should be implemented.

The National Institute of Oceanography and Fisheries (NIOF), the oldest institute in the Middle East dealing with Ocean and fisheries governance, with the newly established fisheries colleges are responsible for the fisheries management and development based on sound scientific research. NIOF is a governmental research institute affiliated with the Ministry of high education and scientific research and its mandate is to maintain, protect, and promote the different Egyptian aquatic environments.

4.4 FISHERIES LEGISLATION

1. Law No. 124 of 1983 concerning the fishing operations, aquaculture and fish farms regulations. This Law repeals law Number 144 which first appeared in 1960 about fishing. It provides GAFRD with responsibility for the development and management of fishery resources, including aquaculture, issuing fishing licenses, supervising fisheries cooperatives, and producing statistical data.
2. Presidential Decree Number 465 of 1983 for describing and detecting the water surface in which GAFRD administers and controls.
3. Ministerial Decree Number 329 of 1985 for identifying some areas of aquaculture. This Decree implements the third item of article one of the Presidential Decree No. 465 of 1983 on the identification of regions of aquaculture and the provision of article 49 of Law No. 124 of 1983 on licensing areas of establishment of fish farms. There are five annexes to this Decree identifying the different areas that shall be used for fish resources development.
4. Decree Number 174 of 1989 defines the minimum mesh size and length of nets by fishing gear.
5. Ministerial Decree Number 1043 of 1990 for stopover renewing rental agreements about the whole areas surrounded Teraat- El Salam (El-Salam Canal) and finalized all undeveloped construction projects in these areas.
6. Resolution Number 342 of 1992 prohibits the issuing of any new fishing licenses for trawl in either the Red Sea or the Mediterranean Sea from January 1, 1994.

7. Ministerial Decree Number 2654 of 2003 about the leasing period of fish farmers.
8. Ministerial Decree Number 2655 of 2003 stops the use of "testosterone hormone" in hatcheries for its dangerous impact on public health.
9. Ministerial Decree Number 2656 of 2003 indicated that GAFRD should be supervised and controlled for any renting or selling aquaculture farms carried out.

4.5 EGYPTIAN FISHERIES

The reduction of most fish stocks in the world's oceans and the usage of increasingly modernized industrial fishing practices are resulting in heavy harm to valuable marine ecosystems. Our marine and inland fisheries contribute a great deal to food security and therefore play a significant role in the economy. Egyptian fisheries are one of the primary sources of wealth in the country. If we provide them due attention and adequate care, we can improve our national incomes and resolve most of our problems mainly food security and employment.

Fish production in Egypt comes from three primary resources; marine fisheries (the Mediterranean and Red seas), inland fisheries (lakes, lagoons, the Nile River, irrigation, and drainage canals), and aquaculture. By the following of Egyptian fish production through the last three decades (1990–2018), it was found that the bulk of catch comes from inland waters (Nile and lakes). The inland waters provide Egypt with about 55% of the total landing in the 90s decreased to only 14% in 2018 (Fig. 4.3). While the Egyptian marine fisheries' contribution to the total fish production was about 24% of it in the 90s, it seriously decreased to only 5% in 2018 (Mehanna, 2020a,b). In contrast, aquaculture started in a low percentage contributing about 21% of the total landing in the 90s increased to become the primary source of fish production in Egypt forming 81% of the total fish production in 2018 (GAFRD, 1990–2018; Mehanna, 2020a,b).

Even with the enhanced fish production in Egypt ($\approx$1,900,000 tons in 2018: 19% from capture fisheries and 81% from fish farms), the high pressure due to the growing populations and the need to rise consumption of fish per capita obliges us to reorder priorities to achieve the rational exploitation of existing fisheries resources. Egypt, like most of the world's countries, requires focusing on fisheries resources, as renewable productive resources of food that can significantly help in the development and increase national income as well as achieve food security for upcoming generations.

Egyptian marine fisheries have a coastline of about 2400 km and a continental shelf area of about 87.2 thousand km^2. While the inland fisheries have about 10,160 km coastline ($\approx$2140 km for coastal lakes, 1188 km for inland lakes, 5238 km for Nasser lake, and 1532 km for Nile). The fish production from natural resources in 2018 is 373.3 thousand tons with a value of approximately 10.7 million LE. The fishing sector contributes 5.1% of total GDP

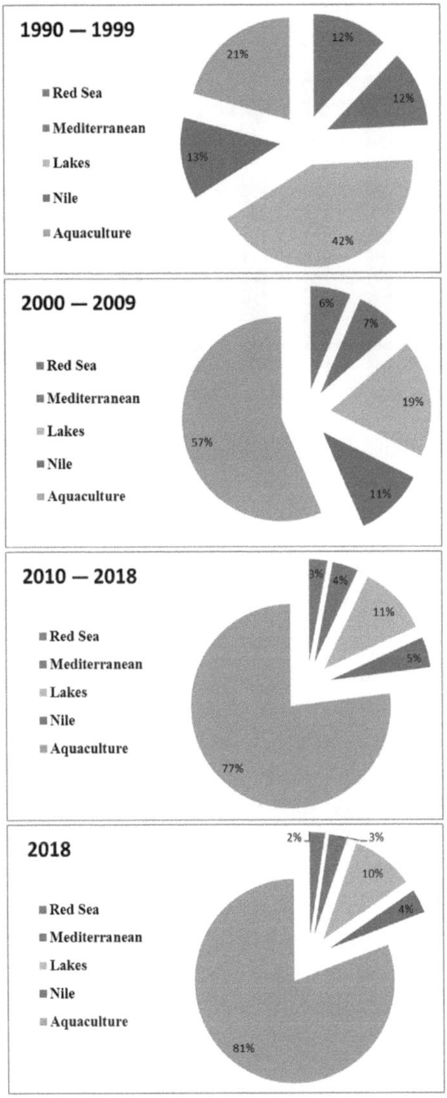

FIGURE 4.3 Egyptian fish production during the last three decades.

and 9.3% of agricultural GDP. The marine fish landings in Egypt from its exclusive Economic Zone differed during the last 10 years (Fig. 4.4), with a reported catch of 127.8 thousand tons in 2009 decreased to 104.6 thousand tons in 2018 (GAFRD, 2018). During the year 2018 Egypt imported upto 350 thousand tons of marine products with a cost of about US $400 million (GAFRD, 2018) to overcome the huge gap between fish production and consumption.

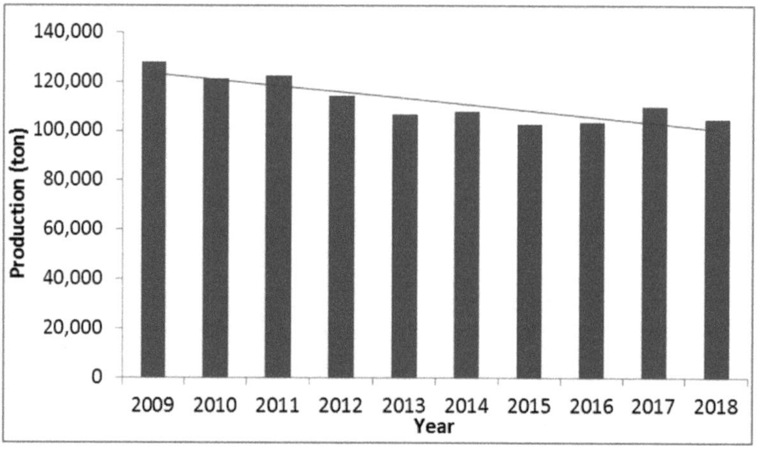

FIGURE 4.4 Marine fish production in Egypt during the last 10 years.

The main gears which exploited the demersal marine finfish, crustacean, and cephalopods resources in Egypt are bottom trawl nets, while pelagics are the purse-seine nets. Small-scale fisheries, especially the hooks, lines, trammel nets, and gill nets, are also used in a wide range. The contribution of different categories of marine catch to the fish production is pelagic fishes 54%, demersal fishes 24%, cartilaginous fishes 3.5%, crustaceans 11.5%, and molluscs 7%. About 1140 trawlers, 370 purse seiners, 3560 small-scale fisheries are operated in the Egyptian marine fisheries. All of them are motorized, besides 1100 sailing boats and 3850 recreational boats are registered in marine fisheries. On the other hand, upto 10,000 wooden boats of small size (4–6 m length) are working in the Nile. About 12,840 registered sailing boats are operated in the lakes (2220 boats of first grade, 11 boats of second grade, and 10,610 boats of third grade) with about 3500 licenses for recreational fishing. It is worth mentioning that, these numbers of fishing boats especially in inland water are underestimated and the real number may be double the registered ones.

It is generally assumed that Egyptian marine and inland fish stocks have severely declined in recent years. Many challenges are facing the development of Egyptian marine fisheries and are responsible for the severe decline of these resources. Excessive fishing effort, high level of exploitation, destructive fishing practices (such as the use of small meshed gears and fishing for juveniles), increasing of tourism and industrial expansion which destroyed the habitat, pollution, and global warming are reported as challenges by different authors (Mehanna, 2005, 2009a,b, 2010a,b, 2011, 2019,a,b; Mehanna and Haggag, 2010, 2011,a,b; El-Serafy et al., 2015; Zaahkouk et al., 2017; Mohamed et al., 2020). Also the number of small-scale vessels and fishers is very high, and, management of these fisheries remains challenging to develop and implement (Mehanna, 2020a,b). Another critical issue is the legislation. Though jurisdictive reforms have improved the regulatory agenda, such reforms application has remained inadequate and suffering from the absence of effective monitoring, control, and surveillance (MCS).

On the other hand the most common challenges facing inland fisheries development (Mehanna, 2007a,d, 2008, 2014, 2020c) are:

1. Degradation, drought, and filling up for housing purposes caused the severe decline of the lakes' areas. Most Egyptian lakes decreased to less than half of their original areas such as Manzala, Borollus, Port Fouad, Timsah, Edku, and Mariut lakes.
2. Severe pollution due to the discharge of untreated effluents and drains from industrial effluents, agricultural drains, raw sewage from the expanding unserviced area, and other sources such as navigation and submersed aquatic weeds control.
3. The spread of aquatic plants and weeds which cover huge areas in the lakes, especially the coastal ones.
4. Overexploitation, illegal, and destructive fishing methods, and illegal harvesting of fish fry to feed fish farms.
5. The silting and blockage of Boughazes (the linkage between the coastal lakes and lagoons with the sea). This makes the spawning migration difficult for fishes and changes the water characters due to stop the water circulation and exchange between the Mediterranean Sea and coastal lakes. Consequently the salinity decreased and anthropogenic nutrients increased, which together have a serious impact on the catch composition from these lakes and human health.
6. Eutrophication and loss of habitat.
7. The low awareness of fishers about environmental issues and the significance of fisheries regulatory measurements.

Some regulatory measurements were taken to protect and conserve the declined stocks in our marine fisheries. The GAFRD, the Governmental body that responsible for the development of Egyptian fisheries, has taken the following regulations based on the scientific researches done by the Egyptian scientists:

1. Controlling the fishing effort by stopping issuing any new fishing licenses and preventing vessels from changing their habitual fishing area and changing the type of gear used. This action was taken as a result of many studies and scientific projects which concluded that most Egyptian fisheries are overexploited and need to reduce the fishing mortality by about 30%–50% of its current value for most commercial species (examples for Mediterranean species: *Solea aegyptiaca* by Mehanna, 2007b; *Sparus aurata* by Mehanna, 2007c; *Mullus surmuletus* by Mehanna, 2009a; *Mullus barbatus* by Mehanna, 2009b; *Boops boops* by Mehanna, 2010a; *Pagellus erythrinus* by Mehanna, 2010b; *Sepia officinalis* by Mehanna and Haggag, 2011a; *Sardinella aurita* by Mehanna and Salem, 2012; *Merluccius merluccius* by Mehanna, 2019a).
2. The increase in engine power is not allowed as the increasing it reduces the time to reach the fishing ground and consequently increases the days of actual fishing.

3. Promoting the usage of new fishing gear and equipment to improve and enhance the fishing vessels' capability to go too deep in waters and fish offshore.
4. Regulating the mesh sizes applied according to regional agreements and based on estimated critical lengths of target species.
5. They working at improving and updating the fishery law as it is not changed since the 80s.
6. Proposing closed seasons to protect the spawning areas and conserve young fishes in the Red Sea and the Mediterranean based on studies undertaken by the NIOF.
7. Improving the system for collecting and compiling fisheries statistics to ensure the quality and accuracy of fishery data.
8. Constructing marine hatcheries to cover the excessive demand for fish fry and juveniles for aquaculture.

Despite all these regulations the fisheries resources in Egypt are still suffering from the continuous decline due to most of these regulations are not enforced and due to the illegal behaviors of many fishers. This continuous decline has stressed the necessity for an effective fisheries management plan. To improve fish production in Egypt the attention of the government and all authorities in charge must be forwarded to develop, control, and conserve our natural marine and fresh-water fish resources.

4.6 EGYPTIAN MARINE FISHERIES

4.6.1 Red Sea Fisheries

The Red Sea is globally well-known for its exceptional and attractive marine and coastal environments, the high diversity of species inhabiting it, the high grade of endemism, and the importance of these resources for human development and as a portion of the area's cultural heritage. It extends from Lat. 30°N at Suez to Lat. 12°S at Bab-Elmandab with an average width of 280 km and an average depth of 490 m. The Red Sea is a semienclosed, elongated warm water body of about 2150 km long. Its maximum width is about 355 km^2, its greatest depth is 2000 m, and its surface area is approximately 458,620 km^2 (Head, 1987). Red Sea connected to the Indian Ocean by a narrow and shallow sill at Bab-Elmandab and to the Mediterranean Sea in the North by the Suez Canal, which was dug in 1869. The Bab-Elmandab and the Suez Canal allow only limited exchange of surface water between the Red Sea and the neighboring water bodies. About 2.33% of its area is protected and contains 3.8% of the globe's coral reefs (Sea Around Us, 2007). The fauna of the Red Sea biologically belongs to the Indo-Pacific region. It is distinguished by rich, diversified fauna and flora (Morcos, 1970; Sheikh-Eldin, 1988; Mehanna, 1993; Tesfamichael and Mehanna, 2012; Osman, 2015, 2018; Mohammad, 2016; El-Mahdy, 2017; Farrag et al., 2018; Mehanna et al., 2019; Mohamed et al. 2020). Information regarding the fish communities of the Red Sea is limited. About 1200 species of fishes are known to live in the Red Sea (Ormond and Edwards, 1987). The bulk of these species inhabits coral reefs where they constitute a dominant component of the fish fauna (Zekeria, 2003). There are marked differences among the different regions of the Red Sea in fish species richness, assemblage compositions and species' abundance (Shepperd et al., 1992).

The Egyptian sector (Fig. 4.5) of the Red Sea is about 1080 km in length extending from Suez city in the north to Mersa Halayeb in the south. Numerous fishing grounds are identified along with the Red Sea's Egyptian

FIGURE 4.5 The Egyptian Red Sea with the main landing sites.

sector, yielding an average catch of about 45 thousand tons annually (Mehanna, 1993; Mohamed et al., 2020; Osman et al., 2020). Egyptian sector of the Red Sea is classified into three main fishing grounds: Gulf of Suez, proper Red Sea (from Hurghada to Halayeb involving Foul Bay), and Gulf of Aqaba. Many landing sites are distributed along the Egyptian Red Sea, they are Attaka, El-Tor, Hurghada, Safaga, Quseir, Berenis, Shalateen, and Abo Ramad. Also Egypt signed many treaties with its neighbors to fish outside its territories (Fig. 4.6). The most typical fishing gears operated in the Red Sea by the Egyptian fishermen are bottom trawling and purse seining (the industrial fishery), hand lines, long lines, and gillnets (artisanal fishing), and a variety of gears used by some of the traditional coastal communities. The industrial fishing fleets operate essentially in the Gulf of Suez and its neighboring areas, and Foul Bay. Semiindustrial fishing fleets are concentrated near Ataka, Salakhana, Hurghada, and El-Tor. Small-scale fisheries are common in Safaga, Quseir, Shalateen, and Abo Ramad.

Bottom trawls are one of the leading fishing gears in the Egyptian marine fisheries yielding the highest earnings among all the other fishing techniques. It constituted about 20% of the Egyptian Red Sea fisheries forming about 54% of the gross revenue due to the high price of some demersal caught by it like shrimp and cephalopods. The purse-seine forms about 69% of the Egyptian Red Sea fish production, while the artisanal catch contributed about 11% of the total Red Sea fish production (GAFRD, 1999–2018).

Based on stock assessment and other fisheries studies carried out on the Egyptian Red Sea since 70s, 90% of the commercial species assessed are fall in the range full to overexploited (Latif, 1974; Rafail, 1970, 1972; Mehanna, 1993; Salem, 1971; Sanders and Kedidi, 1981; Sanders and Kedidi, 1984a; Sanders and Kedidi, 1984b; Sanders and Morgan, 1989, Mehanna, 1993, 1997, 1999a,b, c, 2001, 2002, 2005, 2011; Mehanna and Abdallah, 2002; El-Gammal and Mehanna, 2002, 2006; El-Mahdy, 2017). The main challenges facing the rational

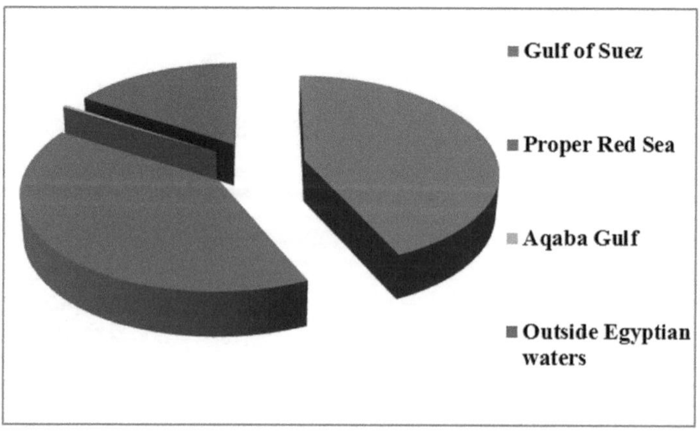

FIGURE 4.6 Fish production from the different fishing grounds on the Red Sea.

exploitation and fisheries management in the Egyptian Red Sea are the conflicts among different uses of the area, fisheries strategies remain development-driven, and without attention to biological, socioeconomic, and the standards of the sustainable environment, socioeconomic data are not collected. Consequently the involvement of small-scale fisheries in human well-being, food security, and poverty mitigation and avoidance is poorly understood and the influences of prospective management measures can't be assessed, information on shared and transboundary stocks is still missing or inadequate, the multispecies—multifleet nature of our marine fisheries has not been taken into consideration, and finally, the integration of different stakeholders in the management process (comanagement) is minimal causing many difficulties in implementing of any management regulations.

4.6.2 Gulf of Suez

The length of the Gulf of Suez is about 250 km starting at Suez city in the north (lat 29° 56′ N) ending at Shadwan Island in the south (lat 27° 36′ N). The Gulf width varies between 20 and 40 km, while the depth range is about 80–90 m at its mouth, and about only 20–30 m along most of its length. The Gulf of Suez is one of the most productive fishing grounds along with the Egyptian sector of the Red Sea as well as in Egypt. The importance of the Gulf as a fishery resource can be attributed to the shallowness and sandy bottom which make it appropriate for trawling. Also it is characterized by the presence of a great diversity of highly economic fish and invertebrate species (Sanders and Morgan, 1989). Three main fishing methods are operated in the Gulf; trawling "caught demersal fishes," purse seining "caught pelagic fishes" as well as small scale fisheries especially long and hand lines "caught demersal and pelagic fishes" (Mehanna, 1993; Mehanna and El-Gammal, 2007). There are four main landing sites along the Gulf, one to the east of the Gulf and three to the west. These are Ataka and Salakhana at Suez city, Ghareb at Ghareb city, and El-Tor at El-Tor city (Fig. 4.7). The fishing in the Gulf of

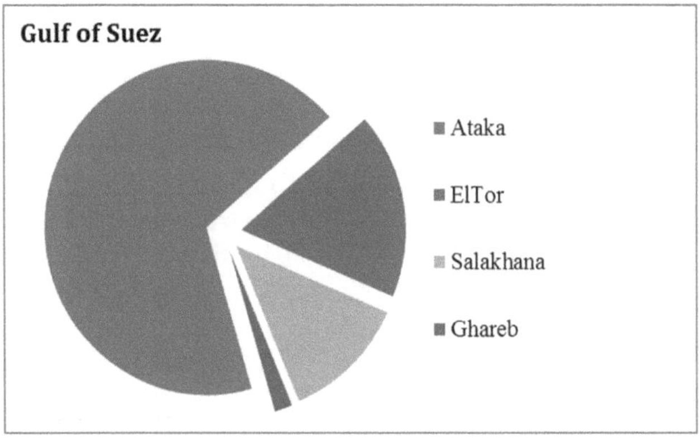

FIGURE 4.7 Contribution of different fishing grounds to fish production in the Gulf of Suez.

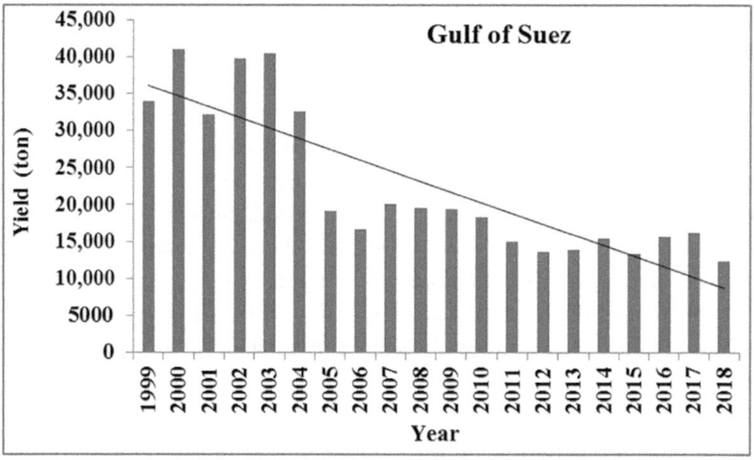

FIGURE 4.8 Fish production from the Gulf of Suez from 1999 to 2018.

Suez is seasonal, starting in September ending in May and all fishing activities are ceased 3 months and sometimes 4 months each year. Generally, there is a dramatic decrease in annual fish production from the Gulf during the period from 1999 to 2018 due to the overfishing, illegal fishing practices in the Gulf, pollution, industrial and tourism expansion activities, and illegal harvesting of fish fry (Fig. 4.8).

4.7 TRAWL FISHERY

The number of fishing vessels operating in the Gulf of Suez is 78 vessels with lengths diverse between 20 and 30 m. Each vessel is motorized by a main engine of 200–600 hp, with the majority (68 vessels) of 400–600 hp. All vessels are provided with mechanized winches. Some of them are equipped with echo-sounders. Trawl net is of the Mediterranean type; the net length varied between 20 and 30 m with an average mesh size of 1.5 cm in the cod-end. The sweep length of the gear ranged between 200 and 250 m. The fishing trip has taken about 5–10 days with a crew number of about 10–15 persons. The trawl fishery is seasonal, generally from October to May (Mehanna and El-Gammal, 2007). The fishing effort (number of fishing days) at the beginning of the fishing season, the first 3 months of the fishing season, forms 43% of the total seasonal fishing effort. The contribution of trawl fishery to the total fish production from the Gulf is about 21%. Trawl catch achieves 50% of the gross revenue of the Gulf fisheries due to the high prices of fish, cephalopod, and crustacean species caught by trawling in the local markets. The most economically important fish families represented in the catch of the trawl are given in Fig. 4.9. The fishes belonging to the family Synodontidae (lizard fishes) came in the first grade contributing about 29.9% of the total trawl catch, family Lutjanidae (stripped snapper) came in the second degree, constituting about 14.9%, followed by the large shrimp (family: Penaeidae) which constituted

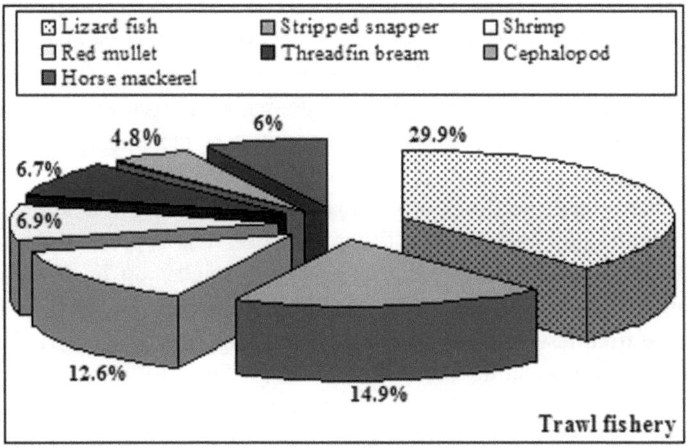

FIGURE 4.9 Catch composition of trawl fishery in the Gulf of Suez.

upto 11%, red mullet (family: Mullidae) contributed by about 6.9% and the fishes belonging to the family Nemipteridae (threadfin bream) forming about 6.7%. Horse mackerel and scads, cuttlefish, and small shrimp are starting 6%, 4.8%, and 1.6% of the total trawl catch, respectively. On the other hand the unsorted species or those of lesser importance are grouped in the "others" group, representing 18.2% of the total trawl catch. It is worth mentioning that, the large shrimp is the most economically important species in the trawl fishery where it sold at high prices fluctuating between 300 and 500 LE per kg, representing about 40% of the gross revenue from the trawl fishery.

4.8 PURSE-SEINE FISHERY

The number of fishing vessels fluctuated between 56 and 83 vessels that worked inside the Suez Gulf. The length range of these vessels varied from 12.5 to 30 m. All are motorized and provided with engines of 150–600 hp. The bulk of the purse-seine fleet (63 vessels) is 400–600 hp. The purse-seine units are worked at night using lighted dinghies. This artificial light makes fishes concentrate before setting the seine net. All fishing activities stop for roughly 10 days each month during the full moon days. The seine net has a length varied between 200 and 300 m with depth oscillated between 50 and 80 m, and the seine nets are dragged manually. The number of crew reaches 25–30 persons per vessel in the trip. The purse-seine fishery is also seasonal usually from October through May. The duration of the fishing trip is short at the beginning of each season. It takes 2–5 days because the most of fishing activities are carried out comparatively near the landing site of Ataka at Suez city. Later in the fishing season the fishing trip takes more than 10 days.

Pelagic fishes, caught by purse seine in the Gulf of Suez, represent upto 70% of the total annual fish production from the Gulf (GAFRD, 2018) forming about 35% of the gross revenue of the Gulf; hence they contribute to the main fishery in Egypt. They are mostly small, highly mobile, schooling fish, live in the surface, and near-surface waters. They occupy an important place in the marine food webs; acting as predators on phytoplankton, zooplankton (planktonic invertebrates; chiefly copepods and other small crustaceans, fish larvae), and other smaller shrimps, cephalopods and fishes, while serving as prey for many larger fish, marine mammals and seabirds that happen higher up in the food web. Consequently, they play an essential role in the marine ecosystems by transforming energy from lesser trophic levels to more complex trophic levels. The most common fish species in the purse-seine landing (Fig. 4.10) are those belonging to the family Carangidae (the horse mackerel and scads), the round herring and sardines (family: Clupeidae), slimy and Indian mackerel (Scombridae),

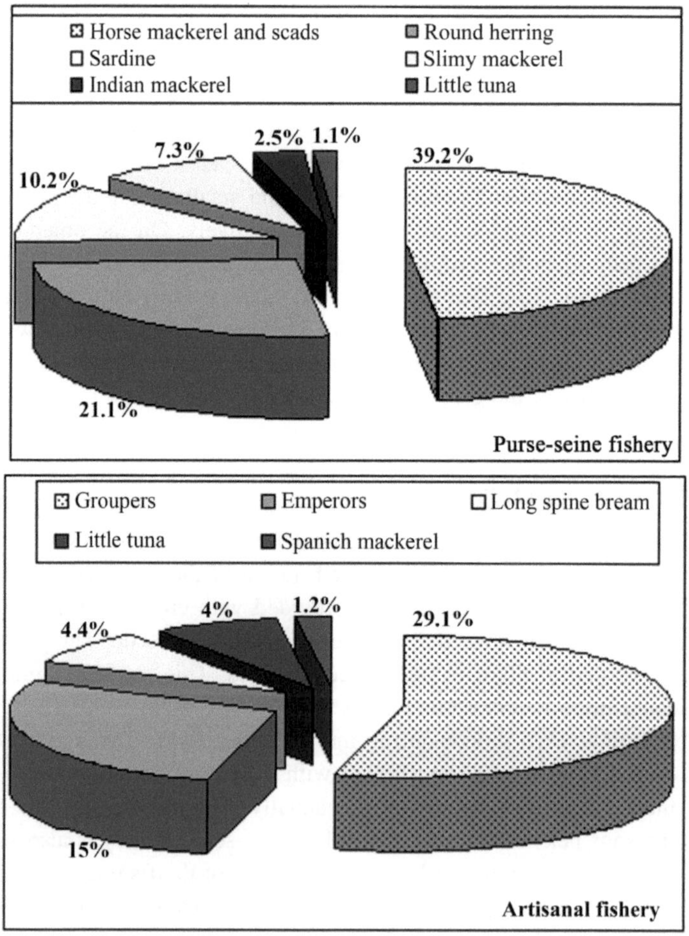

FIGURE 4.10 Catch composition per fishing gear from the Gulf of Suez.

Anchovies (Engraulidae), barracudas (Sphyraenidae), silversides (Atherinidae), and little tuna (Scombridae). The species of lesser importance or unsorted species, as well as the unidentified species, were gathered in the "others" group, which constitute about 18% of the total purse seine catch.

4.9 ARTISANAL OR SMALL SCALE FISHERY

A total number of varied between 93 and 178 fishing boats are used small-scale fisheries especially hooks and lines. These artisanal boats are of length range fluctuated between 10 and 15 m and motorized by inboard engines of horsepower varied from 50 to 200 hp. The crew number ranged between 2 and 10 persons with a fishing trip duration of around 10 days. Several and different fishing gears are used by the local fishers like long and hand lines, gill and trammel nets, and beach-seine. The small-scale or artisanal fishery in the Suez Gulf was all over the year till the fishing season 1996/97, where it converted to seasonal mostly from September to June. The artisanal landing constituted about 9% of the total fish landed from the Gulf of Suez, forming upto 15% of the gross profits of the Suez Gulf fishery because of the high prices of the fish species fished by the small scale fishing gears. The dominant fish families in the artisanal catch were the groupers (Family: Serranidae), the Emperor (Family: Lethrinidae), long spine bream (Family: Sparidae), the little tuna, and the Spanish mackerel (Family: Scombridae). More than 100 species of fishes belonging to upto 20 fish families were grouped without any sorting or identification in the "others" group (Mehanna, 1999c). This "others" group constituted about 46.3% of the total artisanal landings indicating that the current system of collecting and recording the fishery statistics lacks accuracy and reality.

4.9.1 The Proper Red Sea

This sector covers the area from Hurghada in the north until the Halayb triangle in the south. Many fishing grounds were identified in this sector, which are Hurghada, Safaga, El-Quseir, Ras Banas, Berenice, Hemira, Shalatein, Abo Ramad, and Halayb (Fig. 4.11).

From all these fishing grounds the Foul Bay fishing area is the most important and has five fishing landing sites Ras Banas, Berenice, Hemira, Shalateen, and Abo Ramad. Foul Bay lies on the Red Sea's west side near our boundaries with Sudan and enclosed between the Ras Banas (23° 55′ N, 35° 47′ E) in the north and Abo Dara in the South (22° 41′ N, 36° 11′ E). Its length is about 150 km and its width varied between 20 and 65 km with a depth of about 200 m (El-Sharkawy, 1984; Mehanna, 2011; Mehanna et al., 2016; Mohamed et al., 2020). The Bay fisheries give back 34% of the total fish production from the Egyptian Red Sea. Three main fishing techniques operate in the Bay; trawlers, purse-seiners, and artisanal or small-scale fisheries particularly long and hand lines. The catch composition is more or less similar to that recorded from the Gulf of Suez.

The artisanal fishery in the proper Red Sea contributed approximately 36% of its total fish production (GAFRD annual reports). This contribution is

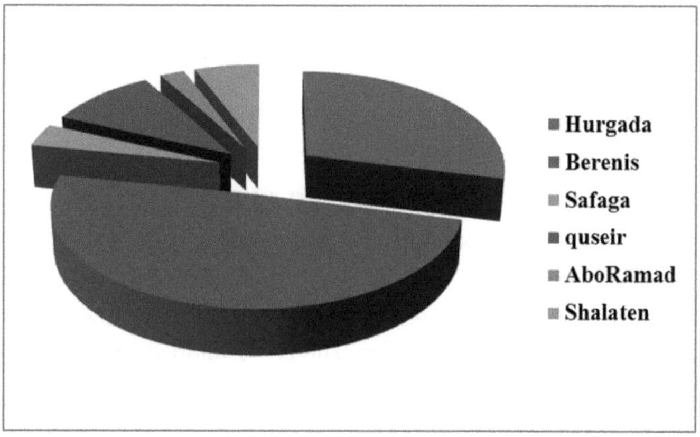

FIGURE 4.11 Contribution of different fishing grounds to fish production in the proper Red Sea.

representing about 55% of the gross profits of that fishery owing to the high value of the fish species fished by the artisanal fishery (Mehanna, 2009, 2011; Osman, 2017; Mohamed et al., 2020). The artisanal fishing (gill, trammel net, and longline) has vessel lengths ranged from 5 to 17 m made from wood. The horsepower of the vessels used gill, and trammel nets are varied between 25 and 80 hp and between 9.9 and 80 hp for the long line vessels. The crew number fluctuated between 3 and 12 persons for gill and trammel net and between 3 and 17 persons for longline. The trip duration is 7–10 days. The fishing season for artisanal fishery started in September and ended in June of each year. The net consists of one layer with narrow mesh sizes in gill nets and consists of three layers in the case of trammel nets. The first and third layers of the trammel net have wider openings than the second or middle layer. The length of both gill and trammel net is varied from 60 to 100 m while line length varied from 50 to 2500 m. Needlefish, squirrelfish, goatfish, rabbitfish, grouper, emperors, and barracuda are the main target species.

In recent years, many challenges were identified in the Foul Bay fisheries like the domination of small-sized fish in trawl and purse-seine landings (growth overfishing indications), decline in the total landings of the Bay throughout the last decades (interviews), increase in the number of fishing units during the previous two decades (interviews), illegal netting around coral reefs, and the damage of the coastal areas as a result of tourism expansion and intensive human activities there (Mohamed et al., 2020).

4.10 GULF OF AQABA

The Gulf of Aqaba is about 170 km in length with an average depth of 650 m and an average width of 16 km. The mean annual fish production does not exceed 200 tons (Fig. 4.12), and the main species are groupers, emperors, red mullet, and mollusks. The main landing sites are Dahab and Nwabaa and the

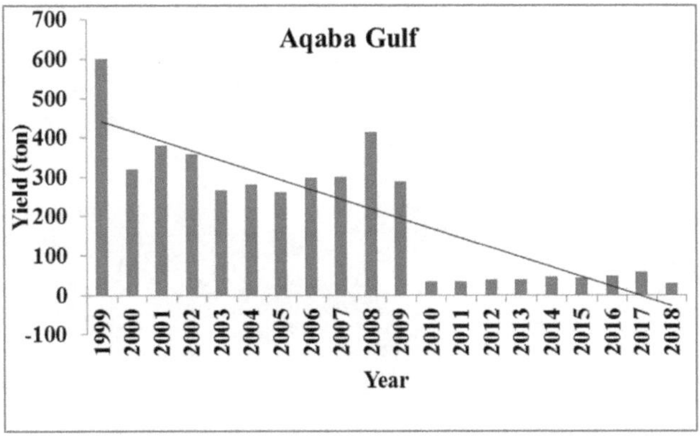

FIGURE 4.12 Fish production from the Gulf of Aqaba from 1999 to 2018.

only fishing method is the artisanal fishing methods. In contrast to the Gulf of Suez and the proper Red Sea fisheries which are heavily studied, the studies on the Gulf of Aqaba fishery are very limited. As most fish species are overexploited in the Gulf of Suez and Foul Bay (Mehanna, 2011; Osman, 2015; Mohamed, 2016; El-Mahdy, 2017; Mehanna et al., 2017, 2018, 2019, 2020), the Gulf of Aqaba fishery is still unexploited. So, we should focus on this valuable resource and explore the most productive areas and how to exploit it rationally.

4.11 MEDITERRANEAN SEA FISHERIES

The Mediterranean Sea (Fig. 4.13) is almost entirely enclosed by land: Europe and Anatolia to the north, North Africa to the south, and the Levant to the east. The sea is connected to the Atlantic Ocean and is sometimes considered a part of the Atlantic Ocean, although it is often identified as a completely different water body.

The Mediterranean sea, as an enclosed sea, is heavily affected by many actions such as intense fishing activity, pollution, and tourism, as well as the invasion of alien species. The bioinvasion of the Mediterranean involves nearly all the major marine taxonomic groups. Fish invade the Mediterranean via three pathways: the Strait of Gibraltar, the Suez Canal, and direct human activities such as shipping, aquaculture, and fish trade. (Mehanna, 2015, 2018a,b; Mehanna and Hassanein, 2020). But the most significant change in the Levantine marine environment took place after the opening of the Suez Canal. Fisheries in the Mediterranean are multispecies, presenting difficulties for using particular gear types.

From the fisheries point of view the Mediterranean sea is considered a fishery data-poor area because the number of stocks that has been assessed as very low in accordance with its biodiversity and the number of exploited fish species (Demirel et al., 2020). Although the Mediterranean in some studies is considered as a part of the Atlantic the stocks assessment are routinely and regularly carried out. The number of these stocks in the Northeast Atlantic is

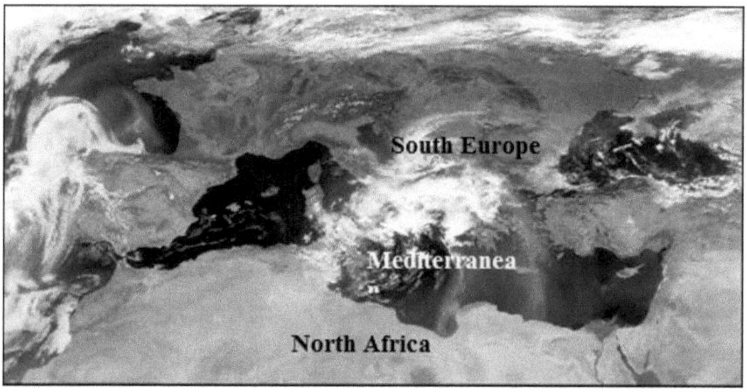

FIGURE 4.13 Mediterranean sea.

higher (Cardinale et al., 2013) than those in the Mediterranean and Black seas (Lleonart, 2015; Stergiou et al., 2016). This situation is mainly a result of the multispecies and multifleets nature of the Mediterranean fisheries (Lleonart and Maynou, 2003). Also, Northern European countries have more advanced capacities for fisheries management. So the Mediterranean and Black seas are classified as fisheries data-poor areas, resulting from two facts: the first is the very few numbers of stocks that empirically assessed and the second one is that the landings and biological data are unavailable (Pilling et al., 2008). Since 2010 this situation has been slightly improved as more assessment studies were done for more fish stocks (Colloca et al., 2013). But the quality of these data is a critical issue and must still be improved upon, especially for the eastern Mediterranean regions, which are incredibly data-poor (Tsikliras et al., 2015).

Recently, many governmental efforts are exerted to conserve and develop fish production in Egypt. Also numerous fisheries research are focusing on the evaluation of different fish stocks to establish target reference points for fisheries management like the ratio of the current fishing mortality F to that which produces the maximum sustainable yield (MSY) F_{msy} and current biomass B/B_{msy} that can achieve efficient management toward sustainable fisheries. On the international level, both the European Union Common Fisheries Policy (EU CFP) and the General Fisheries Commission for the Mediterranean (GFCM) are working hard together along with North Africa and Black sea countries to manage the commercial fish stocks in Mediterranean and Black seas to attain the MSY MSY.

The Egyptian coast of the Mediterranean sea is about 1100 km extending from El-Arish in the East to El-Sallum in the West (Fig. 4.14). This distance is divided as 600 km between El-Sallum and Alexandria with 10–100 m deep, 300 km from Alexandria to Port Said and 200 km from Port Said to El-Arish. A maximum of 70 m long is recorded in front of the Delta. According to GAFRD the Mediterranean sea‘s Egyptian coast is divided into four main fishing grounds namely, western fishing ground which covers Alexandria and El-Mex, Abu-Qir, Rasheed or Rosetta, El-Maadiya, and

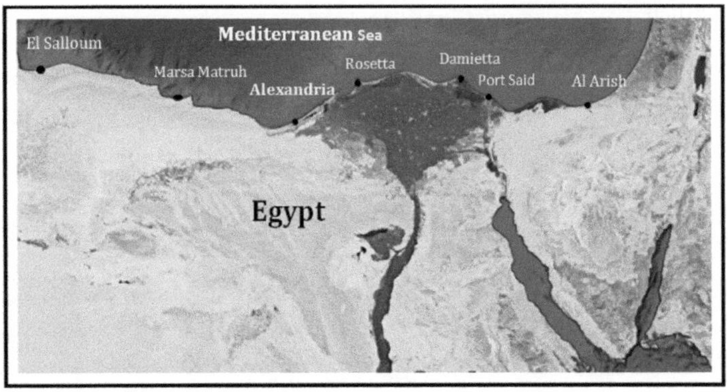

FIGURE 4.14 Egyptian Mediterranean coast with the main landing sites.

Mersa Matrouh landing sites. The Eastern fishing ground has two important landing sites: Port Said and El-Arish, then Demietta and Nile Delta fishing grounds (GAFRD, 2018). Also there are several small landing sites in between these big ones. The western continental shelf is wider than the eastern one, which is very narrow and does not exceed 8 km wide, except in front of the Nile Delta where the 200 m depth contour lies 60 km offshore. There are six coastal lakes and lagoons along the Egyptian coast of Mediterranean Sea with water ranging from brackish to hypersaline. These lakes and lagoons are from east to west: Bardawil lagoon, PortFouad, Manzalah, Borollus, Edku, and Mariut lakes. These lagoons and lakes act as the nursery and sometimes the spawning grounds of several commercially important species inhabiting the Mediterranean. The main fishing grounds used by the Egyptian fleet in the Mediterranean sea are the continental shelf in front of the Nile Delta and may extend to the eastern of Port Said and rarely to the western side of Alexandria. Limited grounds for trawling are available on the western coast due to the spreading of rocky bottoms there (Mehanna, 2009; Mehanna et al., 2011; El-Serafy et al., 2015; Mehanna et al., 2015). The mean annual fish production from this vast area did not exceed 60 thousand tons (GAFRD, 2009–2018 Fig. 4.15). Inshore fisheries widely exist with artisanal fishers along the coast. About 21.3%–33.1% of the catch landed at Alexandria and Damietta respectively (GAFRD, 2009–2018) (Fig. 4.16). There is a closed fishing season for 45 days each year since 2006 (from 1st May to mid-June) in the Egyptian Mediterranean.

4.12 BOTTOM TRAWL FISHERY

A total of 1150 registered bottom trawl vessels are operated on the Mediterranean coast of Egypt, with total length varies between 15 and 30 m. These vessels are motorized with a primary engine of 50–800 hp for each. About 85% of these vessels having engines of 100–250 hp and bottom otter trawl type. All trawlers provided with mechanized winches with crew varied

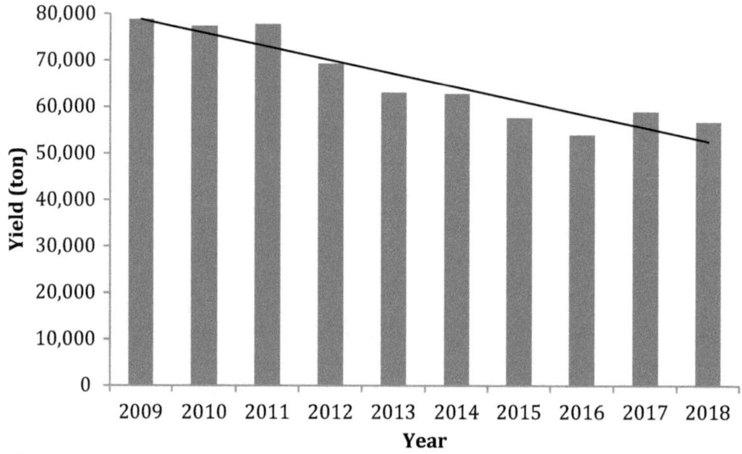

FIGURE 4.15 Fish production from the Egyptian Mediterranean from 2009 to 18.

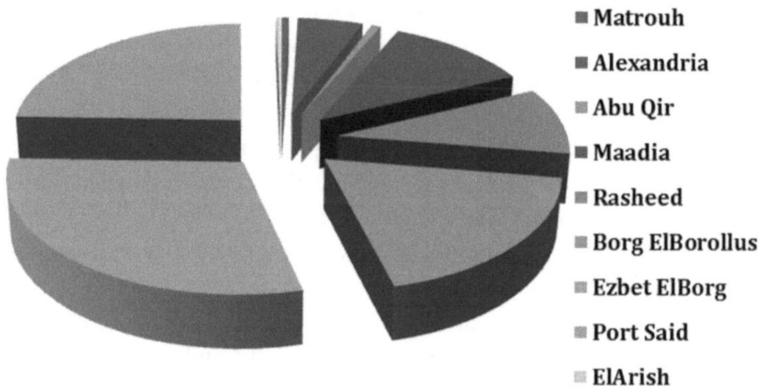

FIGURE 4.16 Fish production by different fishing areas along the Egyptian Mediterranean coast.

in number between 7 and 10 persons. Some trawlers are equipped with GPS and echosounders and use a trawl net of old Italian type with some modifications. The lengths of trawl nets used in these vessels are about 70–110 m, and the average cod-end mesh ranged between 1.5 and 2 cm. The trawl fleet has different sizes to exploit diverse fish stocks and multiple fishing areas based on the depth and distance from the fishing port. The catch of trawlers is composed of many target species such as shrimp, cuttlefish, crabs, sparid fishes, red mullet, lizard fish, triglid fish, flatfish, hake, rabbitfishes, snappers, and grouper (Fig. 4.17). Many other commercial species are also caught as bycatch (Hashem, 1972; Rizkalla, 1992; Mehanna, 2007b,c, 2014, 2018; El-Serafy et al., 2015; Mehanna et al., 2015; Hassanein, 2017). Trawl occupy the first grade in the fishing methods in Egyptian Mediterranean water and contribute about 38% of the total techniques used.

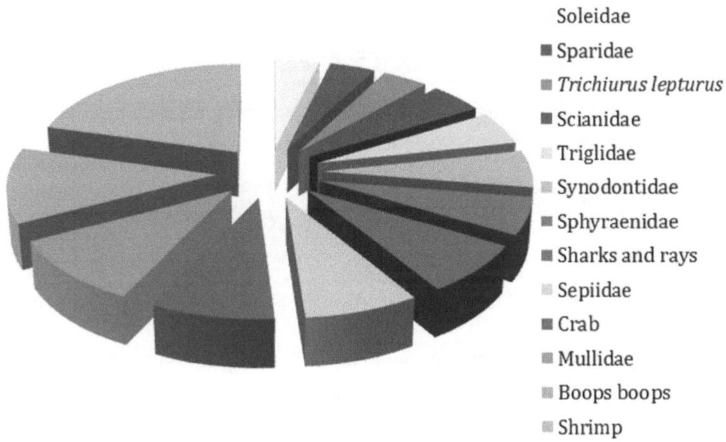

FIGURE 4.17 Catch composition of trawl fishery in the Egyptian Mediterranean sea.

4.13 PURSE SEINE FISHERY

Purse seine fishing method, which is locally called chancholla, is of great importance in Egypt, where it is used for catching the pelagic species. About 238 registered purse seine vessels, with length ranged between 15 and 25 m. These vessels are motorized with engines of 50–500 hp, with about 68% of them having engines of 100–200 hp (GAFRD, 2018). Purse seine fleet in the Egyptian Mediterranean waters can be divided into small purse-seine boats that working through the day in shallow areas without artificial lights, and large purse-seine boats that are working during the night with the assistance of small boats provided with artificial lights to gather and concentrate the fishes before setting the net. Generally, fishing activities during the night ceased for about 10 days each month when the moon is full. The length of the seine net fluctuates between 200 and 400 m with depth varied from 40 to 60 m. The nets are hauled manually, and the crew number varies between 25 and 30 individuals per boat. The catch of purse-seine fishery is composed mainly of Sardines (*Sardina pilchardus*, and *Sardinella aurita*), anchovies (*Engraulis encrasicolus*), carangid species (*Trachurus trachurus* and *T. mediterraneus*), mullets, and little tuna (Faltas, 1983; Sanders et al., 1984a; Mehanna et al., 2008; Mehanna and Haggag, 2010; Mehanna et al., 2011; Mehanna and Salem, 2012). The purse seine fishery contributes about 8% of the fishing techniques in the Egyptian Mediterranean waters (Fig. 4.18).

4.14 ARTISANAL OR SMALL-SCALE FISHERIES

The artisanal or small-scale fisheries in the Egyptian Mediterranean are represented by about 1800 boats, with a length range of 7–15 m, and are provided with small outboard or inboard engines of 8–150 hp (GAFRD, 2010–2018). The duration of the fishing trip is about 1–5 days, and the crew number ranges from 2 to 8 persons per boat. This fishery sector uses several fishing methods includes long and hand lines, gill nets, crab nets, and trammel nets.

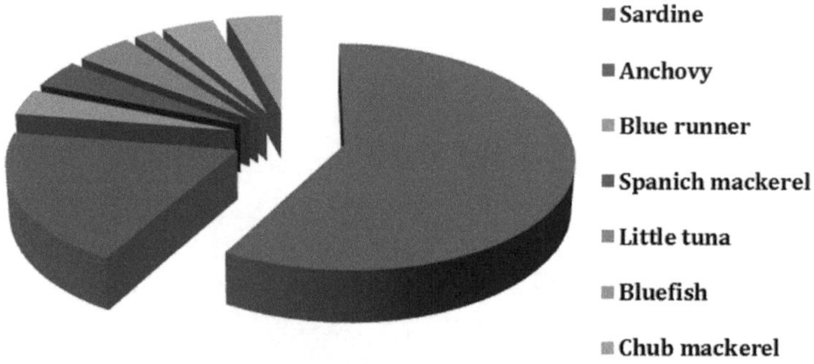

FIGURE 4.18 Catch composition of purse-seine fishery in the Egyptian Mediterranean sea.

The small-scale fishery targets both pelagic and demersal fish species according to their seasonality. The hooks and lines constitute about 36% of the total methods, while the different kinds of nets are 18% of the whole fleet.

4.15 SUSTAINABILITY OF EGYPTIAN MARINE FISHERIES

The basic standard that administrates the sustainable development of fisheries is that it should be conducted in a way that does not lead to overfishing or for those fish stocks that are already overfished. The given fishery should be operated toward the maximization of probability degree that ensures the recovering of the stock(s). Also fishing activities should be managed to minimize their effect on the population structure, productivity, ecosystem functions, and ecosystem biological diversity. The severe decline in marine resources fish production during the last decade raises concern about the sustainable development of Egyptian marine fisheries. The fundamental problems facing sustainable development and lead to the decline of marine fish production in Egypt are:

1. Overfishing, pollution, and habitat degradation affected productivity and the carrying capacity of the fishing areas.
2. Climate change, invasive species, and IUU fishing.
3. Weak governance: laws, policy, and enforcement.
4. Competition between several sectors on the marine space (marine resources and services) and the conflict among users.
5. Damage the benthic ecosystem and benthos by extensive sweeping of the same fishing ground by bottom trawl nets which frequently destroying the food web of commercially vulnerable species.
6. Conflicts among different categories of fishers particularly between the artisanal and mechanized fishermen.
7. Harvest ad postharvest losses.
8. Lack of Participatory Fisheries Management.

9. Market access barriers.
10. High operating cost (fuel, ice, spare parts, maintenance, food, etc.).
11. No integrated-governance framework.

With all these challenges, ensuring sustainability is a difficult but necessary task (Modayil, 2005). So, some suggested options are as follows:

1. Implementing the Code of Conduct for Responsible Fisheries and the Ecosystem Approach to Fisheries.
2. Reducing the negative impacts of fishing on the environment through technological and community-based management solutions.
3. Fighting illegal, IUU fishing IUU.
4. Develop and implement international rules relating to fisheries activities comprising bycatch management and discards reduction as well as decrease the fish loss and supply chain effectiveness.
5. Improve the understanding of fisheries' socioeconomic issues taking into account market access and value chain dynamics, the fisheries' resources status, and rights of different users.
6. Develop a governance framework for the integrated use of marine resources includes: Legislation and policies, Institutional arrangements, Training, and capacity building, sustainably utilize the ecosystem goods and services.
7. Improve the trawl selectivity, reduce the fishing effort, and protect the spawning and nursery grounds through the implementation of MPA.
   a. Prohibition the trawling, trammel nets, and gillnets in the shallow coastal areas to reduce the portion of undersized specimens, protect juveniles and increase or at least maintain the spawning stock biomass SSB.
8. Establishing an accurate fisheries database with continuous updating.
9. Increasing awareness of fishers about illegal fishing gears, bycatch risk, environmental issues, and the importance of fisheries regulatory measures.
10. Recognizing small-scale fisheries as a fundamental contributor to poverty alleviation and food security for securing sustainable small-scale fisheries.

Nowadays, there is an increasing necessity for coordination and collaboration among all interested parties at all available levels for more sustainable and effective fisheries management and better protection. Also more efforts for capacity development are required, particularly those strengthening the environment policy, institutional preparations as well as cooperative processes that empower fisher's communities, civil societies, and public bodies.

REFERENCES

Cardinale, M., Dorner, H., Abella, A., Andersen, J.L., Casey, J., Doring, R., et al., 2013. Rebuilding EU fish stocks and fisheries, a process under way? Marine Policy 39, 43–52. Available from: https://doi.org/10.1016/j.marpol.2012.10.002.

Central Agency for Public Mobilization and Statistics, 2019. Central Agency for Public Mobilization and Statistics, annual statistical report 2019.

Colloca, F., Cardinale, M., Maynou, F., Giannoulaki, M., Scarcella, G., Jenko, K., et al., 2013. Rebuilding Mediterranean fisheries: a new paradigm for ecological sustainability. Fish and Fisheries 14, 89–109. Available from: https://doi.org/10.1111/j.1467-2979.2011.00453.x.

Demirel, N., Zengin, M., Ulman, A., 2020. First large-scale eastern mediterranean and black sea stock assessment reveals a dramatic decline. Frontiers in Marine Science 7, 103. Available from: https://doi.org/10.3389/fmars.2020.00103.

El-Gammal, F.I., Breika, M.I., 1994. Stock assessment of the Red Sea fish resources. Report submitted to the Academy of Scientific Research & Technology. 188pp.

El-Gammal, F.I., Mehanna, S.F., 2002. Purse seine fishery in the gulf of Suez with special reference to sardine fishery. Asian Fisheries Science 15, 83–89.

El-Gammal, F.I., Mehanna, S.F., 2006. Management of invertebrate fisheries in the Gulf of Suez, Red Sea. Journal Agricultural Marine Science Sultan Qaboos University 11 (1), 37–45.

El-Mahdy, S.M., 2017. Biological Studies, Population Dynamics and Stock Assessment of *Acanthopagrus bifasciatus* From the Red Sea. Egypt. (Ph.D. thesis). Assiut University.

El-Serafy, S.S., El-Gammal, F.I., Mehanna, S.F., Abdel-Hamid, N.H., Farrag, E.F.E., 2015. Age, growth and reproduction of the tub gurnard, *Chelidonichthys lucerna* (Linnaeus, 1758) from the Egyptian Mediterranean waters off, Alexandria. International Journal of Fisheries & Aquatic Sciences 4 (1), 13–20.

El-Sharkawy, S.M., 1984. Physics Investigations Foul Bay Red. Sea. Technical Report, 1 Part 1, 78p.

Faltas, S.N., 1983. Study of Purse Seine fishery in Egyptian Mediterranean Waters with Special Reference to the Biology of Sardine in the Catch. (MSc. thesis). Faculty of Science, Alex University, 194 pp.

FAO, 2012. The State of Food Insecurity in the World. Rome. 61 pp. <http://www.fao.org/docrep/016/i3027e/i3027e00.htm>.

FAO, 2020. The State of World Fisheries and Aquaculture 2020. Sustainability in Action. FAO, Rome. Available from: https://doi.org/10.4060/ca9229en.

Farrag, M.M., Osman, A.G.M., Mehanna, S.F., Osman, Y.A., 2018. Fisheries status of the common species of family Mullidae in the Southern Red Sea, Hurghada, Egypt. Egyptian Journal of Aquatic Biology and Fisheries 22 (5), 249–265.

GAFRD, 1990–2018. Publications of the General Authority for Fish Resources Development Annual Fishery Statistics Report. Ministry of Agriculture, Egypt.

Hashem, M.T., 1972. Bottom trawling survey for Abu-Qir Rosetta region during 1969–1970. Bulletin of the National Institute of Oceanography and Fisheries 2, 1–22.

Hassanein, E.M., 2017. Biological Studies, Stock Assessment and Fisheries Management of Some Species of Family Mullidae From the Mediterranean Sea at Alexandria. (Ph.D. thesis). Faculty of Science, Assiut University.

Head, S.M., 1987. Corals and Coral Reefs of the Red Sea International Union for Conservation of Nature and Natural Resource. IV. Series, pp. 128–149.

Latif, A.F., 1974. Marine fisheries of the Gulf of Suez. Journal of the Marine Biological Association of India 16 (1), 24–34.

Lleonart, J., 2015. Mediterranean fisheries stocks, assessments and exploitation status. IEMED Mediterranean yearbook 2015. Strategics Sectors | Economy & Territory 2015, 276–281.

Lleonart, J., Maynou, F., 2003. Fish stock assessments in the Mediterranean: state of the art. Scientia Marina 67, 37–49. Available from: https://doi.org/10.3989/scimar.2003.67s137.

Mehanna, S.F., 1993. Rational Exploitation of Kuruma Shrimp *Penaeus japonicus* Bate, 1888 in the Gulf of Suez. (MSc. thesis). Faculty of Science, Zagazig University, p 236.

Mehanna, S.F., 1997. A study of the Biology and population dynamics of *Lethrinns mahsena* (Forsskal, 1775) in the Gulf of Suez, Faculty Science of Zagazig University Egyptian. PhD Thesis.

Mehanna, S.F., 1999a. An assessment and management of the coral reef fish stocks in the Gulf of Suez. Egyptian Journal of Aquatic Biology and Fisheries 3 (2), 103–114.

Mehanna, S.F., 1999b. Stock assessment of the horse mackerel *Trachurus indicus* in the Gulf of Suez, Egypt. Indian Journal of Fisheries 46 (4), 327–335.

Mehanna, S.F., 1999c. Population dynamics of the round scad *Decapterus macrosoma* (Bleeker, 1951) in the Gulf of Suez, Egypt. Egyptian Journal of Aquatic Biology and Fisheries 3 (2), 55–68.

Mehanna, S.F., 2001. Population dynamics and fisheries management of the Indian mackerel *Rastrelliger kanagurta* in the Gulf of Suez, Egypt. Journal of KAU Marine Sciences 12, 217–229.

Mehanna, S.F., 2005. Population dynamics of the areolate grouper *Epinephelus areolatus* from the Egyptian sector of Red Sea. In: Proceedings of the Twelfth International Conference of Union of Arab Biologists, November, 2005, El-Hodeida University, Yemen.

Mehanna, S.F., 2007a. Lake Bardawil fisheries: current status and future sight. Journal Egyptian - German Society of Zoology 51 (D), 91–105.

Mehanna, S.F., 2007b. Stock assessment and management of the Egyptian sole Solea aegyptiaca Chabanaud, 1927 (Osteichthyes: Soleidae), in the Southeastern Mediterranean (Port Said region), Egypt. Turkish Journal of Zoology 31, 379–388.

Mehanna, S.F., 2007c. A preliminary assessment and management of gilthead bream *Sparus aurata* in Port Said fishery, Southeastern Mediterranean, Egypt. Turkish Journal of Fisheries and Aquatic Sciences 7 (2), 123–130.

Mehanna, S.F., 2007d. Fisheries assessment of Lake Nasser, Egypt with special reference to tilapia fishery. Journal of Egyptian Academic Society for Environmental Development 8 (4), 27–35.

Mehanna, S.F., 2008. Northern Delta Lakes, Egypt: Constraints and Challenges. Tropentag 2008, Hohenheim University, Germany, 7–9 October.

Mehanna, S.F., 2009a. Growth, mortality and spawning stock biomass of striped red mullet *Mullus surmuletus*, in the Egyptian Mediterranean waters. Mediterranean Marine Science 10 (2), 5–17.

Mehanna, S.F., 2009b. Stock assessment of *Mullus barbatus* in GSA 26. In: GFCM Report of the Working Group on Stock Assessment of Demersal Species. Ancona, Italy. 19–23 October.

Mehanna, S.F., 2009c. Stock assessment of *Merluccius merluccius* in GSA 26. In: GFCM report of the working group on stock assessment of Demersal species. Ancona, Italy, 19–23 October.

Mehanna, S.F., 2010a. Stock assessment of *Boops boops* in GSA 26 Egyptian waters (2010). In: Report of the SCSA Working Group on Stock Assessment of Demersal Species, Istanbul, Turkey. 18–23 October. GFCM: SAC13/2011/Inf.19.

Mehanna, S.F., 2010b. Stock assessment of *Pagellus erythrinus* in GSA 26 Egyptian waters. In: Report of the SCSA Working Group on Stock Assessment of Demersal Species, Istanbul, Turkey. 18–23 October. GFCM: SAC13/2011/Inf.19.

Mehanna, S.F., 2011. Population dynamics and management of snubnose Emperor *Lethrinus bungus (L. borbonicus)* from the Foul Bay, Red Sea. In: Proceedings of the INOC-XI International Symposium 2011, Bogor, Indonesia, pp. 121–129.

Mehanna, S.F., 2013. Sustainable development of Bardawil lagoon fisheries. In: Proceedings of the First Regional Symposium on Sustainable Small-Scale Fisheries 27–30 November, St. Julian's Malta. conference paper.

Mehanna, S.F., 2014. Population dynamical parameters of the Atlantic lizardfish *(Synodus saurus)* from the Mediterranean waters of Egypt. In: Proceedings of the International Congress Estuaries & Coastal Protected Areas, ECPA 2014, 4–6 November 2014, Turkey.

Mehanna, S.F., 2015. The positive impacts of the Lessepsian migration in Egypt. In: Proceedings of the MedCoast Annual Conference, Bulgaria, 5–7, October.

Mehanna, S.F., 2018a. Population dynamics of two Lessepsian migrant goatfish species; *Upeneus pori and Upeneus moluccensis* from the southeastern Mediterranean, Port

Said region, Egypt. In: Proceedings of the First Euro-Mediterranean Conference for Environmental Integration, 22–25 November, 2017, Sousse, Tunisia (Springer).

Mehanna, S.F., 2018b. Impacts of some alien species in Egypt: crayfish as fresh water species and puffer fish as marine species. In: FishForum 2018, FAO: Rome.

Mehanna, S.F., 2019a. An overview on fish production in Egypt and how to achieve its sustainability. In: Proceedings of the Third International Conference for Women in Science, 12–14 March.

Mehanna, S.F., 2019b. Stock assessment of European hake, *Murleccius murleccius* from the Egyptian waters of Mediterranean Sea. In: Proceedings of the Expert Group Meeting on Hake, Rome, 2–7 December.

Mehanna, S.F., 2020a. Challenges faced the small scale fisheries and its sustainable development. In: ICAR – Central Marine Fisheries Research Institute, Research Centre Mangalore, 7–10 January.

Mehanna, S.F., 2020b. An overview of the aquaculture sector in Egypt: current trend and the development strategies. Tropentag 2020, 9–11 September, Germany.

Mehanna, S.F., 2020c. Isopod parasites in Egyptian fisheries and its impact on fish production: Lake Qarun as a case study. Egyptian Journal of Aquatic Biology and Fisheries 24 (3), 181–191.

Mehanna, S.F., Abdallah, M., 2002. Population dynamics of the rabbit fish *Siganus rivulatus* from the Egyptian sector of the Red Sea. Journal of King Abdulaziz University, Marine Science 13, 161–170.

Mehanna, S.F., Abo, M., Elregal, Aid, N.M., 2015. Critical lengths, mortality rates and relative yield per recruit of the common sole *Solea solea* from the Egyptian Mediterranean coast off Alexandria. Egyptian Journal of Aquatic Biology and Fisheries 19 (2), 13–20.

Mehanna, S.F., El-Gammal, F.I., 2005. Stock assessment of the round herring, *Etrumeus teres* (DeKay, 1842) from the Egyptian sector of Red Sea. Indian Journal of Fisheries 52 (4), 377–383.

Mehanna, S.F., El-Gammal, F.I., 2007. Gulf of Suez fisheries: current status, assessment and management. Journal of King Abdulaziz University, Marine Science 18, 3–18.

Mehanna, S.F., El-Sherbeny, A.S., Ahmed, M.H., El-Kafrawy, S.B., Khalil, M.T., 2020. Dynamical parameters and relative yield per recruit of the Japanese anchovy *Engraulis japonicus* (Schlegel, 1846) in the Gulf of Suez, Red Sea, Egypt. Egyptian Journal of Aquatic Biology and Fisheries 24 (1), 399–405.

Mehanna, S.F., Fattouh, S.A., Farrag, E.F., El-Sherbeny, A.S., 2011. Spatio–temporal variations in the distribution and abundance of demersal fishes in the Egyptian Mediterranean waters. In: Proceedings of the Fourth International Conference on Fisheries and Aquaculture, 3–5 October, Cairo, Egypt.

Mehanna, S.F., Haggag, H.M., Sallam, N.A., 2008. Population dynamics and management of *Sardinella aurita* (Osteichthyes: Clupeidae), in the Southeastern Mediterranean, Egypt. Journal of the Egyptian Academy of Environmental Development 9 (1), 65–75.

Mehanna, S.F., Haggag, H.M., 2010. Port Said Fisheries: current status, assessment and management. In: Proceedings of the Third International Conference on Fisheries and Aquaculture, 29 November–1 December, Cairo, Egypt. <http://www.cabdirect.com>.

Mehanna, S.F., Haggag, H.M., 2011a. Stock assessment of the common cuttlefish, *Sepia officinalis* in the Southeastern Mediterranean, Egypt. In: Proceedings of the Fourth International Conference on Fisheries and Aquaculture, 3–5 October, Cairo, Egypt.

Mehanna, S.F., Haggag, H.M., 2011b. Maximum sustainable yield of the flatfishes (Family: Soleidae) from Port Said waters, Mediterranean sea, Egypt. In: Proceedings of the Fourth International Conference on Fisheries and Aquaculture, 3–5 October, Cairo, Egypt.

Mehanna, S.F., Hassanien, E.M., 2020. Lessepsian Migration with the First Record of the Red Sea Goatfish, *Parupeneus forsskali* (Fourmanoir & Guézé, 1976) in the Coastal Waters of Egyptian Mediterranean Sea. In: Advances in Agriculture and Fisheries Research Vol. 1. Available from: https://doi.org/10.9734/bpi/aafr/v1.

Mehanna, S.F., Mahmoud, U., Mohammed, A.S., 2016. Fishery status of *Carangoides bajad* and *Caranx melampygus* (Family Carangidae) from Shalateen Fishing Area, Red Sea, Egypt, based on yield per recruit (Y/R) analysis. Egyptian Journal of Aquatic Biology and Fisheries 20 (2), 61–68.

Mehanna, S.F., Mohammed, A.S., Mohsen, S., Abdel-Maksoud, Y., 2018. Stock assessment and management of rabbitfish *Siganus rivulatus* from the Southern Red Sea, Egypt. In: Proceedings of the Second International Conference, Faculty of Science, Ain Shams University (Sustainable Innovations and Sustainable Development), 23–26 October.

Mehanna, S.F., Osman, A.M., Farrag, M.M., Osman, Y.A., 2017. Age and growth of three common species of goatfish exploited by artisanal fishery in Hurghada fishing area, Egypt. Journal of Applied Ichthyology 34 (4), 917–921.

Mehanna, S.F., Osman, Y.A., Khalil, M.T., Hassan, A., 2019. Age and growth, mortality and exploitation ratio of *Epinephelus summana* (Forsskål, 1775) and *Cephalopholis argus* (Schneider, 1801) from the Egyptian Red Sea coast, Hurghada fishing area. Egyptian Journal of Aquatic Biology and Fisheries 23 (4), 65–75.

Mehanna, S.F., Salem, M., 2012. Population dynamics of round sardine *Sardinella aurita* in El-Arish waters, southeastern Mediterranean, Egypt. Indian Journal of Fundamental and Applied Life Science 1 (4), 286–294.

Modayil, M.J., 2005. How do we ensure sustainability in the open access marine fisheries of India? In: Souvenir Sustain Fish 2005, 16–18 March 2005, Cochin, India, pp. 55–57.

Mohamed, A.S., Mehanna, S.F., Osman, Y.A., El-Mahdy, S.M., 2020. Age, growth and population parameters of the spiny squirrelfish, *Sargocentron spiniferum* (Forsskål, 1775) from Shalateen fishing area, Red Sea, Egypt. Egyptian Journal of Aquatic Biology and Fisheries 24 (2), 469–480.

Mohammad, A.S., 2016. Biological Studies, Stock Assessment and Fisheries Management of *Carangoides bajad* (Forsskal, 1775) and *Caranx melampygus* (Cuvier, 1833) from the Red Sea, Egypt. (Ph.D. thesis). Faculty of Science, Assiut University.

Morcos, S.A., 1970. Physical and chemical oceanography of the Red Sea. Oceanography and Marine Biology – An Annual Review 8, 73–202.

Ormond, R.F.G., Edwards, A., 1987. Red Sea fishes. In: Edwards, A.J., Head, S.M. (Eds.), Red Sea. Pergamon Press, Oxford, pp. 252–287.

Osman, Y.A., 2015. Population Dynamics and Stock Assessment of the Most Important Species of Family Scaridae (Parrot) in Hurghada, Red Sea. (MSc. thesis). Fishery and Aquaculture Technology, Faculty of Science, Port Said University.

Osman, Y.A., 2017. Fisheries management of the common species of family Mullidae in Hurghada, Red Sea, Egypt. Al-Azhar University, Assiut Branch. PhD Thesis.

Osman, Y.A., Mehanna, S.F., El-Mahdy, S.M., Mohamed, A.S., Mahe, K., 2020. Age precision and growth rate of *Rhabdosargus haffara* (Forsskål, 1775) from Hurghada fishing area, Red Sea, Egypt. Egyptian Journal of Aquatic Biology and Fisheries 24 (2), 341–352.

Pilling, G.M., Apostolaki, P., Failler, P., Floros, C., Large, P.A., Morales-Nin, B., et al., 2008. Assessment and management of data-poor fisheries. In: Payne, A., Cotter, J., Potter, T. (Eds.), Advances in Fisheries Science: 50 Years on From Beverton and Holt. Blackwell Publishing, Hoboken, NJ, pp. 280–305. Available from: https://doi.org/10.1002/9781444302653.ch12.

Rafail, S.Z., 1970. Studies of populations and exploitation status of Egyption Red Sea abundant sardines. Bulletin of the National Institute of Oceanography and Fisheries 1, 129–148.

Rafail, S.Z., 1972. Studies of Red Sea fisheries by light and purse-seine near Al-Ghardaqa. Bulletin of the National Institute of Oceanography and Fisheries 2, 25–49.

Rizkalla, S.I., 1992. Studies on the Fishery Biology of Sea Breams (Genus: Pagellus) in the Egyptian Mediterranean Waters. (Ph.D. thesis). Faculty of Science, Alexandria University, p. 209.

Salem, S.A., 1971. Biological Studies on Family Lethrinidae from the Red Sea. (MSc. thesis), Cairo University, Egypt.

Sanders, M.J., Kedidi, S.M., 1981. Summary Review of the Red Sea Commercial Fisheries, Catches and Stock Assessments Including Maps of Actual and Potential Fishing Grounds. Cairo, Project for the Development of Fisheries in the Red Sea and Gulf of Aden, FAO/UNDP, RAB/77/008/19, p. 49.

Sanders, M.J., Kedidi, S.M., 1984a. Catches, Fishing Effort, Catches per Fishing Effort and Fishing Locations for the Gulf of Suez and Egyptian Red Sea Fishery for Reef Associated Fish During 1979–1982, UNDP/FAO. RAB/81/002/22, p. 54.

Sanders, M.J., Kedidi, S.M., 1984b. Catches, Fishing Effort, Catches per Fishing Effort and Fishing Locations for the Gulf of Suez and Egyptian Red Sea Fishery for Reef Associated Fish During 1979–1982, UNDP/FAO. RAB/83/002/02, p. 85.

Sanders, M.J., Morgan, G.R., 1989. Review of the Fisheries Resources of the Red Sea and Gulf of Aden. FAO Fisheries Technical Report. 304, p. 138.

Sea Around Us, 2007. A Global Database on Marine Fisheries and Ecosystems. Fisheries Centre, University British Columbia, Vancouver, Canada. <http://www.seaaroundus.org/ lme/Summary Info. aspx? LME=33>.

Sheikh-Eldin, M.Y., 1988. Biological Studies on Certain Marine Teleosts. (MSc. thesis), Ain Shams University, Egypt.

Shepperd, C., Price, A., Roberts, C., 1992. Marine Ecology of the Arabian Region: Patterns and Processes in the Extreme Tropical Environments. Academic Press, Cambridge.

Stergiou, K.I., Somarakis, S., Triantafyllou, G., Tsiaras, K.P., Giannoulaki, M., Petihakis, G., et al., 2016. Trends in productivity and biomass yields in the Mediterranean large marine ecosystem during climate change. Environmental Development 17 (Suppl.1), 57–74. Available from: https://doi.org/10.1016/j.envdev.2015.09.001.

Tesfamichael, D., Mehanna, S.F., 2012. Red Sea fisheries of Egypt: heavy investment and their consequences. In: Tesfamichael, D., Pauly, D. (Eds.), Catch Reconstruction for the Red Sea Large Marine Ecosystem by Countries (950–2010), vol. 20. Fisheries Centre Research Reports, Vancouver.

Tsikliras, A.C., Dinouli, A., Tsiros, V., Tsalkou, E., 2015. The Mediterranean and Black Sea fisheries at risk from overexploitation. PLoS One 10, e0121188. Available from: https://doi.org/10.1371/journal.pone.0121188.

Zaahkouk, S.A., Khalaf-Allah, H.M., Mehanna, S.F., El-Gammal, F.I., Makkey, A.F., 2017. Studies on age, growth, and mortality rates for management of the redspot emperor, *Lethrinus lentjan* (Lacepède, 1802) in the Egyptian sector of Red Sea. Egyptian Journal of Aquatic Biology and Fisheries 21 (1), 63–72.

Zekeria, Z.A., 2003. Butterflyfishes of the southern Red Sea: Ecology and population dynamics. (Ph.D. thesis), University of Groningen, Groningen, the Netherlands, p. 120.

Chapter | five

Life Cycle Impact of Industrial Aquaculture Systems

Paula Gabriela da Silva Pires[1], Ines Andretta[1], Maria Sara Cabrera Mendéz[1], Marcos Kipper[2], Naglezi de Menezes Lovatto[3] and Bruno Bianchi Loureiro[3]

[1]Universidade Federal do Rio Grande do Sul, Porto Alegre, Rio Grande do Sul, Brazil, [2]Elanco Animal Health, São Paulo, São Paulo, Brazil, [3]Universidade Federal de Santa Maria, Santa Maria, Rio Grande do Sul, Brazil

5.1 INTRODUCTION

Traditionally, fisheries were the primary source of seafood for human consumption. However, with the majority of fish stocks now caught at maximum capacity or unsustainable levels, in recent decades, there has been a progressive transition from seafood fisheries to aquaculture (FAO, 2016). Aquaculture has been playing a pivotal role in increasing food production for human nutrition and food security (Béné et al., 2016). Global per capita fish consumption reached 20.5 kg in 2018 (FAO, 2020). The rapid development of aquaculture has been considered part of the "blue revolution" (Simpson, 2011).

According to data from FAO (2016), aquaculture is one of the food production sectors with the most expressive growth in the world and offers little more than 50% of all fish consumed in human eat. As a result the impact of aquaculture on the environment and the understanding of its sustainability have been a concern since the early 1990s (Folke and Kautsky, 1992; Naylor et al., 2000; Perdikaris et al., 2016). However, aquaculture continues to be associated with a series of impacts on the environment (Naylor et al., 2000; Diana, 2009; Ottinger et al., 2016). The growth of aquaculture production leads to a more significant environmental burden, such as eutrophication of water bodies, the entry of invasive species, parasites and diseases for wild populations, greater use of antibiotics, and effluent emissions contribute to

Sustainable Fish Production and Processing. DOI: https://doi.org/10.1016/B978-0-12-824296-4.00011-6

climate change (Henriksson et al., 2015a,b). These impacts may generate costs for the whole society and problems for the farmers themselves due to production and, consequently, financial losses (Neiland et al., 2001).

Sustainability requires a human lifestyle within limits imposed by nature, or in other words, we must live within the capacity of natural capital (Valenti et al., 2018). In the concept of sustainability, production systems must persist throughout human generations. Every future generation must inherit a stock of natural resources equal to or greater than that inherited by the previous generation (WCED, 1987).

It is not enough to just monitor the development of aquaculture, it is necessary to measure sustainability to assess the strengths and weaknesses of each current aquaculture system. The economic, environmental, and social dimensions of sustainability must be considered. Thus, comparing measures with variables of a very different nature is mandatory. Among the methods to assess sustainability in the aquaculture system is the life cycle assessment (LCA) (Aubin et al., 2009; Santos et al., 2015; Medeiros et al., 2017).

LCA methodology can quantify and compare the environmental impacts of productive systems throughout their life cycle (ISO, 2006). Since the mid-2000s, LCA has also been applied to aquaculture products (Avadí and Fréon, 2013). This method has the advantage of its ability to address a wide range of environmental issues and to include the entire life cycle of a product, thus avoiding changing problems. It presents a scientific and quantitative approach, in addition to the ability to allow comparisons of products or services, using a standard functional unit (Nemecek et al., 2016). Considering the relevance of the topic, this chapter will systematically analyze and summarize LCA studies focusing on aquaculture that can mitigate the environmental impact of production systems.

5.2 WORLD MARKETS AND TRADE OF AQUACULTURE SYSTEMS

The per capita food fish consumption increased at an average rate of 1.5% a year, against 1.1% a year for total meat consumption in the same period. Consumption depicted in 1961 was 9.0 kg (equivalent to live weight) and reached 20.3 kg in 2017; it is estimated that in 2018 the per capita fish consumption at 20.5 kg (FAO, 2020).

According to FAO (2020), in 2018 world aquaculture fish production reached 82.1 million tons. The aquaculture is dominated by finfish (54.3 million tons—47 million tons from inland aquaculture and 7.3 million tons from marine and coastal aquaculture), mollusks, mainly bivalves (17.7 million tons), and crustaceans (9.4 million tons).

The global aquaculture production showed an annual growth of 10% in 1990, falling to 5% over 2018 (FAO, 2020). Presently aquaculture exceeded wild-capture fisheries as the primary source of seafood for human consumption (FAO, 2016). While almost all aquaculture production is destined for human consumption, the amount of fish produced in 2018 estimated that 88% was used for human consumption. The rest (12%) was used to feed aquatic and terrestrial organisms (in the form of fish meal and oil) and in the pharmaceutical industry (FAO, 2020).

Seafood protein represents an essential nutrient in many countries, especially where total protein intake levels are low. Capture fisheries and aquaculture make vital contributions to food security due to the composition of high-quality proteins, essential amino acids, polyunsaturated fatty acids, vitamins, and minerals (Godfray et al., 2010; Guillen et al., 2019; FAO, 2020). Due to the diversity of aquatic life the nutritional composition of seafood differs significantly between species and way of processing and commercialization. However, seafood is a versatile food product; a wide variety of species can be prepared in many different ways.

In addition to these interactions, if we compare to other commodities, the quantities of seafood products, produced worldwide, that are traded internationally is very high. This mainly occurs mostly due to globalization and the geographical discrepancy between aquaculture production and demand for seafood products (Guillen et al., 2019). The globalization of the seafood market makes it possible to find seafood from all over the world (Watson et al., 2016, 2017).

Thus the discussion about the long-term sustainability of aquaculture when compared to captured seafood resources (mainly small and medium pelagic) should be placed in a systematic context of the global market. One must still consider the dependencies between the demand for seafood, fisheries capture, aquaculture, livestock, and the feed industry (Tacon and Metian, 2009; Guillen et al., 2019).

Considering the food industry, we must consider the species derived from extractive fishing that is used in the production of ingredients for the manufacture of feed, which can be used directly in human food. The main species used for the manufacture of fish meal and fish oil are anchovies, menhaden, or herring.

Carnivorous species are the ones that most benefit from the use of these ingredients in their diets due to the high quantitative and qualitative requirement of protein and lipids in the diets. This is an essential factor in the aquaculture market since most fish species derived from intensive aquaculture monoculture are carnivorous species, with high commercial value and demand for population consumption.

5.3 ENVIRONMENTAL IMPACTS OF AQUACULTURE SYSTEMS

Presently, 90% of global wild-fish stocks are overfished or fished at capacity and, the seafood extractable from the wild has probably reached a ceiling or is declining. Future expansion of seafood will come from aquaculture. Despite the rapid development of aquaculture, adverse social and environmental effects can occur, including habitat destruction, overfishing for feed ingredients, and social displacement (Willett et al., 2019; FAO, 2020). Table 5.1 (adapted from Fernandes et al., 2001) summarizes current environmental concerns from aquaculture.

Sustainable food systems and the EAT-Lancet Planetary Health guidelines recognize that fish consumption from sustainable fisheries and aquaculture is highly recommended (Willett et al., 2019). Linked to the increased consumption of seafood is the increase in the production of feed for aquaculture. Research in sustainable aquaculture feeds rapidly developing; however, development and implementation remain overdue.

The future environmental footprint of seafood depends on the species farmed, what they eat, and where aquaculture takes place. Aquaculture will not solve the challenges posed by feeding about 10 billion people healthy diets but could help steer the production of animal source proteins toward reduced environmental effects and enhanced health benefits.

On the other hand, intensive aquaculture can significantly adversely impact an economic, social, and environmental nature. Under the optics of the environment the production of aquatic organisms, mainly fish, creates considerable amounts of effluents, primarily those from food residues and feces. Black (2001) and Fernandes et al. (2001) reported that potentially polluting products and services in aquaculture include three broad categories: (1) aquaculture production, per se, with seasonal discharges; (2) agricultural activities to generate inputs; and (3) medicines, the latter two with periodic releases.

However, the behavior of any waste discharged into the water column, whether sweet or marine, depends on hydrographic conditions, the bottom topography, and the geography of the place in question (Fernandes et al., 2001). Dissolved wastes commonly include dissolved organic carbon, phosphorus, and ammonia, resulting from animal feces or feed remains. Also, lipids released from feed pellets can form a film on the water surface, commonly called an oil film (Black, 2001).

TABLE 5.1 Current Environmental Concerns from Aquaculture Aystems

Potential Direct Impacts	Potential Consequences	Actions
Organic enrichment	Impact on wildlife/habitats	Locational guidelines
Nutrient enrichment	Trigger of toxic blooms	Biomass maximum
Chemicals release	Demise of wild stocks	Maximum feed limit
Spread of diseases		Restricted use of chemicals, management guidelines (including codes of practice/conduct)
Escapees	"Genetic dilution" demise of wild stocks	Improved cage design, management guidelines

In a review study on the nutrient emission in aquaculture Olsen et al. (2008) claim that two main mechanisms are responsible for the nutrient assimilation capacity on water column ecosystems: (1) absorption of nutrients by phytoplankton and subsequent transfer to higher trophic levels and (2) dilution of nutrients mediated by the hydrodynamics of the production sites.

Within this system the movement of water is the most critical factor, essential for sizeable intensive fish farms. The management of impacts in aquatic environments requires monitoring at regional scales because the potential environmental effects will usually become significant at the local scale of fish farming.

Still, among the environmental impacts that aquaculture can have, there are undesirable effects on wild populations such as genetic changes due to unsuitable crossings (Crozier, 2000; Fleming et al., 2000) and disease transfer fugitives (Heggberget et al., 1993). These factors are liable to be controlled within intensive production farms, with monitoring and investment in animal breeding structures. However, they are complicated factors to be maintained in small productions where there is little control over floods, increased water volume, and animal escape to rivers, lakes, and water sources, per example.

Feed production is identified as one of the highest environmental and economic costs in the aquaculture and livestock sectors. Although farmed aquatic animals show more efficient feed conversion when compared to land animals, many farmed marine animals are fed diets that contain significant amounts of fish meal and fish oil. For a long time, this has led to criticism of the aquaculture sector for "using fish to feed fish" and cause environmental problems in addition to possible competition with human food (Bartley et al., 2007). Cashion et al. (2017) point out, importantly a 90% of fish destined for uses other than direct human consumption are food-grade or prime food-grade fish.

Studies to reduce the use of fish meal and fish oil in aquaculture diets, date back more than two decades and are still a priority for intensive and monoculture systems, mainly for species at the top of the food chain. Thus the formulation of diets with ingredients that are residues or byproducts of human food, control over the food offered, use of modern and sustainable ingredients that have high digestibility can significantly reduce the global impacts of aquaculture (Bartley et al., 2007; Smárason et al., 2017a,b).

Smárason et al. (2017a,b) used the LCA methodology to assess the environmental performance of three different types of feed for *Salvelinus alpinus* about the following parameters: global warming, acidification, eutrophication, abiotic depletion, human toxicity potential, marine ecotoxicity potential, and cumulative energy demand. The conventional aquafeed used in fish farming was compared with two new types of feed: based on black soldier fly (BSF) larvae and the ECO feed (composed of larger quantities of plant-based ingredients). As a result of the study, we found that feed production is responsible for causing the most significant environmental impacts. However, the BSF ratio demonstrated the best environmental performance among the three types of allocation. In addition, it can be concluded that by increasing plant-based ingredients at the expense of marine-based components, better environmental performance can be achieved.

Pelletier and Tyedmers (2007) used LCA to assess the benefits, or not, associated with a transition to the use of organic foods in salmon (*Salmo salar*) farming. For that, it was compared the cradle-to-mill gate life cycle energy use, biotic resources, global warming, acidification, eutrophication, and impacts of aquatic ecotoxicity, associated with ingredients production for four hypothetical aquafeeds in conventional and organic salmon farming. As a result the authors observed a decrease in environmental impacts when the diets contained reduced proportions of ingredients of animal origin (fish or poultry). The results demonstrate that current standards for organic salmon aquaculture, which advocates the use of ingredients from organic cultures and flours and oils from fishery byproducts, fail to reduce the environmental impacts of feed production for the set of categories evaluated in this study.

On the other hand a study by Papatryphon et al. (2004) investigated, through LCA modeling and use, the theoretical impact of the substitution of fish meal and fish oil in diets for rainbow trout (*Oncorhynchus mykiss*). The authors concluded that the total substitution of fish meal and fish oil by vegetable sources did not mitigate environmental impacts, despite strongly reducing the dependence on the production of trout with food derived from fishing.

These divergences on the environmental impacts in aquaculture demonstrate how complex the seafood production system, reinforcing the urgency in the assessment and studies of the ecological effects and the processes linked to the sustainability of the productive system.

The environmental impact is dependent on the production system and the target species. In general, it is observed that intensive courses that use carnivorous species tend to cause more significant environmental impacts, possibly due to the greater dependence on technologies and protein of high nutritional quality in their diets. In carnivorous fish, plant sources tend to have a lower rate of absorption and a higher percentage of excretion of nutrients.

Therefore the decrease in the use of ingredients of animal origin (mainly fish meal) and the increase of plant-based ingredients does not ensure the reduction in environmental impact. This occurs because the digestibility of the ingredients that make up the diet, the rate of feed conversion, and lower levels of excretion of nitrogen and phosphorus depend on the ability of the animal organism to digest, absorb, and metabolize the nutrients contained in the food. Because of this, there is a strong tendency to use and studies of omnivorous species (for accepting several ingredients both of animal origin and plant based).

With regard to aquafeeds, we can reflect that the sustainability process in aquaculture depends on local actions ranging from choosing the species to be produced, local and global studies of conventional and nonconventional ingredients for the manufacture of diets, and finally the evaluation of digestibility, feed conversion, and retention rates of body protein.

5.4 TECHNIQUES FOR ENVIRONMENTAL IMPACT EVALUATION

Modern and sustainable agriculture systems need to meet the demand for food without compromising the ability of future generations to satisfy their requirements. The concept of sustainability also refers to the necessity of maximizing the real benefits that society receives from agri-food production. Environmentally sustainable systems are continuously searching for ways to improve the efficiency of resource use, such as water, land, and nutrients. This approach is particularly important in the current context of urbanization and the increasing demand for food.

Food demand is expected to double by the year 2050, mainly because of the global population growth (Tilman et al., 2002). It is also estimated that 70% of this new food will need to be supplied by innovative systems or emerging technologies (Simmons, 2011). In other words the current food production systems will not be able to ensure food security in the coming years, and therefore they need to be reassessed and adapted. This context is particularly important for livestock production systems, which are frequently the target of multiple criticisms. Among the livestock sectors, those adopting highly intensive practices are probably the ones that deserve more concerns, mainly due to their association with high demands for resources followed by high volumes of pollutant excretions.

Building a more sustainable world is a big challenge for our society. Apart from the difficulties faced to change the process to improve its sustainability, just indicating what is more or less sustainable is already a challenging task. The choice of a better product in terms of sustainability begins by accessing many dimensions of its life cycle need, including the environmental, economic, and social attributes of this production chain. Several methodologies can be used in this regard, with some focusing on local impacts and other addressing the global effects of a given life cycle. Both are critical instruments in decision-making strategies, but those evaluating the whole system may provide more insights to move toward more sustainable agri-food systems.

The LCA is an ISO-based standard method widely recognized in the context of global environmental impact analyzes. This tool is frequently indicated as a reference methodology for supply-chain impact assessment. In addition to providing a global overview of the product life cycle the LCA method can encompass a vast number of environmental dimensions (or impact categories) within a single framework or approach.

The LCA studies have been developed in many different fields, including a great variety of perspectives. In the last decades, its use in agriculture and livestock systems increased considerably, as the research in these fields shifted from a productivity-focused view to include sustainability concerns. Despite its relatively recent adoption, it represents an essential change in the analysis of environmental impacts. The LCA-based research can mitigate the ecological damage of food production and provide a better understanding of the real effects, allowing the agri-food products to be compared to other products present in daily life.

The method allows us to examine a given system (e.g., product) and systematically analyze the environmental impacts associated with each step of the life cycle ("from the cradle to the grave" or from the extraction of raw materials to their final disposal). An important assumption in LCA projects is that there is an environmental impact associated with each resource consumed in the system. During the analysis, all significant emissions or resource uses in the life cycle are summarized and expressed concerning the chosen functional unit.

Developing an LCA project demands multidisciplinary and interdisciplinary. To identify the research problem and to build a complete inventory, it is essential to have an in-depth knowledge of the field under evaluation. As in any other research project the data used in the study need to be accurate, and the problem address needs to be understandable and relevant. If well performed, the results obtained from LCA studies allow us to identify critical points in the chain that can (and need) be improved in environmental sustainability. This information is an exceptional facilitator for the decision-making process and can also be used for the strategic planning of governments, companies, or anyone seeking to maximize the environmental performance of agri-food chains.

5.5 IMPORTANT ASPECTS TO BE CONSIDERED WHEN APPLYING LCA TO AQUACULTURE

Aquaculture has become the world's fastest-growing animal production sector due to the significant global demand for aquatic products. In attempts to evaluate the environmental impact of the industry in expansion the LCA has become a frequently used method.

Aquaculture is often claimed to be sustainable (Tlusty and Thorsen, 2016). However, depending on the assessment method the seafood sector is associated with potentially critical environmental damages. For this reason, it is essential to further evaluate current seafood farming practices and identify the production systems that are the most environmentally sustainable.

There is no doubt that LCA is a useful tool to assess aquaculture systems. The method can be used to identify feasible ways to reduce the environmental impacts or compare alternative systems to determine which one has the lowest environmental impacts. This can be done at the microlevel (e.g., feed production), mesolevel (e.g., an entire farm), or macrolevel (e.g., aquaculture sector or country as a whole; Bohnes et al., 2018).

However, like any scientific method, LCA has its limitations, and some of these gaps are relevant or more likely to occur in studies developed on livestock. Such restrictions should be addressed and acknowledged when interpreting results from LCA studies based on animal production chains.

Although LCA addresses water use the water consumption or water footprint is not directly accounted for in many studies developed in the agriculture and livestock field. Factors that might impact human health are also insufficiently covered. Additionally, topics as biodiversity, welfare, social aspects, and also the positive impact of specific activities in the simulations are still little explored, particularly for agri-food chains.

Another significant limitation in LCA studies is related to the difficulties in the choice of the functional unit. For instance, when analyzing fish production, the operating unit can be one animal, a given quantity of final product, or the whole farm. To compare different production systems (e.g., intensive vs extensive systems) the units must be the same. The model choice (attributional or consequential LCA approaches) can also interfere with the results observed in the studies. Also the intention of the functional unit could be related to the scope of the course (e.g., 1 t of live animals or 1 t of meat). In this last case the comparison among studies is even more limited.

Analysis of agri-food production is data demanding. The accuracy of the data used in the study is essential for the quality of the results. Along with the assertiveness the data used in the model need to be representative of the given system. This may not seem like an issue, but specialists in a given product will always think of additional topics to be considered in the simulated scenarios to represent the reality of so diverse production chains correctly. Usually the data describing the foreground system are collected from a specific case, intending to have an updated picture of the system. However, there is no single model to describe fish producers worldwide, and probably neither for some specific regions. A wide variety of techniques could be described from larger-scale production units to small-scale family producers; some engaged in vertically integrated models, other working as independent units. It is also clear that great diversity exists between producers from different locations or technification levels, contributing to the previously stated difficulty in cross-study comparison. The variability of simulated scenarios among studies is one of the causes of variation on the LCA results, highlighting the importance of an in-depth description of the simulated background in the publications.

Significant variability is found among results obtained for a single production system, even with most previous LCA studies focusing on evaluating the impacts considering static scenarios. The more the dynamic nature of agri-food systems is considered in the LCA projects, the more variability will be found in the results. Properly assessing the variability inherent to the food sectors in the LCA studies is a challenging task. However, understanding the livestock chain particularities is essential to improve its global sustainably.

The topics raised here are limitations that require more flexibility when implementing the LCA methodology and using its results to adapt them to different decision-making contexts. Despite these issues, LCA is one of the complete ways to assess the sustainability of a given product, as its entire chain will be evaluated from a global perspective. These studies are very important for each part of the productive chain because the understanding of the chain allows its crucial point to be easily identified and improved.

5.5.1 Methodology Applied During Systematic Review

This systematic review was conducted was developed with the aim of retrieving as many as possible LCA studies on aquaculture, disregarding its specific objectives. The searches were based on the title, abstract, and keywords, and included studies published in scientific journals, conference proceedings, technical bulletins, and written in Portuguese, Spanish, and English. No restrictions or filters were imposed in terms of study design, the origin of study, and publication date.

The database was developed, focusing on the aquaculture systems. The strategy "PICo" was used to identify the population, interest, and context for both types of research. Thus the community was defined as "aquaculture." Also the part was outlined as "life cycle assessment," while the context was delimited as "climate change." A list of main topics and alternate terms for each item was used in the search strategy. The entire process was planned using English words. The online search was conducted in three electronic bibliographic databases, Web of Science, Scopus, and PubMed, to increase the probability of identifying all potentially relevant publications. The literature search was performed in August 2020.

The searches were conducted by combining the following terms: *(crustacea* OR mollusk* OR seafood* OR fisher* OR aquaculture* OR "Atlantic salmon*" OR "rainbow trout*" OR codes OR catfish* OR carp* OR fish* OR panga* OR shrimp* OR crab* OR crawfish* OR oyster* OR Rana* OR "marine farm*" OR tilapia*) AND ("life cycle assessment" OR "carbon emission" OR "carbon footprint" OR "greenhouse gas*" OR "global warming" OR LCA).*

When the search was completed, the results found in each database were exported to a software application used to manage and organize bibliographic references. Duplicate references were identified by the software and excluded. After titles and abstracts were examined by both reviewers and a record was only removed from the database, there was a mutual agreement. The full versions of the papers' preliminary selection were then evaluated. In this step a set of information about each study was analyzed through a checklist. Thus the criteria for final publication selection were: (criterium 1) original LCA studies; (criterium 2) that assessed production systems of aquaculture; (criterium 3) from the cradle to farm, slaughterhouse, or processor gate; and (criterium 4) estimating at least the climate change potential in CO_2-eq.

After selecting the studies, their quality was evaluated (data not presented in this document), and relevant information to describe the proposed theoretical model was transferred to an electronic spreadsheet. Finally, cross-study comparisons were performed.

The initial database comprised a list of 6722 references. The online search reported 773 references in PubMed, 2671 references in Web of Science, and 3278 references in the Scopus database. After the removal of 2513 duplicate references, 4209 studies remained for analysis. From these sets of references, 3951 were removed after assessing the titles. In contrast the other 156 were removed after the abstract reading because they did not fit at least one of the selection criteria. At this step the database comprised a list of 111 references that were individually search for full-text analysis. From this, 62 studies were removed due to any of the selection criteria described previously (criterium 4). At the end of the evaluation procedure, 49 reviews were selected to be part of the final database.

Additionally, a 'snowball' approach was adopted to broaden the search by scrutinizing reference lists of the literature retrieved and other previously published reviews (Henriksson et al., 2012; Clark and Tilman, 2017; Bohnes et al., 2018) were crosschecked for additional studies. Another 21 studies were recovered. Finally, 70 studies were considered eligible for the systematic review. A flow diagram of the summarized search procedure can be seen in Fig. 5.1.

5.5.2 General Description of the Available Literature

The list of LCA studies applied to aquaculture production is presented chronologically in Table 5.2. Research papers were published from 2004 to 2020 in 22 different journals, from which The International Journal of Life Cycle Assessment (14 papers), Journal of Cleaner Production (12 papers), and Aquaculture (7 papers) had more significant participation in the database. The number of scientific publications in this field has increased over the years, with a considerable increase in studies between 2010 and 2020 (75%). The increase of publications in the last decade is mainly due to consumer concern about the environment, and the damages caused by aquaculture.

The studies were developed in Europe (45 papers), Africa (6 papers), North America (10 papers), South America (4 papers), Asia (4 papers), and Oceania (2 papers). A description of this distribution is presented in Fig. 5.2. The greater concern amongst consumers and the demand for production systems with less impact on the environment may also be associated with discrepancies in the geographical distribution of the scenarios assessed in the LCA studies. For example, 64.3% of the reviews were carried out on the European continent, while only 5.7% are from Asia. However, Asia represents approximately 90% of global aquaculture production, while Europe only accounts for 3% of the worldwide output (Bohnes et al., 2018). LCA studies from the Southern Hemisphere and developing countries are less common, despite the significant amount of seafood sourced from these areas.

Of the 70 studies, 74.3% exclusively assessed monoculture, 25.7% polyculture. Aquaculture is highly diverse, with more than 310 different farmed species recorded by FAO in 2008. The overview of the studies revealed 71 different types of species considered in LCA studies. Most references studied European pilchard—*Sardina pilchardus* (7 papers), rainbow trout—*Oncorhynchus mykiss* (7 papers), Nile tilapia—*Oreochromis niloticus* (7 papers), and Atlantic salmon—*Salmo salar* (6 papers) production. Aquacultural LCAs probably focus on this species due to high economic value, such as salmon (Wilfart et al., 2013; Ayer et al., 2016), shrimp—*Litopenaeus vannamei* (Belettini et al., 2018), rainbow trout (Papatryphon et al., 2004; Dekamin et al., 2015), and tilapia (Yacout et al., 2016; Boxman et al., 2016).

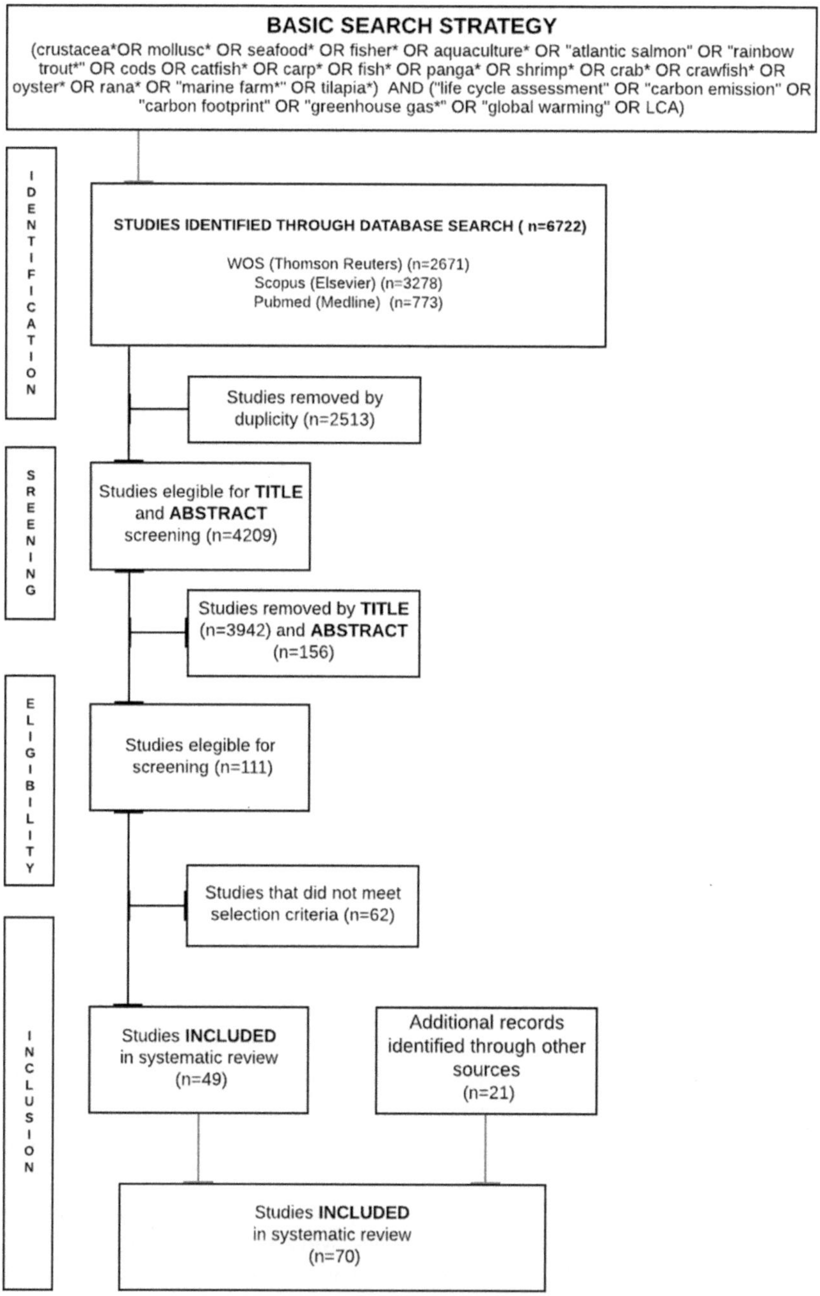

FIGURE 5.1 Prisma flow diagram describing the study selection process for the database focusing on aquaculture production.

TABLE 5.2 Summary of the LCA Studies on Aquaculture Production

Code	Citation	Country	Focus	Functional Unit	System Boundary
1	Papatryphon et al. (2004)	Sweden	Fish meat	1 t of live-weight	Cradle-to-grave
2	Hospido and Tyedmers (2005)	Spain	Fish meat	1 t of frozen unprocessed tuna	–
3	Aubin et al. (2006)	France	Fish meat	1 t of live-weight fish	–
4	Thrane (2006)	Denmark	Fish meat	1 kg of frozen flatfish fillet	–
5	Baruthio et al. (2008)	France	Fish/ Seafood	1 t of product	–
6	Emanuelsson et al. (2008)	Sweden	Seafood	1 kg of frozen shrimp	–
7	Zufia and Arana (2008)	Spain	Fish meat	2 kg of tray tuna with tomato	–
8	Aubin et al. (2009)	France	Fish meat	1 t of live-weight fish	To-farm-gate
9	Ayer and Tyedmers (2009)	Canada	Fish meat	1 t of live-weight fish	Cradle-to-farm-gate
10	D'orbcastel et al. (2009)	France	Fish meat	1 t of fish	–
11	Pelletier et al. (2009)	Canada	Fish meat	1 t of live-weight	Cradle-to-farm-gate
12	Driscoll and Tyedmers (2010)	Canada	Fish meat	1 t of fish	–
13	Iribarren et al. (2010a)	Spain	Seafood	100 t of mussel shell or meat	–
14	Iribarren et al. (2010b)	Spain	Sea product	1 t of sea product	Cradle-to-grave
15	Iribarren et al. (2010a,b)	Spain	Seafood	100 kg of mussel processed	–
16	Pelletier and Tyedmers (2010)	Canada	Fish meat	1 t of tilapia fillet or feed	Cradle-to-farm-gate
17	Vázquez-Rowe et al. (2010)	Spain	Fish meat	1 t of round horse mackerel	Cradle-to-gate
18	Cao et al. (2011)	USA	Seafood	1 t of live-weight shrimp	Cradle-to-farm-gate
19	Iribarren et al. (2011)	Spain	Sea product	1 t of sea product	Cradle-to-grave
20	Ramos et al. (2011)	Spain	Fish meat	1 t of landed round fish	Cradle-to-gate

Continued

TABLE 5.2 continued

Code	Citation	Country	Focus	Functional Unit	System Boundary
21	Vázquez-Rowe et al. (2011)	Spain	Fish meat	645 g of hake or 500 g hake fillet	Cradle-to-grave
22	Ziegler et al. (2011)	Sweden	Seafood	1 kg of shrimp in package	–
23	Iribarren et al. (2012)	Spain	Fish meat	1 t of marine aquafeed	–
24	Jerbi et al. (2012)	Tunisia	Fish meat	1 t of live-weight fish	Cradle-to-gate
25	Kluts et al. (2012)	Netherlands	Fish meat	1 t of fresh striped fish	Cradle-to-gate
26	Samuel-Fitwi et al. (2012)	Germany	Fish meat	1 t of live rainbow trout	Cradle-to-farm-gate
27	Vázquez-Rowe et al. (2012)	Spain	Fish meat	1 t of landed fish	Cradle-to-gate
28	Mungkung et al. (2013)	France	Fish meat	1 t of fresh fish	Cradle-to-grave
29	Pongpat et al. (2013)	Thailand	Fish meat	1 kg of fish or 1 kg of fish meal	Cradle-to-gate
30	Samuel-Fitwi et al. (2013a, b)	Germany	Fish meat	1 t of rainbow trout	–
31	Vázquez-Rowe et al. (2013)	Spain	Seafood	1 kg of barnacles	–
32	Wilfart et al. (2013)	France	Fish meat	1 t of live fish	Cradle-to-gate
33	Almeida et al. (2014)	Portugal	Fish meat	1 kg of sardine	Cradle-to-gate
34	Aubin and Fontaine (2014)	France	Seafood	1 t of mussels "ready-to-cook"	Cradle-to-gate
35	Fréon et al. (2014)	France	Fish meat	1 t of landed anchoveta	Cradle-to-gate
36	Vázquez-Rowe et al. (2014)	Spain	Fish meat	17.26 g of protein or 137.1 g of sardine or 183 g of hake	–
37	Almeida et al. (2015)	Portugal	Fish meat	1 kg of edible product	Cradle-to-gate
38	Aubin et al. (2015)	France	Fish meat	1 t of fish	Cradle-to-gate
39	Avadí et al. (2015)	France	Fish product	1 t of final product or fish product	Cradle-to-gate
40	Dekamin et al. (2015)	Iran	Fish meat	1 t of live-weight fish	Cradle-to-farm-gate

Continued

TABLE 5.2 continued

Code	Citation	Country	Focus	Functional Unit	System Boundary
41	Driscoll et al. (2015)	Canada	Seafood	1 t of live lobster	–
42	Jonell and Henriksson (2015)	Sweden	Seafood	1 t of live-weight	Cradle-to-grave
43	McGrath et al. (2015)	Canada	Fish meat	1 t of salmon	Cradle-to-farm-gate
44	Parker et al. (2015)	Australia	Fish meat	1 t of tuna	–
45	Santos et al. (2015)	Brazil	Seafood	1 t of prawn live-weight	Cradle-to-delivery-market
46	Ayer et al. (2016)	USA	Fish meat	1 t of live-weight	Full growth cycle
47	Besson et al. (2016)	Netherlands	Fish meat	1 t of fish	–
48	Denham et al. (2016)	Australia	Fish meat	1 t of processed fish fillet	–
49	García et al. (2016)	Spain	Fish meat	1 t of sea bream	Cradle-to-grave
50	Liu et al. (2016)	Norway	Fish meat	1 kg of salmon live-weight	–
51	Rodrigues et al. (2016)	Brazil	Fish product	1 kg of product	Gate-to-gate
52	Yacout et al. (2016)	Egypt	Fish meat	1 t of live-weight	Cradle-to-farm-gate
53	Zucaro et al. (2016)	Italy	Seafood	1 t of edible meat or protein	Cradle-to-farm-gate
54	Abdou et al. (2017)	Tunisia	Fish meat	1 t of fish	Cradle-to-farm-gate
55	Badiola et al. (2017)	Spain	Fish meat	1 kg of grown out cod	–
56	Boxman et al. (2016)	USA	Fish meat	1 t of live-weight fish	Cradle-to-gate
57	Chang et al. (2017)	Taiwan	Seafood	1 kg of shrimp	Cradle-to-grave
58	Lourguioui et al. (2017)	Algeria	Seafood	1 t of fresh wet-weight	Cradle-to-gate
59	Smárason et al. (2017a,b)	Iceland	Fish meat	1 kg of live-weight	Cradle-to-gate
60	Verones et al. (2017)	Norway	Seafood	1 kg of live-weight	–

Continued

TABLE 5.2 continued

Code	Citation	Country	Focus	Functional Unit	System Boundary
61	Abdou et al. (2018)	Tunisia	Fish meat	1 t of fish	Cradle-to-gate
62	Astuti and Hadiyanto (2018)	Indonesia	Fish meat	1 kg of amplang	–
63	Avadí et al. (2018)	France	Fish meat	1 t of whole hake on ice	–
64	Belettini et al. (2018)	Brazil	Seafood	1 kg of marine shrimp	Cradle-to-gate
65	Laso et al. (2018)	Spain	Fish meat	1 kg of landed round anchovy	Cradle-to-gate
66	Villanueva-Rey et al. (2018)	Spain	Fish meat	1 t of landed pilchard	Cradle-to-farm-gate
67	Biermann and Geist (2019)	Germany	Fish meat	1 kg of live carp	Cradle-to-farm-gate
68	García et al. (2019)	Spain	Fish meat	1 t of live-weight seabass	Cradle-to-farm-gate
69	Ray and Maguire (2019)	USA	Seafood	1 kg of protein	Cradle-to-gate
70	Abdou et al. (2020)	Tunisia	Seafood	1 t of seafood	Cradle-to-gate

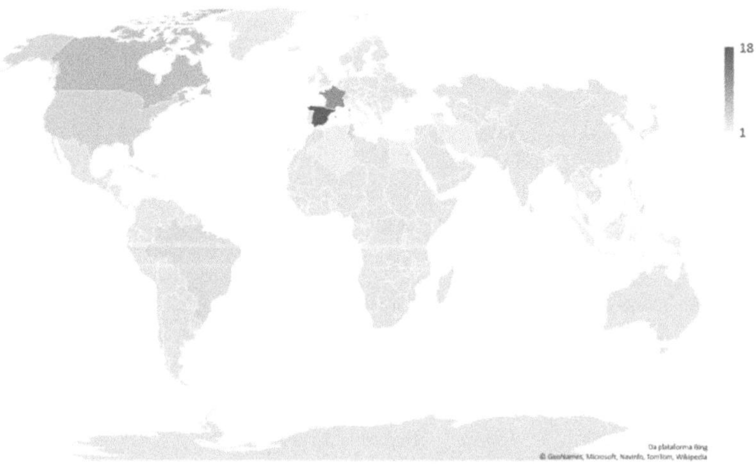

FIGURE 5.2 The number of studies undertaken in each country worldwide (gray = 0 studies).

Most studies used a cradle-to-farm-gate scope, which means that the assessment started with the inputs used in the agricultural phase (e.g., seeds and fertilizers production, land, farm energy use) and included up to the processes related to the animal rearing phase (farm gate). An important limitation of these studies is that information on the packaging, distribution, and consumption processes are not considered in the scopes. Some environmental impacts of postfarm stages may be significant and need to be included in future studies (e.g., distribution to the market). A LCA containing a complete overview of all stages of the production system could demonstrate the specific processes responsible for significant environmental impacts. However, only eight studies included broader system boundaries (cradle-to-grave).

The frequency of different impact categories used in LCA studies is presented in Table 5.3 with detailed characteristics. Global warming, eutrophication, and acidification have been employed with the highest frequency. Other impact categories were observed in a smaller number of studies, such as energy use,

TABLE 5.3 Impact Categories Commonly Used in Aquacultural LCAs and Frequency of Applying Different Impact Categories in on Aquaculture studies

Impact Category	Σ	Characterization Factor	Category Indicator	Equivalency Unit	Interpretation
Climate change	70	GWP	CO_2	kg CO_2-eq	Atmosphere absorption of infrared radiation
Acidification	57	AP	SO_2	kg SO_2-eq	Acid deposition
Eutrophication	56	EP	PO_4	kg PO_4-eq	Nutrient enrichment
Energy use	32	EUP	MJ	MJ	Depletion of nonrenewable energy resource
Ozone depletion	25	ODP	CFC	kg CFC-eq	Stratospheric ozone breakdown
Ecotoxicity	23	Ecotoxicity potential	1,4 DB	kg 1,4 DB-eq	Toxic to flora, fauna and humans
Abiotic resource depletion	20	ADP	Sb	kg Sb-eq	Depletion of nonrenewable resources
Photochemical oxidant	18	POP	C_2H_4	kg C_2H_4-eq	Photochemical smog
Water dependence	11	WD	m^3	m^3 water-eq	Water input relative to fish biomass production

AP, Acidification potential; *ADP*, abiotic resource depletion potential; *BDP*, biotic depletion potential; *EP*, eutrophication potential; *EUP*, energy use potential; *GWP*, global warming potential; *ODP*, ozone depletion potential; *POP*, photochemical oxidant potential. Category indicators: *CO_2*, carbon dioxide; *PO_4*, phosphate; *SO_2*, sulfur dioxide; *MJ*, mega Joules; *NPP*, net primary productivity; *Sb*, antimony; *1,4 DB*, 1,4 dichlorobenzenes; *CFC*, chlorofluorocarbon; *C*, carbon.

Source: *Adapted from Cao, L., Diana, J.S., Keoleian, G.A., Lai, Q., 2013. Role of life cycle assessment in sustainable aquaculture. Reviews in Aquaculture 5, 61–71.*

ozone depletion, ecotoxicity, abiotic resource depletion, photochemical oxidant, and water dependence.

The characterization of the LCA studies on seafood production in terms of the functional unit, global warming potential, and scope contribution in relation to the estimated total impact is presented in Table 5.4. The environmental effects calculated for aquaculture systems life cycles considerably different over the evaluated scenarios. An essential part of this variability is caused by the species, heterogeneity in the definition of functional units, and is also due to the amplitude of scope considered in the studies. The available unit is particularly important for comparative assessments. Using an identical unit of measurement allows a fair comparison (Bohnes and Laurent, 2019). The different forms of the way to define the functional unit reduces the possibility of comparison between studies, and there a risk that the choice of the operating unit might change the results of the study (Cao et al., 2013).

However, several other attributes considered in the simulations are also responsible for the variability among studies. In this particular, results may be partially explained by production system (Aubin et al., 2006, 2009; Cao et al., 2011), regional characteristics (Pelletier et al., 2009), feed production (Papatryphon et al., 2004), and culture comparison (Baruthio et al., 2008).

5.5.2.1 IMPORTANCE OF PRODUCTIVE SYSTEMS

The production systems can be classified according to the intensity and the use of technologies. For example, from low intensity, such as extensive and semiintensive, to intensive systems with different farming technologies. However, all aquaculture production systems generate wastes, require energy, use water, and change the land cover.

In extensive aquaculture is aquatic organisms are placed into an appropriate environment until being harvested (there is little or even no human interference). Consequently the systems with low intensity generally have a lower production compared to the system with more intensity. In semiintensive system, water quality may be manipulated by flushing new water into the system or by using aerators to increase the rate of growth for the organism. The intensive system uses regular aeration and adding new freshwater, feed and chemical supplements, and high density. Finally, as the level of farming intensity accelerates, the production per unit area increases considerably, and in some cases, the feed-conversion efficiency decreases, costs rise, and more waste is generated. In developing countries, aquaculture mostly takes place in both semiintensive and intensive systems, while it remains intensive in developed countries (Diana, 2009).

Semiintensive aquaculture is a way to decrease environmental problems associated with intensive systems. An example is the semiintensive shrimp production in China, which presents a great production but uses commercial feed and fertilizer-based natural food. The system receives criticism because needs high material and energy inputs, and increases the effluent discharge, and consequently increases environmental burdens (Cao et al., 2013). Intensive shrimp systems have

TABLE 5.4 Characterization of the LCA Studies on Seafood Production in Terms of Functional Unit, Global Warming Potential, and Scope Contribution in Relation to the Estimated Total Impact

Code	Species	System	Scope	Climate Change (kg CO_2-eq)	Acidification (kg SO_2-eq)	Eutrophication (kg PO_4-eq)	Comments
1	Rainbow trout	Raceway fish farm	Feed comparison	1120–1560	5.88–6.69	36.10–50	Carnivorous
2	Tuna	Capture	Delivery comparison	1600–2200	20–29	3.40–4.50	–
3	Turbot	Various	System comparison	6017–10,647	48.28–54.96	61.16–76.96	Carnivorous
4	Flatfish	Capture	Fillet product	20.86	0.16	–	–
5	Prawn, tilapia, milkfish, crab	Ponds	Mono and Polyculture comparison	3553–5108	25–48	109–172	Omnivorous
6	Pink shrimp	Capture	Methods comparison	7.80–38	–	–	–
7	Tuna	Capture	Product comparison	23.36–23.75	0.10–0.11	0.0151–0.0164	–
8	Rainbow trout, seabass, turbot	Various	System comparison	2753–6017	19.17–48.28	65.91–108.85	Carnivorous
9	Atlantic salmon, Arctic char	Various	System comparison	1900–2770	15.80–22.60	20.10–35.30	Carnivorous
10	Rainbow trout	Various	System comparison	1602–2043	10.50–13.40	17.80–28.50	Carnivorous
11	Atlantic salmon	Net pen	Country comparison	1790–3270	17.10–29.70	41–74.90	Carnivorous
12	Atlantic herring	Capture	Vessel comparison	52–503	–	–	–
13	Mussel	Farming	Product comparison	−132,000 to 559,000	−5770 to 3830	−3520 to 1 − 650	
14	Several	Capture	Sea product comparison	16.21–277	–	–	–

Continued

TABLE 5.4 continued

Code	Species	System	Scope	Climate Change (kg CO_2-eq)	Acidification (kg SO_2-eq)	Eutrophication (kg PO_4-eq)	Comments
15	Mussel	Farming	Sector evaluation	7.42–158.83	−0.0028 to 1.28	−0.02 to 1.01	–
16	Tilapia	Net pen, ponds	Production cycle	787–2290	11.1–28.90	2.53–48.50	Omnivorous
17	Horse mackerel	Capture	Vessel comparison	796–2278	10.10–27.22	1.84–4.97	–
18	Shrimp	Farming	System comparison	2750–5280	19.40–43.90	2.30–63	Omnivorous
19	Several	Capture	Sea product comparison	16.21–338.45	–	–	–
20	Atlantic mackerel	Capture	Fisheries comparison	95–2279	1.04–27.20	0.24–4.97	–
21	Hake	Capture	Vessel comparison	0.00383–0.00574	0.00000477–0.00000762	0.00000898–0.0000143	–
22	Pink shrimp	Farming	Fishery comparison	0–38	–	–	Omnivorous
23	Turbot	Farming	Production cycle	428.84–532.27	4.67–5.24	0.48–1.68	Carnivorous
24	Seabass	Raceway fish farm	System comparison	11,087–17,449	54–70	180–215	Carnivorous
25	Catfish	Farming	System comparison	−167 to 8952	–	65.20–99.50	Omnivorous
26	Rainbow trout	Recirculating system	System comparison	2239.44–13,622.38	10.73–40.72	4.04–60.36	Carnivorous
27	Several	Capture	Vessel comparison	796–8759	10.10–115	1.84–21	–
28	Carp, tilapia	Net cage	System comparison	800–2025	9.90–15.90	70–105	Omnivorous
29	Catfish, tilapia	Floating cage, ponds	Species comparison	0.96–5.69	0.0079–0.0408	–	Omnivorous

30	Rainbow trout	Foreground	System comparison	767–3633	7.70–11.30	59.80–60.40	Carnivorous
31	Goose barnacle	Capture	Production cycle	0.64–11.53	–	–	–
32	Atlantic salmon	Recirculating, ponds	System comparison	1427–8585	10.50–50	16–35	Carnivorous
33	Sardine	Capture	Overall fleet	0.36	0.0024	0.0005	–
34	Bouchot mussels	Farming	Production cycle	−6.56 to 77.45	1.59–2.29	−1.24 to −0.98	–
35	Peruvian anchoveta	Capture	Vessel comparison	62.96–71.51	0.67–0.77	0.0060–0.0072	–
36	European pilchard	Capture	Scenario comparison	0.21–3.36	0.0019–0.0198	0.0000231–0.00236	–
37	Sardine	Capture	Product comparison	0.80–7.70	0.0049–0.038	0.0011–0.012	–
38	Prawn, crab, tilapia, milkfish	Ponds	Production cycle	2057	31.10	101.50	Omnivorous
39	Shrimp	Farming	System comparison	19,800–31,300	9.14–11.20	−31.60–1.44	Omnivorous
40	Rainbow trout	Various	System comparison	1157.28–6102.57	10.03–23.16	8.53–135.73	Carnivorous
41	Lobster	Capture	Industry comparison	5209–9197	61.38–74.74	–	–
42	Ecuadorian tuna	Capture	Production cycle	2623	–	–	–
43	Atlantic salmon	Various	Scenario comparison	0.25–3874	18.90–25.50	–	Carnivorous
44	Tuna	Capture	Fishing cycle	1140	–	–	–
45	Prawn	Farming	System comparison	3807.10–6862.70	21.80–38.10	−8.10 to 175.10	Omnivorous
46	Atlantic salmon	Various	System comparison	2210–2660	22.90–27.60	–	Carnivorous

Continued

TABLE 5.4 continued

Code	Species	System	Scope	Climate Change (kg CO_2-eq)	Acidification (kg SO_2-eq)	Eutrophication (kg PO_4-eq)	Comments
47	Catfish	Recirculating system	Production cycle	1461.10 kg CO_2-eq	8.2 kg SO_2-eq	6.5 kg PO_4-eq	Omnivorous
48	Several	Capture	Supply chain comparison	18,870–92,560	–	–	–
49	Sea bream	Various	Production cycle	124	38.46	81.79	Carnivorous
50	Atlantic salmon	Various	System comparison	3.40–7	–	–	Carnivorous
51	Productive chain	Capture, farming	Product comparison	1.06–2.56	–	–	–
52	Tilapia	Cages, ponds	System comparison	960.70–6126.10	9.80–24.40	6.30–14.10	Omnivorous
53	Snail	Farming	Production cycle	0.68–5.21	0.0087–0.0672	0.000222–0.00171	Omnivorous
54	Seabass, sea bream	Sea cage offshore	Specie comparison	3182.22–3668.65	18.85–21.61	91.03–98.86	Carnivorous
55	Codfish	Recirculating system	System comparison	11.81–27.26	0.05–0.22	0.008–0.05	Carnivorous
56	Tilapia	Recirculating system	Production cycle	8640	54.30	0.75	Omnivorous
57	Arctic char	Land-based system	Production cycle	6.94	–	–	Carnivorous
58	Mussel	Longline farm	Scenario comparison	248–478	1.20–2.36	0.12–0.29	Omnivorous
59	Shrimp	Farming	Feed comparison	1.02–2.22	0.0159 kg	0.0230	Omnivorous

60	Several	Capture	Techniques comparison	0.0384–0.698	0.000109–0.000179	0.00000752–0.0000200	–
61	Seabass, sea bream	Sea cages	Practices comparison	3421–4400	16–22	92–124	Carnivorous
62	Mackerel	Capture	Processing cycle	9.66	0.079	0.02	–
63	Hake	Capture	Fishing cycle	390	–	0.025	–
64	Shrimp	Raceway system	System comparison	1–48	–	–	Omnivorous
65	Several	Capture	Fishing cycle	1.45	0.0289	0.0042	–
66	European anchovy	Capture	Fleet comparison	619.10–1359.10	–	0.0180–0.58	
67	Carp	Ponds	System comparison	4.30–6	–	0.0060–0.0061	Omnivorous
68	Seabass	Offshore farm	Production cycle	7293	44.15	115.23	Carnivorous
69	Oyster	Farming	Production cycle	0.13	–	–	Omnivorous
70	Seafood	Capture	Scenario comparison	13,773.340–21,930.800	155,813–248,100	9320–14,800	–

almost double the environmental impacts, in all the studied, impact categories than semiintensive systems and this is due to an increase in electricity use, feed, and nutrients in effluents (Cao et al., 2011).

The culture type has little influence on the impacts (eutrophication), and this suggests that monoculture and polyculture systems may not have significantly different results (Bohnes et al., 2018). This may be explained by the variety and technology used in each design. At the same time the culture intensity has more influence on the results, such as acidification. This may be explained by the high productivity of intensive systems that manage to balance the higher input of energy and resources (Bohnes et al., 2018).

From an environmental point of view the monoculture of carnivorous fish is not the best form of aquaculture. However, there is high consumer demand for these species. Nevertheless, carnivorous fish species require a high level of protein and oil. The use of these ingredients causes the release of a significant amount of nitrogen and phosphorus in the aquatic environment and consequently, this requires either higher water availability or costly techniques such as recirculating water systems (Aubin et al., 2009).

5.5.2.2 IMPORTANCE OF FEEDING SCENARIOS

The feeding program determines the growth of the organism and determines the quality of the product, and also generates the most significant impact on the environment. In addition, feed is the main production cost in aquaculture. The fish feed can affect practically all assessed environmental impact categories (Konstantinidis et al., 2020) and the most affected are freshwater and marine eutrophication, freshwater and marine ecotoxicity, land use categories. The role of feed as an environmental impact parameter is also reflected in its share in the "global warming" indicator.

As the feedstuffs account for part of the total environmental impact associated with the production, it is expected that strategies that mitigate the effects of the acquisition of feed ingredients will have a positive impact on the final product footprint. The feed is the most significant contributor to most of the environmental effects of fish farms, except eutrophication (Pelletier et al., 2009; Iribarren et al., 2012; García et al., 2016). The environmental impact of using alternative ingredients in aquaculture and some beneficial effects were already reported for global warming potential (Boissy et al., 2011; García et al., 2016).

Significant criteria for ingredient selection are nutrient density and digestibility. The use of highly digestible feed ingredients and smaller nutrient loss and feed wastage minimize any possible adverse environmental and ecosystem impacts (Tacon and Metian, 2009). Also, nutrient impacts from feeds can be negated by integrating production with other cultured species, which can benefit from these nutrient waste streams (Soto, 2009).

Fish meal and fish oils from forage fisheries are the reference of nutritional quality for farmed species. Feed-fed aquaculture production (e.g., shrimp, salmonids, marine finfish, and eels) depends on capturing fisheries for primary dietary sources of protein and lipid. The use of coproducts is an excellent case to be analyzed in this context. Many vegetable materials are sustainable alternatives to marine proteins and lipids. Alternative lipids sources to fish oil can be used in more significant amounts and include vegetable oils and poultry oil. The use of oil from farmed fish offal is also a potential omega-3 source for other farmed fish. Soybean meal is an essential raw material due to its high protein and actual amino acid content. However, in Europe, the use of soybean is faced with the non-GMO problem, and the certified non-GMO soybean is more expensive compared to no-GMO, due to its reduced availability and some logistical constraints. Soybean meal is commonly used for tilapia, the second most-produced species in the world.

Fish feed, especially the raw material used, such as soybean, can be contributed to the most environmental impact in gilthead sea bream (*Sparus aurata*) production, and a significant reduction in possible negative effects can be achieved by increasing feeding efficiency. For example, the contribution to the potential environmental impacts of soybean meal and fish meal is higher when compared to that of wheat and maize gluten meal (García et al., 2016). This fact occurs due to the demand for significant amounts of these specific materials (soybean and fish meal) according to the current aquafeed formulation (Iribarren et al., 2012). At present, there are still a lot of raw vegetable materials underutilized in fish feeds. It is possible to stimulate and promote the exploration of new sources (e.g., fava beans, peas, and lupin), including coproducts from the pharmaceutical and cosmetic industries, agrifood, agrofuels (Le Gouvello and Simard, 2017).

Another critical argument justifying ingredient replacement as a strategy to mitigate the environmental cost is reducing the distance between the sites of feedstuff production and their place of use. Reducing the distances from the producer to the consumer, means fewer transportation needs, which mitigates several associated costs and emissions. In many cases, using locally grown ingredients in the diet formulation is a way to reduce the aquaculture environment loads.

5.6 CONCLUSION

Due to recent improvements in technologies, aquaculture it became the fastest-growing food production in the world. Although this chain has reached very high-performance levels, its environmental sustainability still needs improvement. This aspect is particularly important because the growing demand for food worldwide must be met at an affordable cost and without compromising environmental integrity. Aquafeeds contribute highly to the environmental impacts of aquaculture. In addition, the environmental impact associated with aquaculture is influenced by the different intensity systems used during production. These aspects had already been reported by many studies using LCA all around the world. Several studies also reported that improving feeding practices with ingredient replacement may mitigate the environmental impact of production. Based on this assessment of the shortcomings current, still there is a need for more detailed LCA studies in developing countries (especially in Asia) and species with less economic value. Studies including postfarm processes (e.g., packaging, distribution, and consumption) are suggested to future projects in this research area.

REFERENCES

Abdou, K., Aubin, J., Romdhane, M.S., Le Loc'h, F., Lasram, F.B.R., 2017. Environmental assessment of seabass (*Dicentrarchus labrax*) and seabream (*Sparus aurata*) farming from a life cycle perspective: a case study of a Tunisian aquaculture farm. Aquaculture 471, 204–212.

Abdou, K., Lasram, F.B.R., Romdhane, M.S., Le Loc'h, F., Aubin, J., 2018. Rearing performances and environmental assessment of sea cage farming in Tunisia using life cycle assessment (LCA) combined with PCA and HCPC. The International Journal of Life Cycle Assessment 23, 1049–1062.

Abdou, K., Le Loc'h, F., Gascuel, D., Romdhane, M.S., Aubin, J., Ben Rais Lasram, F., 2020. Combining ecosystem indicators and life cycle assessment for environmental assessment of demersal trawling in Tunisia. International Journal of Life Cycle Assessment 25, 105–119.

Almeida, C., Vaz, S., Cabral, H., Ziegler, F., 2014. Environmental assessment of sardine (*Sardina pilchardus*) purse seine fishery in Portugal with LCA methodology including biological impact categories. International Journal of Life Cycle Assessment 19, 297–306.

Almeida, C., Vaz, S.A., Ziegler, F., 2015. Environmental life cycle assessment of a canned sardine product from Portugal. Journal of Industrial Ecology 19, 607–617.

Astuti, R.S.D., Hadiyanto, H., 2018. Estimating environmental impact potential of small-scale fish processing using life cycle assessment. Journal of Ecological Engineering 19, 65–74.

Aubin, J., Fontaine, C., 2014. Impacts of producing bouchot mussels in Mont-Saint-Michel Bay (France) using LCA with emphasis on potential climate change and eutrophication. In: Proceedings of the 9th International Conference on Life Cycle Assessment in the Agri-Food Sector Environmental, Zurich, Switzerland, pp. 64–69.

Aubin, J., Papatryphon, E., Van Der Werf, H.M.G., Petit, J., Morvan, Y.M., 2006. Characterization of the environmental impact of a turbot (*Scophthalmus maximus*) recirculating production system using Life Cycle Assessment. Aquaculture 261, 1259–1268.

Aubin, J., Papatryphon, E., Van Der Werf, H.M.G., Chatzifotis, S., 2009. Assessment of the environmental impact of carnivorous finfish production systems using life cycle assessment. Journal of Cleaner Production 17, 354–361.

Aubin, J., Baruthio, A., Mungkung, R., Lazard, J., 2015. Environmental performance of brackish water polyculture system from a life cycle perspective: a Filipino case study. Aquaculture 435, 217–227.

Avadí, A., Fréon, P., 2013. Life cycle assessment of fisheries: a review for fisheries scientists and managers. Fisheries Research 143, 21–38.

Avadí, A., Bolanos, C., Sandoval, I., Ycaza, C., 2015. Life cycle assessment of Ecuadorian processed tuna. International Journal of Life Cycle Assessment 20, 1415–1428.

Avadí, A., Adrien, R., Aramayo, V., Freon, P., 2018. Environmental assessment of the Peruvian industrial hake fishery with LCA. International Journal of Life Cycle Assessment 23, 1126–1140.

Ayer, N.W., Tyedmers, P.H., 2009. Assessing alternative aquaculture technologies: life cycle assessment of salmonid culture systems in Canada. Journal of Cleaner Production 17, 362–373.

Ayer, N., Martin, S., Dwyer, R.L., Gace, L., Laurin, L., 2016. Environmental performance of copper-alloy Net-pens: life cycle assessment of Atlantic salmon grow-out in copper-alloy and nylon net-pens. Aquaculture 453, 93–103.

Badiola, M., Basurko, O.C., Gabina, G., Mendiola, D., 2017. Integration of energy audits in the Life Cycle Assessment methodology to improve the environmental performance assessment of Recirculating Aquaculture Systems. Journal of Cleaner Production 107, 155–166.

Bartley, D.M., Brugère, C., Soto, D., Gerber, P., Harvey, B., 2007. Comparative assessment of the environmental costs of aquaculture and other food production sectors: methods for meaningful comparisons. FAO Fisheries Proceedings 10. FAO, Rome, 241 pp.

Baruthio, A., Aubin, J., Mungkung, R., Lazard, J., van der Werf, H.M., 2008. Environmental assessment of Filipino fish/prawn polyculture using life cycle assessment. In: Proceedings of the 6th International Conference on LCA in the Agri-Food Sector, Zurich, Switzerland, pp. 242–247.

Belettini, F., Seiffert, W.Q., Lapa, K.R., Vieira, F.D., Santo, C.M.D., Arana, L.A.V., 2018. Carbon footprint in commercial cultivation of marine shrimp: a case study in southern Brazil. Brazilian Journal of Animal Science 47 (e20160353).

Béné, C., Arthur, R., Norbury, H., Allison, E.H., Beveridge, M., Bush, S., et al., 2016. Contribution of fisheries and aquaculture to food security and poverty reduction: assessing the current evidence. World Development 79, 177–196.

Besson, M., Aubin, J., Komen, H., Poelman, M., Quillet, E., Vandeputte, M., et al., 2016. Environmental impacts of genetic improvement of growth rate and feed conversion ratio in fish farming under rearing density and nitrogen output limitations. Journal of Cleaner Production 116, 100–109.

Biermann, G., Geist, J., 2019. Life cycle assessment of common carp (*Cyprinus carpio L.*) – a comparison of the environmental impacts of conventional and organic carp aquaculture in Germany. Aquaculture 501, 404–415.

Black, K.D. (Ed.), 2001. Environmental Impacts of Aquaculture. Sheffield Academic Press, Sheffield, 214 pp.

Bohnes, F.A., Laurent, A., 2019. LCA of aquaculture systems: methodological issues and potential improvements. International Journal of Life Cycle Assessment 24, 324–337.

Bohnes, F.A., Hauschild, M.Z., Schlundt, J., Laurent, A., 2018. Life cycle assessments of aquaculture systems: a critical review of reported findings with recommendations for policy and system development. Reviews in Aquaculture 11, 1061–1079.

Boissy, J., Aubin, J., Abdeljalil, D., Van Der Werf, H.M.G., Bell, G.J., Kaushik, S.J., 2011. Environmental impacts of plant-based salmonid diets at feed and farm scales. Aquaculture 321, 61–70.

Boxman, S.E., Zhang, Q., Bailey, D., Trotz, M.A., 2016. Life cycle assessment a commercial-scale freshwater aquaponic system. Environmental Engineering Science 34, 299–311.

Cao, L., Diana, J.S., Keoleian, G.A., Lai, Q., 2011. Life cycle assessment of Chinese shrimp farming systems targeted for export and domestic sales. Environmental Science and Technology 45, 6531–6538.

Cao, L., Diana, J.S., Keoleian, G.A., Lai, Q., 2013. Role of life cycle assessment in sustainable aquaculture. Reviews in Aquaculture 5, 61–71.

Cashion, T., Le Manach, F., Zeller, D., Pauly, D., 2017. Most fish destined for fishmeal production are food-grade fish. Fish and Fisheries 18, 837–844.

Chang, C.C., Chang, K.C., Lin, W.C., Wu, M.H., 2017. Carbon footprint analysis in the aquaculture industry: assessment of an ecological shrimp farm. Journal of Cleaner Production 168, 1101–1107.

Clark, M., Tilman, D., 2017. Comparative analysis of environmental impacts of agricultural production systems, agricultural input efficiency, and food choice. Environmental Research Letters 12, 1–11.

Crozier, W.W., 2000. Escaped farmed salmon, *Salmo salar* L., in the Glenarm River, Northern Ireland: genetic status of the wild population 7 years on. Fisheries Management and Ecology 7, 437–446.

Dekamin, M., Veisi, H., Safari, E., Liaghati, H., Khoshbakht, K., Dekamin, M.G., 2015. Life cycle assessment for rainbow trout (*Oncorhynchus mykiss*) production systems: a case study for Iran. Journal of Cleaner Production 91, 43–55.

Denham, F.C., Biswas, W.K., Solah, V.A., Howieson, J.R., 2016. Greenhouse gas emissions from a Western Australian finfish supply chain. Journal of Cleaner Production 112, 2079–2087.

Diana, J.S., 2009. Aquaculture production and biodiversity conservation. BioScience 59, 27–38.

D'orbcastel, E.R., Blancheton, J.P., Aubin, J., 2009. Towards environmentally sustainable aquaculture: comparison between two trout farming systems using Life Cycle Assessment. Aquacultural Engineering 40, 113–119.

Driscoll, J., Tyedmers, P., 2010. Fuel use and greenhouse gas emission implications of fisheries management: the case of the New England Atlantic herring fishery. Marine Policy 34, 353–359.

Driscoll, J., Boyd, C., Tyedmers, P., 2015. Life cycle assessment of the Maine and southwest Nova Scotia lobster industries. Fisheries Research 172, 385–400.

Emanuelsson, A., Flysjö, A., Thrane, M., Ndiaye, V., Eichelsheim, J.L., Ziegler, F., 2018. Life Cycle Assessment of southern pink shrimp products from Senegal. An environmental comparison between artisanal fisheries in the Casamance region and a trawl fishery off Dakar including biological considerations. In: Proceedings of the 6th International Conference on LCA in the Agri-Food Sector, Zurich, Switzerland, pp. 271–279.

FAO (Food and Agriculture Organization of the United Nations), 2016. The State of World Fisheries and Aquaculture 2016. Contributing to Food Security and Nutrition for all. FAO, Rome, 200 pp.

FAO, 2020. The State of World Fisheries and Aquaculture 2020. Sustainability in Action. FAO, Rome.

Fernandes, T.F., Eleftheriou, A., Ackefors, H., Eleftheriou, M., Ervik, A., Sanchezmata, A., et al., 2001. The scientific principles underlying the monitoring of the environmental impacts of aquaculture. Journal of Applied Ichthyology 17, 181–193.

Fleming, I.A., Hindar, K., Mjolnerod, I.B., Jonsson, B., Balstad, T., Lamberg, A., 2000. Lifetime success and interactions of farm salmon invading a native population. Proceedings of the Royal Society B: Biological Sciences 267, 1517–1523.

Folke, C., Kautsky, N., 1992. Aquaculture with its environment: prospects for sustainability. Ocean & Coastal Management 17, 5–24.

Fréon, P., Avadi, A., Chavez, R.A.V., Ahon, F.I., 2014. Life cycle assessment of the Peruvian industrial anchoveta fleet: boundary setting in life cycle inventory analyses of complex and plural means of production. International Journal of Life Cycle Assessment 19, 1068–1086.

García, G.B., Rosique, J.C., Aguado-Gimenez, F., Garcia, J.G., 2016. Life cycle assessment of gilthead seabream (*Sparus aurata*) production in offshore fish farms. Sustainability 8, 1228.

García, G.B., Jimenez, C.R., Aguado-Gimenez, F., Garcia, J.G., 2019. Life Cycle Assessment of Seabass (*Dicentrarchus labrax*) produced in offshore fish farms: variability and multiple regression analysis. Sustainability 11, 3523.

Godfray, H.C.J., Beddington, J.R., Crute, I.R., Haddad, L., Lawrence, L., Muir, J.F., et al., 2010. Food security: the challenge of feeding 9 billion people. Science 327, 812–818.

Le Gouvello, R., Simard, F., 2017. Durabilité des aliments pour le poisson en aquaculture: Réflexions et recommandations sur les aspects technologiques, économiques, sociaux et environnementaux. UICN/Comité français de l'UICN, Gland/Paris, 296 pp.

Guillen, J., Natale, F., Carvalho, N., Casey, J., Hofherr, J., Druon, J.N., et al., 2019. Global seafood consumption footprint. Ambio 48, 111–122.

Heggberget, T.G., Johnsen, B.O., Hindar, K., Jonsson, B., Hansen, L.P., Hvidsten, N.A., et al., 1993. Interactions between wild and cultured Atlantic salmon: a review of the Norwegian experience. Fisheries Research 18, 123–146.

Henriksson, P.J.G., Guinee, J.B., Kleijn, R., De Snoo, G.R., 2012. Life cycle assessment of aquaculture systems – a review of methodologies. International Journal of Life Cycle Assessment 17, 304–313.

Henriksson, P.J.G., Rico, A., Zhang, W., Ahmad-Al-Nahid, S., Newtonr, R., Phan, L.T., et al., 2015a. Comparison of Asian aquaculture products by use of statistically supported life cycle assessment, Environmental Science and Technology, 49. 14176–14183.

Henriksson, P.J., Heijungs, R., Dao, H.M., Phan, L.T., De Snoo, G.R., Guinee, J.B., 2015b. Product carbon footprints and their uncertainties in comparative decision contexts. PLoS One 10, 121–221.

Hospido, A., Tyedmers, P., 2005. Life cycle environmental impacts of Spanish tuna fisheries. Fisheries Research 76, 174–186.

Iribarren, D., Moreira, M.T., Feijoo, G., 2010a. Implementing byproduct management into the life cycle assessment of the mussel sector. Resources, Conservation and Recycling 54, 1219–1230.

Iribarren, D., Moreira, M.T., Feijoo, G., 2010b. Revisiting the life cycle assessment of mussels from a sectorial perspective. Journal of Cleaner Production 18, 101–111.

Iribarren, D., Vazquez-Rowe, I., Hospido, A., Moreira, M.T., Feijoo, G., 2011. Updating the carbon footprint of the Galician fishing activity (NW Spain). Science of the Total Environment 409, 1609–1611.

Iribarren, D., Moreira, M.T., Feijoo, G., 2012. Life cycle assessment of aquaculture feed and application to the turbot sector. International Journal of Environmental Research 6, 837–848.

ISO, 2006. ISO 14044:2006—Environmental Management—Lifecycle Assessment—Requirements and Guidelines. ISO, Geneva.

Jerbi, M.A., Aubin, J., Garnaoui, K., Achour, L., Kacem, A., 2012. Life cycle assessment (LCA) of two rearing techniques of sea bass (*Dicentrarchus labrax*). Aquacultural Engineering 46, 1–9.

Jonell, M., Henriksson, P.J.G., 2015. Mangrove-shrimp farms in Vietnam—comparing organic and conventional systems using life cycle assessment. Aquaculture 447, 66–75.

Kluts, I.N., Potting, J., Bosma, R.H., Phong, L.T., Udo, H.M.J., 2012. Environmental comparison of intensive and integrated agriculture aquaculture systems for striped catfish production in the Mekong Delta, Vietnam, based on two existing case studies using life cycle assessment. Reviews in Aquaculture 4, 195–208.

Konstantinidis, E., Perdikaris, C., Gouva, E., Nathanalides, C., Bartzanas, T., Anestis, V., et al., 2020. Assessing environmental impacts of sea bass cage farms in Greece and Albania using life cycle assessment. International Journal of Environmental Research 14, 693–704. Available from: https://doi.org/10.1007/s41742-020-00289-8.

Laso, J., Vazquez-Rowe, I., Margallo, M., Crujeiras, R.M., Irabien, A., Aldaco, R., 2018. Life cycle assessment of European anchovy (*Engraulis encrasicolus*) landed by purse seine vessels in northern Spain. International Journal of Life Cycle Assessment 23, 1107–1125.

Liu, Y., Rosten, T.W., Henriksen, K., Hognes, E.S., Summerfelt, S., Vinci, B., 2016. Comparative economic performance and carbon footprint of two farming models for producing Atlantic salmon (*Salmo salar*): land-based closed containment system in freshwater and open net pen in seawater. Aquacultural Engineering 71, 1–12.

Lourguioui, H., Brigolin, D., Boulahdid, M., Pastres, R., 2017. A perspective for reducing environmental impacts of mussel culture in Algeria. International Journal of Life Cycle Assessment 22, 1266–1277.

McGrath, K.P., Pelletier, N.L., Tyedmers, P.H., 2015. Life cycle assessment of a novel closed-containment salmon aquaculture technology. Environmental Science and Technology 49, 5628–5636.

Medeiros, M.V., Aubin, J., Camargo, A.F.M., 2017. Life cycle assessment of fish and prawn production: comparison of monoculture and polyculture freshwater systems in Brazil. Journal of Cleaner Production 156, 528–537.

Mungkung, R., Aubin, J., Prihadi, T.H., Slembrouck, J., Van Der Werf, H.M.G., Legendre, M., 2013. Life cycle assessment for environmentally sustainable aquaculture management: a case study of combined aquaculture systems for carp and tilapia. Journal of Cleaner Production 57, 249–256.

Naylor, R.L., Goldburg, R.J., Primavera, J.H., Kautsky, N., Beveridge, M.C.M., Clay, J., et al., 2000. Effect of aquaculture on world fish supplies. Nature 405, 1017–1024.

Neiland, A.E., Soley, N., Varley, J.B., Whitmarsh, D.J., 2001. Shrimp aquaculture: economic perspectives for policy development. Marine Policy 25, 265–279.

Nemecek, T., Jungbluth, N., I Canals, L.M., Schenck, R., 2016. Environmental impacts of food consumption and nutrition: where are we and what is next? International Journal of Life Cycle Assessment 21, 607–620.

Olsen, L.M., Holmer, M., Olsen, Y., 2008. Perspectives of Nutrient Emission from Fish Aquaculture in Coastal Waters: Literature Review with Evaluated State of Knowledge. FHF Project No. 542014. The Fishery and Aquaculture Industry Research Fund, Oslo, 87 pp.

Ottinger, M., Clauss, K., Kuenzer, C., 2016. Aquaculture: relevance, distribution, impacts and spatial assessments—a review. Ocean and Coastal Management 119, 244–266.

Papatryphon, E., Petit, J., Kaushik, S.J., Van Der Werf, H.M.G., 2004. Environmental impact assessment of salmonid feeds using life cycle assessment (LCA). Ambio 33, 316–323.

Parker, R.W.R., Vazquez-Rowe, I., Tyedmers, P.H., 2015. Fuel performance and carbon footprint of the global purse seine tuna fleet. Journal of Cleaner Production 103, 517–524.

Patcharaporn Pongpat, P., Tongpool, R., 2013. Life Cycle Assessment of Fish Culture in Thailand: Case Study of Nile Tilapia and Striped Catfish. International Journal of Environmental Science and Development 4, 608–612.

Pelletier, N., Tyedmers, P., 2007. Feeding farmed salmon: is organic better? Aquaculture 272, 399–416.

Pelletier, N., Tyedmers, P., 2010. Life cycle assessment of frozen tilapia fillets from Indonesian lake-based and pond-based intensive aquaculture systems. Journal of Industrial Ecology 14, 467–481.

Pelletier, N., Tyedmers, P., Sonesson, U., Scholz, A., Ziegler, F., Flysjo, A., 2009. Not all salmon are created equal: life cycle assessment (LCA) of global salmon farming systems. Environmental Science and Technology 43, 8730–8736.

Perdikaris, C., Chrysafi, A., Ganias, K., 2016. Environmentally friendly practices and perceptions in aquaculture: a sectorial case-study from a Mediterranean-based industry. Reviews in Fisheries Science & Aquaculture 24, 113–125.

Ramos, S., Vazquez-Rowe, I., Artetxe, I., Moreira, M.T., Feijoo, G., Zufia, J., 2011. Environmental assessment of the Atlantic mackerel (*Scomber scombrus*) season in the Basque Country. Increasing the timeline delimitation in fishery LCA studies. International Journal of Life Cycle Assessment 16, 599–610.

Ray, N.E., Maguire, T.J., 2019. Low greenhouse gas emissions from oyster aquaculture. Environmental Science & Technology 53, 9118–9127.

Rodrigues, E.R.N., Medeiros, D.L., Mendonça, A.A., Marcolin, C.R., Albinati, R.C.B., Franke, C.R., 2016. Life cycle analysis of national imported fish in the state of Bahia (Brazil). Boletim do Instituto de Pesca 42, 792–800.

Samuel-Fitwi, B., Nagel, F., Meyer, S., Schroeder, J.P., Schulz, C., 2013a. Comparative life cycle assessment (LCA) of raising rainbow trout (*Oncorhynchus mykiss*) in different production systems. Aquacultural Engineering 54, 85–92.

Samuel-Fitwi, B., Schroeder, J.P., Schulz, C., 2013b. System delimitation in life cycle assessment (LCA) of aquaculture: striving for valid and comprehensive environmental assessment using rainbow trout farming as a case study. International Journal of Life Cycle Assessment 18, 577–589.

Santos, A.A.O., Aubin, J., Corson, M.S., Valenti, W.C., Camargo, A.F.M., 2015. Comparing environmental impacts of native and introduced freshwater prawn farming in Brazil and the influence of better effluent management using LCA. Aquaculture 444, 151–159.

Simmons, J., 2011. Making safe, affordable and abundant food a global reality. Range Beef Cow Symposium. p. 300.

Simpson, S., 2011. The blue food revolution: making aquaculture a sustainable food source. Scientific American 54–61.

Smárason, B.Ö., Ögmundarson, Ó., Árnason, J., Björnsdóttir, R., Davíðsdóttir, B., 2017a. Life cycle assessment of Icelandic Arctic char fed three different feed types. Turkish Journal of Fish and Aquatic Sciences 17, 79–90.

Smárason, B., Ögmundarson, Ò., Árnason, J., Björnsdóttir, R., Davíðsdóttir, B., 2017b. Life cycle assessment of Icelandic arctic char fed three different feed types. Journal of Fisheries and Aquatic Sciences 17, 79–90.

Integrated mariculture: a global review. In: Soto, D. (Ed.), FAO Fisheries and Aquaculture Technical Paper No. 529. FAO, Rome, 183 pp.

Tacon, A.G., Metian, M., 2009. Fishing for feed or fishing for food: increasing global competition for small pelagic forage fish. Ambio 38, 294–302.

Thrane, M., 2006. LCA of Danish fish products – new methods and insights. International Journal of Life Cycle Assessment 11, 66–74.

Tilman, D., Cassman, K.G., Matson, P.A., Naylor, R., Polasky, S., 2002. Agricultural sustainability and intensive production practices. Nature 418, 671–677.

Tlusty, M.F., Thorsen, Ø., 2016. Claiming seafood is "sustainable" risks limiting improvements. Fish and Fisheries 18, 340–346.

Valenti, W.C., Kimpara, J.M., Preto, B.L., Moraes-Valenti, P., 2018. Indicators of sustainability to assess aquaculture systems. Ecological Indicators 88, 402–413.

Vázquez-Rowe, I., Moreira, M.T., Feijoo, G., 2010. Life cycle assessment of horse mackerel fisheries in Galicia (NW Spain): comparative analysis of two major fishing methods. Fisheries Research 106, 517–527.

Vázquez-Rowe, I., Moreira, M.T., Feijoo, G., 2011. Life Cycle Assessment of fresh hake fillets captured by the Galician fleet in the Northern Stock. Fisheries Research 110, 128–135.

Vázquez-Rowe, I., Moreira, M.T., Feijoo, G., 2012. Inclusion of discard assessment indicators in fisheries life cycle assessment studies. Expanding the use of fishery-specific impact categories. International Journal of Life Cycle Assessment 535–549.

Vázquez-Rowe, I., Moreira, M.T., Feijoo, G., 2013. Carbon footprint analysis of goose barnacle (*Pollicipes pollicipes*) collection on the Galician coast (NW Spain). Fisheries Research 143, 191–200.

Vázquez-Rowe, I., Villanueva-Rey, P., Hospido, A., Moreira, M.T., Feijoo, G., 2014. Life cycle assessment of European pilchard (*Sardina pilchardus*) consumption. A case study for Galicia (NW Spain). Science of the Total Environment 475, 48–60.

Verones, F., Bolowich, A.F., Ebata, K., Boutson, A., Arimotos, T., Ishikawa, S., 2017. A case study of life cycle impacts of small-scale fishing techniques in Thailand. Cogent Environmental Science 3, 1387959.

Villanueva-Rey, P., Vazquez-Rowe, I., Arias, A., Moreira, M.T., Feijoo, G., 2018. The importance of using life cycle assessment in policy support to determine the sustainability of fishing fleets: a case study for the small-scale xeito fishery in Galicia, Spain. International Journal of Life Cycle Assessment 23, 1091–1106.

Watson, R.A., Green, B.S., Tracey, S.R., Farmery, A., Pitcher, T.J., 2016. Provenance of global seafood. Fish and Fisheries 17, 585–595.

Watson, R.A., Nichols, R., Lam, V.W.Y., Sumaila, U.R., 2017. Global seafood trade flows and developing economies: insights from linking trade and production. Marine Policy 82, 41–49.

WCED, 1987. Our Common Future. Oxford University Press, Oxford.

Wilfart, A., Prudhomme, J., Blancheton, J.P., Aubin, J., 2013. LCA and emergy accounting of aquaculture systems: towards ecological intensification. Journal of Environmental Management 121, 96–109.

Willett, W., Rockström, J., Loken, B., Springmann, M., Lang, T., Vermeulen, S., et al., 2019. Food in the Anthropocene: the EAT – Lancet Commission on healthy diets from sustainable food systems, The Lancet, 393. pp. 447–492.

Yacout, D.M.M., Soliman, N.F., Yacout, M.M., 2016. Comparative life cycle assessment (LCA) of Tilapia in two production systems: semi-intensive and intensive. International Journal of Life Cycle Assessment 21, 806–819.

Ziegler, F., Emanuelsson, A., Eichelsheim, J.L., Flysjo, A., Ndiaye, V., Thrane, M., 2011. Extended life cycle assessment of southern pink shrimp products originating in Senegalese artisanal and industrial fisheries for export to Europe. Journal of Industrial Ecology 15, 527–538.

Zucaro, A., Forte, A., De Vico, G., Fierro, A., 2016. Environmental loading of Italian semi-intensive snail farming system evaluated by means of life cycle assessment. Journal of Cleaner Production 125, 56–67.

Zufia, L., Arana, L., 2008. Life cycle assessment to eco-design food products: industrial cooked dish case study. Journal of Cleaner Production 16, 1915–1921.

PART 2

Fish and Seafood processing & By-Products

Chapter | six

Industrial Thawing of Fish Blocks

Christoph Josef Backi
Department of Chemical Engineering, Norwegian University of Science and Technology (NTNU), Trondheim, Norway

6.1 INTRODUCTION

Bacteriological and enzymatic changes cause fish and fish products to spoil rapidly (Huis in't Veld, 1996). Thereby enzymatic or autolytic changes start directly after the fish has died and resulted in phenomena like rigor mortis, see Mørkøre et al. (2008), Skjervold et al. (2001), and Stroud (1969), and weakening/softening of the tissue. Bacteriological changes in fish typically succeed enzymatic changes and are well described in the works of Huss (1995) and Gram and Dalgaard (2002), for example. Hence shelf life extension is vital to guarantee a high-quality product, which is safe for consumption. Preservation techniques, such as freezing and cooling, should be enforced along the whole production chain (from catch to processing and consumption). These cooling and freezing techniques reduce the rate of spoilage while the fish's physical properties won't be affected to a too large degree (Gould, 1996). Shelf life can be extended in ranges of multiple days to several months if the techniques are applied correctly, see, for example, the works of Hall (1997), Barros-Velázquez et al. (2008), and Dawood et al. (1986). Long trawls or seasonal fluctuations (e.g., for white fish) make cooling and freezing necessary steps in the production chain in order to maintain acceptable and safe quality levels. Standard procedures imply freezing-headed and gutted (H/G) fish into blocks. The typical size of these blocks would be a thickness of 10 cm and a length and height of 50 cm, respectively. This leads to a weight of roughly 25 kg, which a worker should still be able to handle and carry. Nevertheless a standard is not defined correctly, leading to various dimensions in the industry.

Freezing is often conducted in plate freezers, which are the preferred method onboard fishing vessels, since they are quite robust, also in bad weather conditions. But again, air blast freezing is put in practice, mostly in

Sustainable Fish Production and Processing. DOI: https://doi.org/10.1016/B978-0-12-824296-4.00001-3

factories onshore. For all freezing techniques, it holds that with reduced thickness of the product, freezing occurs faster. Freezing, or being frozen, in that sense, means that the temperature inside the product's center (core) is $-18°C$ at the warmest.

Understanding the process of thawing implies understanding the process of freezing. Simple thermodynamic relations can describe freezing: An amount of heat or energy, respectively, is withdrawn from the product to be frozen. This amount of energy is defined as $dh = c(T)(T_{\text{init}} - T_{\text{des}}) + h_{\text{latent}}$, where T_{init} is the product's initial temperature, T_{des} denotes the desired temperature, $c(T)$ is the temperature-dependent specific heat capacity, and h_{latent} indicates the latent heat of fusion. Latter defines an amount of energy that must be removed to undergo phase change—in this case, for water, from liquid to solid. Be advised that the equation above depicts the theoretical amount of energy to be removed in order to obtain an evenly distributed desired final temperature T_{des} throughout the freezing product. It is a steady-state calculation and does neither consider temperature dependencies concerning time and space nor any sort of energy losses.

In a series of studies, models incorporating spatial and time-dependencies for the temperature dynamics inside fish blocks were introduced, see Backi et al. (2014a, 2015, 2016a) and Backi and Gravdahl (2013). These models are based on nonlinear partial differential equations, which are based on the well-known heat equation. Thereby the parameters representing the fish's physical properties were modeled as thermodynamic alloys, meaning that parameters like density, specific heat capacity, and thermal conductivity were represented as fractions of essential components (e.g., water, fat, ash, protein, and carbohydrates). In addition, these parameters were modeled temperature-dependent see Harðarson (1996) and Backi (2015), which led to the heat equation becoming nonlinear. A more in-depth look into the modeling of physical properties of different foods can be found in the work of Pham (2014). In order to model the latent heat of fusion a technique called the apparent heat capacity method was utilized, which overestimates the specific heat capacity in a region around the freezing point and hence artificially slows down heat transfer in this particular region. Further literature on modeling the adequate heat capacities for freezing and thawing of food includes the work of Schwartzberg (1976) and Lind (1991). More theoretical studies of heat transfer with phase-change phenomena during thawing operations include the work of Leung et al. (2005), for example.

Since it is almost impossible to process frozen-headed and gutted fish further (e.g., to cut high-quality fillets), it is vital to thaw the fish before further processing. This means that a significant amount of heat (or energy, respectively) has to be provided to the fish to avoid damage while processing in a (partially) frozen state. This amount of heat/energy must cause the water in the fish block's inner domain to change phase from solid to liquid form. It must be mentioned that there exist products, which do not necessarily have to be thawed before further processing. The most known of these products are fish sticks, which are produced by freezing fillets of fish to blocks in plate freezers and then cut fish sticks out of the frozen fish fillet blocks.

It is essential to know that, in general, the freezing process runs faster than the thawing process. The reason for this is that frozen water conducts heat better compared to liquid water. When frozen from the boundary a layer of frozen water will form and accelerate heat transport from the inside toward the outside of the product. It means that the cold front will move faster toward the knowledge of the product when moving through the frozen layer compared to the liquid layer. For thawing, this is somewhat reversed since, when thawed from the boundary, the outer layers of liquid water will decelerate heat transfer toward the inner domain and act as a resistance to the heat flow.

One can find various thawing techniques, which differ in terms of quality and practicality for different fish species or products, respectively. In general, thawing methods are dividable into two principal groups, namely, methods, which supply heat to the product through the (outer) boundary layers, and techniques generating and distributing heat all over the product's (inner) spatial domain. The thawing process's ultimate goal should be the complete thawing of the product. Complete thawing means that the core temperature (i.e., the temperature at the point, which is hardest to affect and furthest away from the outer boundary) is higher than the freezing point (depending on species or product). For fish blocks, it is essential to avoid single fish still being attached to the block or to one another in order to allow further processing optimally. The amount of energy needed to thaw a fish block from initial to a particular final temperature is dictated by the laws of thermodynamics. Hence, reducing energy consumption is only possible by making the thawing process themselves more energy efficient. However, this aspect is not represented in this work.

Generally the quality of the thawed product is maximized for slow thawing. Nilsson (1994) was able to show a reduction of drip loss for slower proceeding thawing processes. However, this relation must not necessarily hold for all applications and thawing methods since other problems like tissue drying and lipid oxidation might arise.

Maximal and minimal temperatures define the speed at which thawing can be conducted. Naturally the lower bound on the temperature is defined by the melting/freezing point of the particular product. One simply cannot thaw a product with temperatures below its freezing point. However, products can be tempered that way. The upper bound on the temperature will heavily influence the thawed product's quality: It will already suffer if the product's temperature exceeds 20°C. Temperatures above 30°C will cause the fish to start cooking, as, for example, stated in Jason (1974). These temperatures do not appear very realistic, however. In literature, there exist various standards for the thawing medium's maximum temperature, which includes both air and water, respectively, as well as that of the product: Thawing media temperatures span from upper values between 12°C and 25°C, whereas product temperatures are mentioned to be between 0°C and 7°C, see for example, Archer et al. (2008). Roiha et al. (2018b) report about quality attributes of prerigor frozen Atlantic cod for thawing conducted in water with added air bubbles at two different temperatures, namely, 10°C followed by −0.5°C and constant 10°C, respectively. They conclude that the two controlled thawing temperatures preserved quality in a good way, meaning that drip loss, cooking yield, total volatile bases nitrogen, texture, microbiological analyses, and sensory evaluation are not affected to a too large degree.

Alizadeh et al. (2007) performed a cross-comparative study between three thawing methods (pressure-assisted thawing, air blast thawing, and immersion thawing) and two freezing methods (pressure shift freezing and air blast freezing). It becomes apparent that the quality after thawing is already affected by the chosen freezing process and procedure. Not even the most advanced thawing process can improve the product's quality after a nonoptimal, badly executed freezing process. That means that damages and quality degradation that have already been introduced during freezing cannot be repaired or reduced by the thawing method. At best the quality level can be preserved by the thawing method.

Only a few studies can be found, which touch on quality changes caused by different thawing methods for varying freezing methods. Therefore the author proposes to conduct research comparing product quality indicators using different thawing methods after utilizing standardized freezing processes. This allows the optimization of thawing methods with respect to product quality. In addition to the techniques themselves, several other important factors should be considered, like the product's freshness before freezing, the frozen storage period, the storage temperature, and very importantly, if and to what degree the storage temperature fluctuated during storage. Latter is a huge problem since varying storage temperatures can cause the formation of larger ice crystals compared to storage at constant temperatures. This can compromise quality factors, such as, for example, the water holding capacity, as reported in Johnston et al. (n.d.).

There are only very few scientific studies concerning the operation of contact thawers, see, for example, Roiha et al. (2018a). In that study, thawing times for water thawing with added air bubbles and contact thawing were comparable, whereas thawing in water without air bubbles was substantially slower. The author of the present chapter expects contact thawing to improve controllability and ability to monitor the temperatures of the thawing process since plate freezer technology, which is the reversed process to contact thawing, is relatively mature and can be applied rather easily. For theoretical studies of contact thawing/plate freezing technology with applications to automatic control and estimation, see the works of Backi et al. (2014b, 2016b).

Besides the broad overview of thawing methods in Backi (2017), other reviews exist, such as the works of Otero and Sanz (2003), Le Bail et al. (2002), Li and Sun (2002), and Carver (1975). These, however, only address subsets of the thawing methods presented in Backi (2017) and this work.

The scope of this work is not to present thawing times for different thawing methods for standardized sizes of fish or fish blocks. This lies in the fact that there are no standardized sizes, which leads to a lack of data in the literature concerning thawing times. The present chapter merely collects thawing methods, divides them into functional groups, and briefly introduces their technological backgrounds and specifications. However, if available among the cited literature, examples for thawing times are provided.

Additionally, articles reporting about the calculation of thawing times for different shapes of foodstuff are included in the literature. These are, for example, Cleland et al. (1987a,c), who introduce simple formulae for thawing and freezing time prediction for multidimensional shapes. Furthermore the authors of Cleland et al. (1987b) used numerical methods to estimate freezing and thawing times, again for multidimensional shapes, whereas Cleland et al. (1987d) provide experimental data for thawing and freezing times for multidimensional shapes as well.

6.2 THAWING METHODS ACTING THROUGH THE BOUNDARY

The thawing methods presented in this section affect the product only via its boundary layers. This implies that heat/energy will only be introduced by a heat source at these boundaries, meaning that in order to reach the center and thaw the product thoroughly, the heat front has to propagate all the way through the product.

6.2.1 Air Thawing

The thermodynamic principle behind air thawing is heat transfer via convection, either natural or forced convection. Hence air thawing can come in several shapes and is generally considered a cheap solution. Forced convection denotes the exchange of heat with exterior interaction (e.g., forced airflow), while natural convection is defined as heat transfer without any further interaction. However, since air is the thawing medium, these methods increase the probability of lipid oxidation and undesired surface drying. Latter is especially a problem with forced convection, while lipid oxidation may be problematic for fatty fish in particular. These quality-decreasing factors can be minimized by the use of humid air, which generally contains a higher fraction of water vapor and less oxygen, as, for example, reported in the works of Jason (1974), Mannapperuma and Singh (1988), and Archer et al. (2008).

6.2.1.1 THAWING IN STILL AIR

With this method the product is thawed by nonmoving, static air. The rate of heat exchange is relatively low, making the thawing time extremely large. Reported thawing times are upto 20 hours for a 100 mm thick block of frozen whole cod thawed in the air of 15°C, see Jason (1974). Accelerating measures are putting the product on a grate, enabling the air also to reach and act on its bottom side. Separation of single fish from a block leads to a larger total surface area and yet accelerated thawing since the area of attack is increased. The separation should be undertaken cautiously, avoiding excessive force not to damage the fish's fillet while it might still be sticking to the rest of the block. Since the rates of heat exchange are relatively low and thus thawing times are quite high, the product is exposed to air for a long time, which implies the danger of high rates of lipid oxidation as well as surface drying. Using humid air with high fractions of water vapor is likely to reduce drying of the surface and lipid oxidation, as mentioned before.

6.2.1.2 AIR BLAST THAWING

Air blast thawing is commonly used in industrial practice. The general principle is the same compared to thawing in the still air as reported above. The difference is that the air is forced to move around and over the product's surface, which is an example of forced convection. Naturally, air blast thawing provides higher rates of heat exchange compared to thawing in still air, where the airspeed is a vital parameter to influence the heat transfer coefficient. Characteristic speeds are around 6–10 m/second as reported in, for example, Footit and Lewis (1999), but the air velocity field inside the thawing chamber is quite irregular. It can be compared to that inside air blast freezing chambers. Alonso et al. (2011) provide a simulation study to investigate the flow velocity field inside a freezing tunnel. Adequate airspeeds might be substantially smaller than the already mentioned 6–10 m/second and hence a focus should be put on the value for the minimum airspeed. Reported thawing times are roughly 4–4.5 hours to thaw a 100 mm thick block of frozen whole cod in humid air at 20°C with a speed of 8 m/second, as reported in Jason (1974).

In industry, both, continuous and batch thawers can be found, but the latter type is most commonly used in practice since their costs are substantially lower. For one type the product to be thawed is put on trays and stacked on trolleys. A uniformly distributed airflow is crucial inside the thawing chamber. Therefore not only the distance between trays but also between trolleys is vital for optimal airflow. To thaw the product uniformly the direction of airflow should be adjustable inside the thawing chamber. Hence a reversible fan should be installed, which should be equipped with nozzles providing heated water. A thawing process of this kind is depicted in fig. 2 in Jason (1974). However, several different constructions exist in the industry, but the basic principles remain the same.

For the described setup the danger of lipid oxidation is usually reduced compared to thawing in still air since thawing times are shorter, and water vapor is added. Nevertheless a slight threat of surface drying remains present. Additionally, if the air temperature is not tightly controlled, overheating of the product or parts of it can occur. Hence, in practice, the product is usually removed from the thawing chamber before complete thawing. The final stage will be conducted in a cooling chamber, where thawing occurs covered in ice, which reduces drying and lipid oxidation to a minimum.

6.2.1.3 IMPINGEMENT THAWING

Impingement denotes the process of directing an air stream toward the product's surface through small nozzles. This technique is commonly used and established in freezing and cooling applications, see the works of Salvadori and Mascheroni (2002), Soto and Bórquez (2001), and Dirita et al. (2007), as well as food processing in general, presented in, for example, Sarkar and Singh (2004). Applying this technique to thawing, however, is still subject to research activity. There exist studies looking into mathematical aspects of impingement thawing, for example, Anderson and Singh (2006a) and determining effective heat transfer, as in Anderson and Singh (2006b). Nevertheless, experimental studies concerning the thawing of fish and fish products are hard to find. Using forced air streams appears to provide some advantages compared to conventional thawing if surface drying can be avoided, especially when it comes to thawing times.

6.2.2 Water Thawing

Water thawing, next to air thawing methods, is widely used in industrial practice. Another name for this technique is immersion thawing, since, during operation, the product is immersed inside a container usually filled with water.

One thing that has to be ensured for the most efficient heat transfer is that the product remains fully covered with water since the density of ice is slightly lower than that of liquid water, and hence the product will rise to the water surface. Naturally the dangers of drying and lipid oxidation that are potential threats to quality in air-based thawing methods, only pose a minor problem for water thawing, if at all.

In industry, there are many different techniques relying on the same principle, namely conduction-driven heat exchange between the product and the thawing medium, in which the product is immersed. In theory, and often in practice, most of the immersion thawing methods are considerably faster due to the larger heat transfer coefficient compared to methods driven by convective heat transfer alone. In order to ensure optimal thawing results with respect to thawing times the thawing medium should be regenerated and exchanged. The reason for this lies in the fact that during the thawing process, the thawing medium will cool down while the thawing product will gradually heat up. This reduces the temperature difference between product and medium, which is the driving force for heat transfer. A smaller temperature difference between medium and product will ultimately lead to an increase in thawing time. In addition to the longer thawing times, bacteria might become enriched if the thawing medium is not regenerated/exchanged, leading to potential contamination and quality reduction of the thawed product. In principle, one can find two kinds of immersion thawing operations, namely, batch and continuous thawers.

An example of an immersion batch thawer is a rectangular tank (typically $0.3-1\ m^3$), which will be filled with water together with the thawing product. One mode of practical application is to consistently feed (heated) water through a hose into the tank, as discussed in MacCallum and Ellis (1964) and

Haugland (2002). Thawing times are reported to be around 3–4.5 hours for 102–115 mm thick fish blocks with a constant water temperature of 15.5°C, as stated in MacCallum and Ellis (1964). In a comparative study investigating freeze–thaw cycles of white shrimp and black tiger shrimp, Sriket et al. (2007) report about water thawing with running water of 27°C, which sounds unusually high. Improvement of heat exchange can be achieved by pumping air into the bottom of the tank. The resulting air bubbles rising to the water top will enhance heat transfer. Over time, single fish will separate from a block, which leads to an increased heat exchange area and further accelerates the thawing process. A picture of such a batch thawing process in a tank enhanced with water bubbles is shown in Fig. 6.1.

The second big group of water thawing operations is continuous immersion thawers, which can be found in many different constructive shapes. One construction is a large cylindrical tank filled with water. The tank is divided into compartments (like cake slices), into which the thawing products (here fish blocks) are filled. These compartments can revolve around the cylinder's rotational axis, and, after one full turn is fulfilled, the products are completely thawed. Ideally, in the final stage, only single thawed fish are present in the compartment. The volumetric efficiency of this construction, which is shown in Fig. 6.2, is very high.

Other constructions include helical screws in big tanks. The frozen product gets filled into the tank at one end, from which it is conveyed to the other end by the helical screw—fully thawed. Yet another construction follows a similar principle: a big tank with a conveyor belt at the bottom. Thawed fish will sink to the bottom (since its density is typically larger than that of water) and is transported out of the tank by the conveyor belt.

FIGURE 6.1 Experimental thawing setup for two blocks of frozen-headed and gutted cod in a tank with air bubbles in order to increase heat transfer—the aluminum frame ensures immersion of the fish blocks in water.

FIGURE 6.2 A thawing tank in a cylindrical shape as built by Melbu Systems AS, Melbu, Norway. Image included with kind permission of Melbu Systems AS.

Kissam et al. (1981) report about a water thawing method equipped with low-frequency acoustics of 1500 Hz with 60 W power. Their claim is a significant reduction of thawing times of up to 82% compared to conventional operation, since the low-frequency acoustics do not aggressively apply heat/energy to the frozen product but rather stimulate heat transfer with the thawing medium. Leung et al. (2007) investigated heat transfer and fluid flow in water thawing processes in a theoretical study utilizing experimental data. Based on their study the authors developed a simple water thawing device with improved efficiency.

6.2.3 Vacuum Thawing

The working principle behind vacuum thawing lies in the decreasing boiling temperature of water for a decrease in pressure. In addition, the condensation of water vapor on a colder surface releases the so-called heat of condensation. This means that a specific boiling temperature at a defined pressure can be chosen to be the thawing temperature of the product. Water vapor is released into the vacuum chamber and will condense on the product's surface while releasing the heat of condensation. In order to operate this method optimally, tight pressure control is crucial to avoid heat damage, since the boiling temperature will instantly increase if the pressure increases toward ambient pressure. When it comes to thawing speed, vacuum thawing is roughly equal to immersion thawing or air blast thawing. A 65 mm thick block of shrimps and a 50 mm thick block of plaice are reported to be thawed in less than 80 minutes in the work of Jason (1974).

Vacuum heat thawing for blocks of headed and gutted whiting and shrimps is reported in Carver (1975). A liquid-ring pump is used to remove air from the thawing chamber, in which water in a container is heated electrically or by steam and vaporizes at low temperatures due to the low pressure. The vacuum thawing process itself is relatively clean since the water vapor condenses only on the product's surface, and ideally, the chamber walls remain dry. According to the author the operation is hygienic, and bacteria do not pose severe a threat while the cleaning effort is negligibly small.

6.2.4 Contact Thawing

Contact thawers are related to plate freezers, which have been used in the industry for a long time. The product is placed between plates, which are flown through by an evaporating refrigerant taking the heat off from the product.

Contact thawers represent the reversed process, where thawing media provide heat to the product, which is again placed between two plates. It is preferable to use thawing media that have relatively high specific heat capacities, such as water or glycol–water mixtures. However, even electrical heating seems possible using heating pads.

There are two advantages to contact thawing operations. Firstly the product is not in direct contact with the thawing medium, and secondly the thawing temperature can be controlled in a better manner than other thawing processes. The advantage lies in the fact that heat transfer dynamics in this kind of construction are well understood from plate freezer technology, see for example, Backi et al. (2016a). A theoretical study on the optimal operation of contact thawers is presented in Backi et al. (2016b), where advanced feedback control methods, in particular model predictive control, are applied. Furthermore, for monitoring purposes, another theoretical study by Backi et al. (2014b) developed so-called observer techniques in order to estimate the inner-domain temperature field inside the fish block—in that case for freezing, but the results are transferable to thawing processes. A broader study of observer techniques for plate freezing and contact thawing is included in Backi (2015). Nevertheless a disadvantage can be the unloading of these devices, which is believed to be quite time consuming, since single, thawed fish will have to be taken out of the contact thawer, for example, by hand. One could think about different constructions incorporating conveyor belts underneath the contact thawer. The thawing chambers can be opened on the bottom such that the thawed fish will fall down on the conveyor and are transported to further processing stages.

6.3 METHODS ACTING IN THE INNER DOMAIN

Besides solely acting on the outer layers of the frozen product (like the methods introduced above) the second major group of thawing methods also acts directly on the product's inner domain/interior. It must be noted that generally, these methods are faster compared to the thawing methods presented above. Although their principles are well understood, their scale-up seems often a challenge, what makes them mostly available on small scales only.

Thermal runaway, or often called local overheating, is a significant problem, especially in dielectric methods, whereas high-pressure methods are mostly limited to (professional) kitchen use for high-quality products and academic research.

6.3.1 High-Pressure Thawing

High-pressure thawing relies on the principle that the freezing temperature of water decreases as the pressure is increased. This effect appears up to a pressure of about 210 MPa, whereas, for a further increase in pressure, the freezing point rises again. At ambient pressure (roughly 0.1 MPa) the freezing temperature of the water is around 0°C, whereas, at 210 MPa, it is around −21°C, see Fig. 6.3. The freezing temperature does not depend on the pressure only but also on any dissolved substances lowering or increasing the freezing point, like salt for example.

In order to obtain high pressures inside the processing chamber, it will become filled with a pressure-transmitting liquid, which is often water with potential additives. The pressure increase itself is achieved by hydraulic pressure intensifiers or hydraulic cylinders moving into the processing chamber reducing its volume, as described in, for example, Mújica-Paz et al. (2011) and Torres and Velazquez (2005). Chevalier et al. (1999) state that, ideally, the products should be put in vacuum-sealed packaging, since the effect of high-pressure operations on, for example, texture and color are quite extensive.

In general, there exist two different methods for high-pressure thawing: Pressure-assisted thawing and pressure-induced thawing. The principle behind pressure-assisted thawing is thawing at constant pressure, while the temperature

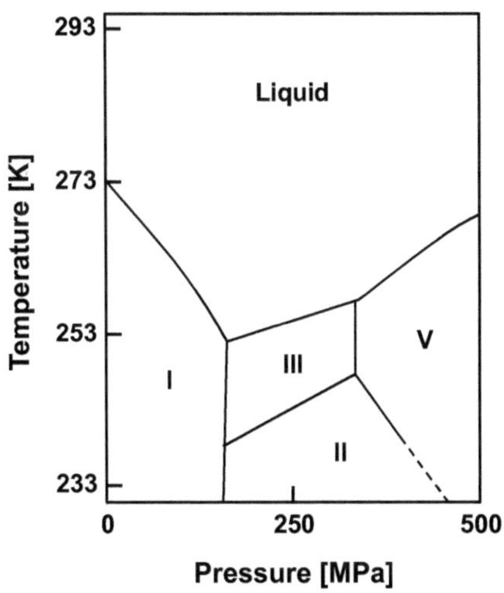

FIGURE 6.3 Pressure–temperature phase diagram for different ice formations of type I, II, III, and V and liquid water.

difference between the product and the pressurized liquid inside the chamber is the driving force. A thawing medium (e.g., water), which surrounds the chamber, typically controls the temperature inside the chamber. This method corresponds roughly to the reversed pressure-assisted freezing process, as described in, for example, Fernández et al. (2006).

For pressure-induced thawing the thawing process is realized by changing the pressure along the melting curve of ice I, see Fig. 6.3. Hence the driving force for temperature increase is pressure decrease. In Otero and Sanz (2003) and Norton and Sun (2008) the differences between the two techniques are well explained. Li and Sun (2002) provide a review over applications for high-pressure freezing and thawing for foodstuff, whereas Le Bail et al. (2002) deliver another review concerning high-pressure freezing and thawing both for foodstuff, and also for microscopic preparation purposes. Furthermore, Chourot et al. (1997) present results for numerical modeling of a high-pressure thawing processes with application to water thawing.

In high-pressure freezing applications the refrigerant/freezing medium must be present around the pressurized chamber such that the temperature inside the chamber and hence of the product can be reduced. In pressure thawing, this is theoretically no requirement since the pressure-transmitting liquid inside the chamber will theoretically heat up during pressurization. Nevertheless a controlled temperature should be applied to the outside of the chamber in order to control the temperature inside the pressurized chamber thoroughly.

The construction's sizes are limited to small scales, however. Reported volumes in academic experiments reach from 0.053 dm^3 as in Cheftel et al. (2000), 0.6 dm^3 as reported by Schubring et al. (2003), to around 3.4 dm^3 in Ousegui et al. (2008), 4.32 dm^3 in the work of Rouillé et al. (2002), and 4.57 dm^3 in Zhu et al. (2004).

In industrial pressure processing applications, for example, for pasteurization, typical volumes are reported to be 100–500 dm^3, see Mújica-Paz et al. (2011).

6.3.2 Electrical Methods

Electrical methods utilize electric fields, which generate heat inside and hence throughout the whole spatial domain of the product. In summary, Zhao et al. (2000) report about methods, such as dielectric thawing and electrical resistance thawing. Another critical method not mentioned in the study is high-voltage electrostatic fields (HVEFs).

Mertens and Knorr (1992) discuss several (electrical) methods for nonthermal food preservation, such as oscillating magnetic field pulses, high-electric field pulses, microwaves, linear induction electron accelerators, intense light pulses, and high-pressure treatments.

6.3.2.1 HIGH-VOLTAGE ELECTROSTATIC FIELD

Thawing by HVEF denotes a method, where the product is put on a metal plate acting as a cathode while electrodes are mounted at some distance (above the product). When the electric circuit is closed, a corona will form between the cathode and the electrode. This method is more like a thawing method using still air and not an inner domain method. Nevertheless HVEF seems to accelerate thawing, as reported in, for example, Hsieh et al. (2010). In this comparative study, chicken thigh meat was thawed at −3°C and 4°C with and without HVEF. The authors conclude that thawing times are reduced and quality factors, for example, water holding capacity and cooking loss, are improved using HVEF.

The study of He et al. (2013) comes to a similar conclusion, namely, that thawing times are reduced utilizing HVEF, in that case, for pork tenderloin thawed at room temperature (around 20°C). A further finding is that thawing times seem to be inversely proportional to the applied voltage. However, the authors did not find improvements for quality factors, such as pH, thawing loss, and cooking loss, in that particular study.

In a series of studies, tuna fish was thawed using still air at 20°C combined with HVEF by Mousakhani-Ganjeh et al. (2016a,b). In these studies, reported thawing times using HVEF are faster compared to the control group without HVEF.

Bai et al. (2011) investigated optimal parameters for HVEF thawing, especially the voltage applied and the space between cathode and electrode. They conclude that for specific areas between cathode and electrode as well as for specific applied voltages, thawing rates and energy consumption are at optimal levels. This, however, must not necessarily hold for all kinds of products/foodstuffs. In another study, Li et al. (2017) investigated the HVEF treatment of cubes of carp. They conclude that the fastest thawing times were achieved utilizing −12 kV HVEF thawing and that at the same time water loss and microbial counts were minimized, compared to air thawing, water thawing, and HVEF thawing at −6 kV. Other benefits of HVEF thawing include an enhanced activity of adenosine monophosphate deaminase, reduced activity of acid phosphatase, and delay of the degradation of inosine monophosphate to hypoxanthine.

More critical studies include Otero et al. (2016), who state that more research should be conducted regarding the use of magnetic fields since, according to the authors, the benefits are not proven, at least for freezing processes. Nevertheless, more research is needed for HVEF thawing applied to fish and fishery products since there is a lack of research in the literature.

6.3.2.2 DIELECTRIC THAWING

Dielectric thawing is a collective term denoting all sorts of thawing methods, which utilize high-frequency waves to thaw foodstuff. Most common are methods based on radio-frequency waves and microwaves, which do not differ in their principle, but in their technical specifications, namely, the frequency band used. Only rough classifications concerning the particular frequency bands can be found in literature, but it generally holds that radio frequencies are in the MHz range while microwave frequencies are in the GHz range. A second important factor is the supplied voltage and hence the applied electrical power to the process. Both methods can not only be used for thawing frozen products but also for cooking foodstuff as well, as stated in McKenna et al. (2006).

A requirement for dielectric thawing to be applicable is that the product can be polarized by an electric field and is, in fact, an electric insulator. The theoretical principle behind these methods is that in an oscillating electric field, molecules start to rotate and ultimately collide with one another, which generates heat inside the product. The thawing impact, which is defined by the penetration depth of the waves, depends highly on the applied frequency. For foodstuff, it is known that the thawing effect is inversely proportional to the frequency, which implies that radio-frequency methods should naturally be more suited for large-scale applications compared to microwave methods, see the work of Wang et al. (2003). To find out which frequencies are best suited for a specific kind of foodstuff, one must find the frequency that maximizes the imaginary permittivity of this foodstuff (also called the Debye resonance). But in practice the known value for water is the first right choice since most foods consist of water to a large extent. In Bengtsson et al. (1963), measurements of dielectric properties for frozen and thawed fish and meat are presented.

Common problems applying dielectric methods are local overheating of the product itself due to irregular shapes and entrapped air. But even overheating and burning of packaging material have been reported due to multiple Debye resonance frequencies, as stated in Zhao et al. (2000).

For microwave techniques, microwave radiation is utilized to thaw frozen products. The principle is in fact the same as for private-use microwave ovens. The frequency range lies in between that for radio waves (30 kHz–300 MHz) and infrared light (300 GHz–385 THz). Local thermal runaway can occur since the penetration depth depends on the frequency and tissue composition of the product. That means that limitations in penetration depth will lead to thawing, which is restricted to the product's outer boundary layers. Hence the center and the inner layers will remain frozen with the consequence that melting for these layers will occur by regular thermal conduction. The heat transfer gradient for microwaves is much bigger than for conduction, and, in addition, liquid water absorbs more energy than frozen water—hence the term thermal runaway, since the product's outer layers will heat up much faster than its interior.

Private-use microwave ovens with frequencies of typically 2.45 GHz and power consumption of 1–2 kW have a penetration depth of not more than a few centimeters. This might suffice for private use but seems unpractical for large-scale industrial applications. This is why the industry applies lower frequencies of exemplary around 915 MHz, whereas the supplied power is increased to, for example, 10 kW. Faghri (1994) conducted an experimental and numerical study concerning heat transfer in microwave thawing. They conclude that the numerical results resemble temperature curves obtained by conventional convective thawing. However, the authors also report about local overheating and large temperature gradients as a result, both in experiments and simulations. In Taher and Farid (2001), another experimental and theoretical study is presented, this time for cyclic microwave operation applied to frozen minced beef. The authors state that their experiments and simulations showed good agreement. They claim furthermore that with controlled surface temperatures, thermal runaway might be minimized or even avoided.

Radio-frequency utilizes low-frequency waves with large wavelengths. Typically, these frequencies for thawing applications include 13.56, 27.12, and 40.68 MHz, as reported in Piyasena et al. (2007). In theory, this implies more considerable penetration depths compared to microwaves. Combined with an appropriate level of power, radio-frequency waves are allegedly better and safer when it comes to an overall uniform temperature distribution as well as overheating. Nevertheless, thermal runaway might still pose a threat to the quality of the thawed product. Marra et al. (2008) provide a useful review over advances in radio-frequency thawing together with aspects of its mathematical modeling. Uyar et al. (2015) conducted a computational study on radio-frequency thawing of food products and come to the conclusion that it should be more applicable for industrial scale compared to microwave thawing. However, the authors report about nonuniform temperature distribution leading to local overheating. Llave et al. (2015) present another computational study for radio-frequency thawing but mention that their theoretical findings should be verified in additional studies. The application of radio-frequency methods for cooking rather than thawing is presented in Rincon et al. (2015). They conclude that the moisture content was highest for lower temperature treatments, as can be expected. In the work of Bedane et al. (2018) a staggered through-field electrode radio-frequency system was used to thaw frozen chicken breast meat. They conclude that radio-frequency thawing was around 23 times faster compared to thawing in a refrigerator at 4°C and provided improved quality aspects, namely, minimized drip loss and better textural properties.

A comparative experimental study between microwave and radio-frequency tempering of frozen blocks of shrimp was conducted by Palazoğlu and Miran (2017), who conclude that thermal runaway was mostly a problem for microwave tempering and can lead to even bigger problems when fully thawing the blocks. They state furthermore that tempering with the radio-frequency operation brings temperatures close to 0°C without excessively compromising on quality.

6.3.2.3 ULTRASOUND THAWING

Ultrasound thawing uses frequencies ranging from the kHz range to the low MHz range and has been reported for the thawing of food in Miles et al. (1998). However, disadvantages may include local overheating as already introduced for radio-frequency and microwave thawing. The even larger wavelengths (compared to radio-frequency and microwaves) should theoretically imply more considerable penetration depths. In their work, Miles et al. (1998) conclude that frequencies of about 500 kHz will deliver optimal thawing results concerning overheating and penetration depth. However, surface overheating became a problem with lower and higher frequencies (compared to 500 kHz) applied, but this is most likely caused by the supplied power rather than the frequency.

Gambuteanu and Alexe (2015) report about an 87% reduction of thawing times for pork applying a frequency of 25 kHz and a power of 0.6 W/cm^2 compared to the control groups, which were thawed using an immersion or air thawing.

6.3.2.4 FAR-INFRARED THAWING

Far-infrared thawing (FIR thawing) has some similarities compared to dielectric thawing. As stated further above the frequency spectrum of infrared waves lies above that of microwaves, and hence FIR thawing can be used as an alternative to microwave thawing, as reported in Sakai and Hanzawa (1994). The authors claim that since the absorption coefficient of the FIR waves is almost the same for ice and water, thermal runaway does not pose a big problem or is nearly reduced to zero. Nevertheless, since penetration depths for the FIR waves are theoretically even smaller than for microwaves, local overheating at the outer layers still poses a threat to thick/bulky foodstuffs.

6.3.2.5 ELECTRICAL RESISTANCE THAWING

When an electric current passes through a resistor, heat is generated. This is the principle behind electrical resistance thawing. Other common names for this method are Joule heating or Ohmic thawing. Its application is not only limited to thawing but also cooking of foodstuff, as reported in the work of McKenna et al. (2006). It holds that the lower the resistance to the current, the faster the thawing process will proceed. Since the specific resistance of frozen fish muscle is substantially large at very low temperatures (e.g., 25×10^6 Ω/cm at −30°C), it is an inferior conductor of electricity. But frozen fish is an excellent conductor of heat at these temperatures, making it applicable for conventional thawing methods. Therefore, at first, the product should be thawed using a traditional method before being applied to electrical resistance thawing in order to finish the thawing. Since the specific electrical resistance for fish muscle at 0°C is 800 Ω/cm, electrical resistance thawing could be used to remove the latent heat of fusion, see for example, Jason (1974), Li and Sun (2002), and de Alwis and Fryer (1990). According to Li and Sun (2002), frozen tuna beef and eggs can be thawed rapidly in the temperature region between −3°C and +3°C by HVEF incorporating a patented method (United States Patent 5034236 and European Patent 0409430). Liu et al. (2017) conducted an experimental study with respect to the electrical conductivity of frozen tuna. They conclude that the temperature, fat content, moisture content, and applied frequency have a significant influence on the electrical conductivity and that it changed rapidly for temperatures above −7°C. These findings fit well with the mentioned values for the specific resistance at different temperatures mentioned above.

Electrical resistance thawing has two significant advantages compared to dielectric thawing. Almost all energy provided to the frozen product is converted into heat, and in theory there exist no limitations concerning the penetration depth. However, local overheating, like for almost all other methods mentioned earlier, poses a constraint on the operation. A further restriction is that the product should be in regular shapes (e.g., rectangular) to be placed between two parallel metal plates (compare contact thawing, where the same applies for best efficiency).

6.4 PRODUCT QUALITY ASPECTS

Quality of fish is a concept, which is hard to describe since it is influenced by a lot of factors and parameters. In Ólafsdóttir et al. (1997), quality is introduced as a function of a number of different factors. Thereby, among these factors, the potentially most important one is freshness, since it is a concept, which also the end consumer can understand and grasp. Freshness again is a function of many other factors as well, such as microbiology, adenosine triphosphate levels, lipids. Nevertheless, for high-quality scores, freshness is not sufficient, but only a necessary condition. Other major quality aspects next to freshness are for example hygienic aspects and nutritional properties, which are affiliated to sensory properties. Regarding consumer safety, however, freshness constitutes the most important condition. One, and probably the major rule in food production is the safety of the foodstuff. That means that a product should not lead to any diseases or sicknesses among the consumers and hence should be safe for consumption. If this is ensured, other quality aspects can be taken into account, for example, the appearance of the product. However, spoilage can occur even without the presence of pathogenic bacteria, meaning that eating a spoiled fish might not necessarily lead to the consumer becoming sick. Many publications cover quality and freshness of fresh, chilled fish. Dawood et al. (1986) present results on quality changes due to delayed icing of rainbow trout, while Barros-Velázquez et al. (2008) conclude that, in order to maintain quality, slurry ice should be used rather than flake ice for storage of fresh fish. However, in industry flake ice is a widespread standard.

Generally, product quality is a relatively complex topic with a wide variety of different parameters, which might be assessed after thawing, such as sensory properties, texture, pH, water and fat contents, color, drip loss, and water holding capacity, see again the work of Ólafsdóttir et al. (1997). In the present chapter the focus mainly lies on texture, color, and drip loss, since these factors are discussed in most of the cited literature.

Nevertheless, other relevant quality factors, such as flavor, odor, lipid degradation, and chemical composition, are rather substantial as well. In their study about the quality of meat, Leygonie et al. (2012) focused on physical quality attributes such as lipid oxidation, color, pH, moisture loss, tenderness, microbial count, and protein denaturation after both freezing and thawing. However, a study of this extent is missing for fish products.

In literature, some studies exist, which concern the quality of fish after thawing, and thereby also differentiate between the freezing methods that have been used to freeze the fish. For quality, it is important not only to take the thawing method alone into account, but also the freezing method. For meat products, more studies comparing different freezing and thawing methods are available, compared to fish products, such as Zhao et al. (1998) and Farag et al. (2011), who investigated frozen beef, as well as Ngapo et al. (1999), whose study concerned frozen pork meat.

In their study, Sriket et al. (2007) conclude that severe changes in the physicochemical properties of white shrimp and black tiger shrimp become worse for an increased number of freeze–thaw cycles. Since the effects of thawing and freezing on these physicochemical properties are known, these results are not surprising. Even a fluctuation in frozen storage temperature can lead to quality degradation, hence freeze–thaw cycles are expected to have an even worse effect on some quality attributes, for example, texture and drip loss.

In Ersoy et al. (2008) the quality of eels was investigated utilizing various thawing methods, namely in a microwave oven for 5 minutes, thawing in 15°C air for 2 hours, water thawing at 14°C for 1.5 hours, as well as thawing at 4°C inside a refrigerator for 4 hours. The effects on the chemical, microbiological, and physical quality parameters of the eel varied with different thawing methods. In conclusion the best overall quality-preserving results were obtained using water thawing. This is not fully surprising, since water thawing seems to provide a good trade-off between rather fast thawing times and good quality preservation.

In a study using a similar methodology for pork meat by Xia et al. (2012), it was concluded that the best quality after thawing could be achieved after thawing in a refrigerator. In contrast, microwave thawing had the most negative outcome. This must, however, not hold for fish and can thus be a peculiarity for pork meat.

In Boonsumrej et al. (2007) the effects of air blast and cryogenic freezing were compared to microwave and refrigerator thawing, respectively. The authors conclude that microwave thawing leads to more enormous thawing losses than thawing in a refrigerator. Furthermore the authors state that the avoidance of freeze–thaw cycles is paramount for quality.

In another study the quality aspects of texture, drip loss, and color were investigated for Atlantic salmon after three different thawing methods and two other freezing methods (Alizadeh et al., 2007). These thawing methods were air blast thawing at 4°C and 8 m/second airspeed, immersion thawing in a stirred water bath at controlled 20°C water temperature and pressure-assisted thawing at steady 20°C in a 3.5 dm^3 vessel. The freezing methods used were air blast freezing and pressure shift freezing, respectively. The authors report that the chosen freezing method influences the mentioned quality factors heavily and that it seems to be more critical for drip loss than the thawing method selected. Compared to air blast freezing, pressure shift freezing was superior in terms of drip loss, but the inverse phenomenon was observable for texture and color. In general, pressure shift freezing, followed by pressure-assisted thawing, gave the worst results in terms of texture and color. However, for drip loss, this was the best combination. Concerning color and drip loss, air blast thawing gave good results, independent of the freezing method. These results are not really remarkable, since the effects of high-pressure treatments on texture and color are enormous.

The authors in Sigurgísladóttir et al. (2000) investigated the effect of thawing on the texture and microstructure of smoked Atlantic salmon. After having been frozen in a cryogenic freezer at −60°C, the product was stored at −20°C for 1 month before being thawed in still air at 4°C for 24 hours in a cold room. Thawing in still air gave drip loss of about 1.3%. In addition, extracellular space was increased for prefrozen salmon compared to salmon smoked without having been frozen. The prolonged rate of thawing is most likely the cause of the drip loss. However, when compared to faster freezing methods, the drip loss still seems relatively small. Therefore a better and more complex thawing method might have led to reduced drip loss.

The thawing of freshwater prawns in a water bath, microwave oven, and a refrigerator after having been frozen in a still-freezer and blast freezer, respectively (both at −29°C), is presented in Srinivasan et al. (1997, 1998). The authors state that the influences of the freezing methods did not differ concerning protein destabilization and that the muscle shear forces increased when frozen in a faster manner and thawed in a slower, respectively. In terms of protein destabilization of freshwater prawns, thawing in a cold-water bath gave the best results, while microwave thawing was least preferable.

For the same thawing methods as mentioned above, Dinçer et al. (2009) state that water thawing gave the best sensory quality for sardine, gilthead seabream, and anchovy after evaluation by a panel of trained experts.

In Díaz-Tenorio et al. (2007), quality attributes for white leg shrimp were investigated after forced convective and cryogenic freezing, followed by thawing at 4°C and 25°C, respectively. The conclusion is that for commercial trade, it is preferable to use forced convective freezing together with slow thawing at lower temperatures.

Nilsson (1994) studied quality factors of frozen rainbow trout thawed by two methods, namely air blast thawing at low airspeeds and immersion thawing. The freezing of the fish occurred with air blast freezing on a spiral belt at −35°C and 3.5 m/second airspeed. Storage conditions of the fish were either −40°C or −18°C for several days up to 18 months. In conclusion, regarding the quality after thawing the freezing method must implicitly be regarded in the evaluation. Generally, better quality in terms of membrane structure was obtained for faster freezing processes and lower storage temperatures. Short thawing times led to reduced membrane deterioration.

In another study by Vyncke (1983), quality factors, such as total volatile bases, volatile acids, pH, and trimethylamine of frozen−thawed and fresh cod fillets, were investigated. The fillets were frozen by air blast freezer while being thawed in a circulating water bath, and in conclusion, no significant differences were found between the frozen–thawed and fresh cod fillets, which is a quite remarkable result.

The works of Rouillé et al. (2002), Schubring et al. (2003), and Zhu et al. (2004) compare methods for thawing at high pressure and conventional thawing at ambient pressure. Thereby quality factors such as drip loss, texture, color, and pH value are being investigated. Rouillé et al. (2002) found that drip loss is reduced for high-pressure freezing and thawing, while Zhu et al. (2004) report that the freezing method is more critical than the thawing method regarding quality. Hardening of tissue and discoloration for pressure thawing methods are stated by Schubring et al. (2003), Zhu et al. (2004), and Tironi et al. (2010). All articles above report faster thawing times for pressure thawing methods when compared to conventional thawing methods.

Murchie et al. (2005) conducted a broad review about quality aspects for high-pressure processing for enzymatic and microbial inactivation for shellfish, whereas the effect of high-pressure treatments on quality factors such as bacterial counts, lipid oxidation, and discoloration is reported by Ohshima et al. (1993). However, the latter two are not covering pressure thawing per se, but high-pressure processing in general.

It must be stated that in literature, there exists a lack of experimental studies investigating the effect of a wide variety of well-controlled thawing methods on the product's final quality. The procedures should be comparable in terms of duration, temperature, and well-defined quality factors. Also, only some of the cited literature cover the topic of investigating possible combinations between different freezing and thawing methods. Thereby, pressure methods, both for freezing and thawing, are the biggest group discussed, see Otero and Sanz (2003), Cheftel et al. (2000), Fernández et al. (2006), Le Bail et al. (2002), and Norton and Sun (2008).

The utilization of nanoparticles in combination with microwave or FIR thawing methods is reported in Cai et al. (2020). They conclude that an increase in total volatile bases nitrogen and pH can effectively be delayed, while the growth of microorganisms is reduced. However, the authors state that more research concerning the effect of the nanoparticles is required in order to receive more quantitative results.

6.5 CONCLUSION

Methods for thawing foodstuff can be divided into two principal groups differing in their respective functional principles. These are on the one hand methods that act solely through the product's boundary, and on the other hand methods that also generate heat inside the product itself. Generally, methods operating via the product's boundary are practically and efficiently applicable in the industry since large volumetric amounts of products can be thawed simultaneously with limited investments in technology. Methods in the second group acting in the inner domain of the product are not (yet) applicable in the industry for large-scale operation. An exception could be electrical resistance thawing, a combination of both functional principles—for example, starting the thawing process in a water bath followed by electrical resistance thawing. Another exception could be dielectric thawing, especially microwave technology, if thermal runaway and hence local overheating can be avoided.

The author believes that a promising future technology can be contact thawing since it provides increased controllability and monitoring of the thawing process. From plate freezer technology, the physical and thermodynamic phenomena are well understood. Hence, problems like local overheating as for dielectric thawing methods do not pose a problem.

In general, when it comes to quality preservation and technical applicability, established methods such as immersion thawing and air blast thawing are preferred over more sophisticated methods (i.e., dielectric and high-pressure methods). As mentioned before, dielectric methods are not controllable in a satisfying manner leading to unwanted disadvantages such as thermal runaway. These disadvantages outweigh the advantages, first of all, the rather fast thawing times.

Combinations of boundary methods with other technological advances, which have proven themselves to accelerate thawings, such as ultrasound or HVEF, seem promising for future large-scale operations. But more research has to be conducted in this direction.

The author would like to encourage researchers to investigate product quality factors after standardized freezing and thawing methods. From these investigations, practitioners will benefit heavily. Also research in the direction of cost-benefit analyses for operational costs and capital expenditures as well as energy usage should be conducted for different thawing methods to obtain comparable results.

When it comes to quality preservation of foods during freezing and thawing processes, amongst all foods, fruits are the benchmark without a doubt. Their physical structure and appearance will be affected to a large degree. Hence, for completion, also two studies investigating freezing and thawing on strawberries and blueberries will be presented in the following.

The effects of freezing and thawing processes on the different quality factor of strawberries have been investigated in Holzwarth et al. (2012). The authors come to the conclusion that the freezing processes did not have a significant effect on the quality parameters color, polyphenol, and ascorbic acid retention. The thawing processes, however, affected these quality factors greatly. Specifically, lower thawing temperatures and hence longer thawing times are reported to have negative effects on pigment and ascorbic acid losses, probably due to increased oxidation. Delgado and Rubiolo (2005) performed another study on the freezing and thawing of strawberries and they conclude that fast freezing is better for structure preservation. In order to reduce water loss and maintain tissue integrity, fast thawing was most practical.

A similar study for blueberries was conducted by Cheng et al. (2020), who conclude that the lower the freezing temperature, the better the freezing characteristics will be. It is important to freeze the blueberries as quickly as possible, however, the difference between liquid nitrogen freezing at −100°C or −80°C was negligible and hence, due to energy-saving reasons, −80°C is recommended. Gradient thawing from −20°C to −5°C and ultimately 4°C is reported to have the best influence on functional nutrition and the preservation of physical properties. Microwave thawing and ultrasonic rapid thawing had the worst outcome.

REFERENCES

Alizadeh, E., Chapleau, N., de Lamballerie, M., Le Bail, A., 2007. Effects of freezing and thawing processes on the quality of Atlantic salmon (*Salmo salar*) fillets. Journal of Food Science 72 (5), 279–284.

Alonso, M.J., Andresen, T., Frydenlund, F., Widell, K.N., 2011. Improvements of air flow distribution in a freezing tunnel using Airpak. Procedia Food Science 1, 1231–1238.

Anderson, B.A., Singh, R.P., 2006a. Useful heat transfer coefficient measurement during air impingement thawing using an inverse method. International Journal of Refrigeration 29 (2), 281–294.

Anderson, B.A., Singh, R.P., 2006b. Modeling the thawing of frozen foods using air impingement technology. International Journal of Refrigeration 29 (2), 294–304.

Archer, M., Edmonds, M., George, M., 2008. Seafood Thawing. Seafish Research & Development.

Backi, C.J., 2015. Modeling, Estimation and Control of Freezing and Thawing Processes—Theory and Applications. Norwegian University of Science and Technology.

Backi, C.J., 2017. Methods for (industrial) thawing of fish blocks: a review. Journal of Food Process Engineering 41 (1), e12598.

Backi, C.J., & Gravdahl, J.T. (2013). Optimal boundary control for the heat equation with application to freezing with phase change. Australian Control Conference. IEEE, pp. 409–414.

Backi, C.J., Bendtsen, J.D., Leth, J., Gravdahl, J.T., 2014a. The nonlinear heat equation with state-dependent parameters and its connection to the Burgers' and the potential Burgers' equation. IFAC Proceedings Volumes 47 (3), 7019–7024.

Backi, C.J., Bendtsen, J.D., Leth, J., & Gravdahl, J.T. (2014b). Estimation of inner-domain temperatures for a freezing process. IEEE Conference on Control Applications. IEEE, pp. 1759–1764.

Backi, C.J., Bendtsen, J.D., Leth, J., Gravdahl, J.T., 2015. Stability properties of a heat equation with state-dependent parameters and asymmetric boundary conditions. IFAC-PapersOnLine 48 (11), 587–592.

Backi, C.J., Bendtsen, J.D., Leth, J., Gravdahl, J.T., 2016a. A heat equation for freezing processes with phase change: stability analysis and applications. International Journal of Control 89 (4), 833–849.

Backi, C.J., Leth, J., Gravdahl, J.T., 2016b. Optimal boundary control of a contact thawing process for foodstuff. IFAC-PapersOnLine 49 (7), 183–188.

Bai, Y., Sun, Y., Li, Z., Kang, D., 2011. Study the optimum parameters of high voltage electrostatic field thawing. Procedia Engineering 16, 679–684.

Barros-Velázquez, J., Gallardo, J.M., Calo, P., Aubourg, S.P., 2008. Enhanced quality and safety during onboard chilled storage of fish species captured in the Grand Sole North Atlantic fishing bank. Food Chemistry 106, 493–500.

Bedane, T.F., Altin, O., Erol, B., Marra, F., Erdogdu, F., 2018. Thawing of frozen food products in a staggered through-field electrode radio frequency system: a case study for frozen chicken breast meat with effects on drip loss and texture. Innovative Food Science and Emerging Technologies 50, 139–147.

Bengtsson, N.E., Melin, J., Söderlind, S., 1963. Measurements of the dielectric properties of frozen and defrosted meat and fish in the frequency range 10–200 MHz. Journal of the Science of Food and Agriculture 14, 592–604.

Boonsumrej, S., Chaiwanichsiri, S., Tantratian, S., Suzuki, T., Takai, R., 2007. Effects of freezing and thawing on the quality changes of tiger shrimp (*Penaeus monodon*) frozen by air-blast and cryogenic freezing. Journal of Food Engineering 80, 292–299.

Cai, L., Dai, Y., Cao, A., Cao, M., 2020. The effects of CS@Fe_3O_4 nanoparticles combined with microwave or far infrared thawing on microbial diversity of red seabream (*Pagrus major*) fillets based on high-throughput sequencing. Food Microbiology 91.

Carver, J.H., 1975. Vacuum cooling and thawing fishery products. Marine Fisheries Review 37 (7), 15–21.

Cheftel, J.C., Levy, J., Dumay, E., 2000. Pressure-assisted freezing and thawing: principles and potential applications. Food Reviews International 16 (4), 453–483.

Cheng, L., Wu, W., An, K., Xu, Y., Yu, Y., Wen, J., et al., 2020. Advantages of liquid nitrogen quick freezing combine gradient slow thawing for quality preserving of blueberry. Crystals 10 (5).

Chevalier, D., Le Bail, A., Chourot, J.-M., Chantreau, P., 1999. High pressure thawing of fish (whiting): influence of the process parameters on drip losses. LWT - Food Science and Technology 32, 25–31.

Chourot, J.-M., Boillereaux, L., Havet, M., Le Bail, A., 1997. Numerical modeling of high pressure thawing: application to water thawing. Journal of Food Engineering 34, 63–75.

Cleland, D.J., Cleland, A.C., Earle, R.L., 1987a. Prediction of freezing and thawing times for multi-dimensional shapes by simple formulae Part 1: regular shapes. International Journal of Refrigeration 10, 156–164.

Cleland, D.J., Cleland, A.C., Earle, R.L., 1987b. Prediction of freezing and thawing times for multi-dimensional shapes by numerical methods. International Journal of Refrigeration 10, 32–39.

Cleland, D.J., Cleland, A.C., Earle, R.L., 1987c. Prediction of freezing and thawing times for multi-dimensional shapes by simple formulae Part 2: irregular shapes. International Journal of Refrigeration 10, 234–240.

Cleland, D.J., Cleland, A.C., Earle, R.L., Byrne, S.J., 1987d. Experimental data for freezing and thawing of multi-dimensional objects. International Journal of Refrigeration 10, 22–31.

Dawood, A.A., Roy, R.N., Williams, C.S., 1986. Effect of delayed icing on the storage life of rainbow trout. International Journal of Food Science and Technology 21 (2), 159–166.

de Alwis, A.A., Fryer, P.J., 1990. The use of direct resistance heating in the food industry. Journal of Food Engineering 11, 3–27.

Delgado, A., Rubiolo, A., 2005. Microstructural changes in strawberry after freezing and thawing processes. LWT - Food Science and Technology 38, 135–142.

Díaz-Tenorio, L.M., García-Carreño, F.L., Pacheco-Aguilar, R., 2007. Comparison of freezing and thawing treatments on muscle properties of Whiteleg shrimp (*Litopenaeus vannamei*). Journal of Food Biochemistry 31, 563–576.

Dinçer, T., Cadun, A., Çaklı, Ş., Tolasa, Ş., 2009. Effects of different thawing methods on the freshness quality of fish. E.U. Journal of Fisheries & Aquatic Sciences 26 (4), 253–256.

Dirita, C., de Bonis, M.V., Ruocco, G., 2007. Analysis of food cooling by jet impingement, including inherent conduction. Journal of Food Engineering 81 (1), 12–20.

Ersoy, B., Aksan, E., Özeren, A., 2008. The effect of thawing methods on the quality of eels (*Anguilla anguilla*). Food Chemistry 111, 377–380.

Faghri, A., 1994. Experimental and numerical study of microwave thawing heat transfer for food Materials. Transactions of the ASME 116, 446–455.

Farag, K.W., Lyng, J.G., Morgan, D.J., Cronin, D.A., 2011. A comparison of conventional and radio frequency thawing of beef meats: effects on product temperature distribution. Food Bioprocess Technology 4, 1128–1136.

Fernández, P.P., Otero, L., Guignon, B., Sanz, P.D., 2006. High-pressure shift freezing versus high-pressure assisted freezing: Effects on the microstructure of a food model. Food Hydrocolloids 20 (4), 510–522.

Footit, R.J., Lewis, A., 1999. The Canning of Fish and Meat. Springer.

Gambuteanu, C., Alexe, P., 2015. Comparison of thawing assisted by low intensity ultrasound on technological properties of pork *Longissimus dorsi* muscle. Journal of Food Science and Technology 52 (4), 2130–2138.

Gould, G.W., 1996. Methods for preservation and extension of shelf life. International Journal of Food Microbiology 33, 51–64.

Gram, L., Dalgaard, P., 2002. Fish spoilage bacteria – problems and solutions. Current Opinion in Biotechnology 13 (3), 262–266.

Hall, G.M. (Ed.), 1997. Fish Processing Technology. 2nd edition Chapman & Hall.

Harðarson, V., 1996. Matvarens termofysiske egenskaper og deres betydning ved dimensjonering av frysetunneler. Universitetet i Trondheim – Norges Tekniske Høgskole.

Haugland, A., 2002. Industrial Thawing of Fish: To Improve Quality, Yield and Capacity (Ph.D. thesis). Norwegian University of Science and Technology, Trondheim, Norway.

He, X., Liu, R., Nirasawa, S., Zheng, D., Liu, H., 2013. Effect of high voltage electrostatic field treatment on thawing characteristics and post-thawing quality of frozen pork tenderloin meat. Journal of Food Engineering 115, 245–250.

Holzwarth, M., Korhummel, S., Carle, R., Kammerer, D.R., 2012. Evaluation of the effects of different freezing and thawing methods on color, polyphenol and ascorbic acid retention in strawberries (*Fragaria* × *ananassa* Duch.). Food Research International 48, 241–248.

Hsieh, C.-W., Lai, C.-H.H.-J., Huang, S.-C., Ko, A.-C., 2010. Effect of thawing and cold storage on frozen chicken thigh meat quality by high-voltage electrostatic field. Journal of Food Science 75 (4), 193–197.

Huis in't Veld, J.H., 1996. Microbial and biochemical spoilage of foods: an overview. International Journal of Food Microbiology 33, 1–18.

Huss, H.H., 1995. Quality and quality changes in fresh fish. FAO Fisheries Technical Paper, 348.

Jason, A.C., 1974. Thawing frozen fish, Torry Advisory Note No 25. Ministry of Agriculture, Fisheries and Food, Torry Research Station.

Johnston, W.A., Nicholson, F.J., Roger, A., Stroud, G.D., n.d. Freezing and refrigerated storage in fisheries. FAO Fisheries Technical Paper, 340.

Kissam, A.D., Nelson, R.W., Ngao, J., Hunter, P., 1981. Water-thawing of fish using low frequency acoustics. Journal of Food Science 47, 71–75.

Le Bail, A., Chevalier, D., Mussa, D.M., Ghoul, M., 2002. High pressure freezing and thawing of foods: a review. International Journal of Refrigeration 25, 504–513.

Leung, M., Ching, W.H., Leung, D.Y., Lam, G.C., 2005. Theoretical study of heat transfer with moving phase-change interface in thawing of frozen food. Journal of Physics D: Applied Physics 38, 477–482.

Leung, M., Ching, W.H., Leung, D.Y., Lam, G.C., 2007. Fluid dynamics and heat transfer in cold water thawing. Journal of Food Engineering 78, 1221–1227.

Leygonie, C., Britz, T.J., Homan, L.C., 2012. Impact of freezing and thawing on the quality of meat: review. Meat Science 91, 93–98.

Li, B., Sun, D.-W., 2002. Novel methods for rapid freezing and thawing of foods—a review. Journal of Food Engineering 54, 175–182.

Li, D., Jia, S., Zhang, L., Li, Q., Pan, J., Zhu, B., et al., 2017. Post-thawing quality changes of common carp (*Cyprinus carpio*) cubes treated by high voltage electrostatic field (HVEF) during chilled storage. Innovative Food Science and Emerging Technologies 42, 25–32.

Lind, I., 1991. The measurement and prediction of thermal properties of food during freezing and thawing—a review with particular reference to meat and dough. Journal of Food Engineering 13, 285–319.

Liu, L., Llave, Y., Jin, Y., Zheng, D.-Y., Fukuoka, M., Sakai, N., 2017. Electrical conductivity and ohmic thawing of frozen tuna at high frequencies. Journal of Food Engineering 197, 68–77.

Llave, Y., Liu, S., Fukuoka, M., Sakai, N., 2015. Computer simulation of radiofrequency defrosting of frozen foods. Journal of Food Engineering 152, 32–42.

MacCallum, W.A., Ellis, D.G., 1964. Water-thawing of frozen cod blocks. Journal of the Fisheries Research Board of Canada 21 (1), 115–131.

Mannapperuma, J., Singh, R.P., 1988. Thawing of frozen foods in humid air. International Journal of Refrigeration 11 (3), 173–186.

Marra, F., Zhang, L., Lyng, J.G., 2008. Radio frequency treatment of foods: review of recent advances. Journal of Food Engineering 91, 497–508.

McKenna, B.M., Lyng, J., Brunton, N., Shirsat, N., 2006. Advances in radio frequency and ohmic heating of meats. Journal of Food Engineering 77, 215–229.

Mertens, B., Knorr, D., 1992. Developments of nonthermal processes for food preservation. Food Technology 46 (5), 124–133.

Miles, C.A., Morley, M.J., Rendell, M., 1998. High power ultrasonic thawing of frozen foods. Journal of Food Engineering 39, 151–159.

Mørkøre, T., Mazo, P.I., Tahirovic, V., Einen, O., 2008. Impact of starvation and handling stress on rigor development and quality of Atlantic salmon (*Salmon salar* L). Aquaculture (Amsterdam, Netherlands) 277, 231–238.

Mousakhani-Ganjeh, A., Hamdami, N., Soltanizadeh, N., 2016a. Effect of high voltage electrostatic field thawing on the lipid oxidation of frozen tuna fish (*Thunnus albacares*). Innovative Food Science and Emerging Technologies 36, 42–47.

Mousakhani-Ganjeh, A., Hamdami, N., Soltanizadeh, N., 2016b. Thawing of frozen tuna fish (*Thunnus albacares*) using still air method combined with a high voltage electrostatic field. Journal of Food Engineering 169, 149–154.

Mújica-Paz, H., Valdez-Fragoso, A., Tonello-Samson, C., Welti-Chanes, J., Torres, J.A., 2011. High-pressure processing technologies for the pasteurization and sterilization of foods. Food and Bioprocess Technology 4, 969–985.

Murchie, L.W., Cruz-Romero, M., Kerry, J.P., Linton, M., Patterson, M.F., Smiddy, M., et al., 2005. High pressure processing of shellfish: A review of microbiological and other quality aspects. Innovative Food Science and Emerging Technologies 6, 257–270.

Ngapo, T.M., Babare, I.H., Reynolds, J., Mawson, R.F., 1999. Freezing and thawing rate effects on drip loss from samples of pork. Meat Science 53, 149–158.

Nilsson, K., 1994. *Quality of Frozen Rainbow Trout. Effects of Different Freezing and Thawing Treatments (Ph.D. thesis)*. Chalmers University of Technology, Gothenburg, Sweden.

Norton, T., Sun, D.-W., 2008. Recent advances in the use of high pressure as an effective processing technique in the food industry. Food and Bioprocess Technology 1 (1), 2–34.

Ohshima, T., Ushio, H., Koizumi, C., 1993. High-pressure processing of fish and fish products. Trends in Food Science and Technology 4 (11), 370–375.

Ólafsdóttir, G., Martinsdóttir, E., Oehlenschläger, J., Dalgaard, P., Jensen, B., Undeland, I., et al., 1997. Methods to evaluate fish freshness in research and industry. Trends in Food Science and Technology 8, 258–265.

Otero, L., Sanz, P.D., 2003. Modelling heat transfer in high pressure food processing: a review. Innovative Food Science & Emerging Technologies 4, 121–134.

Otero, L., Rodríguez, A.C., Pérez-Mateos, M., Sanz, P.D., 2016. Effects of magnetic fields on freezing: application to biological products. Comprehensive Reviews in Food Science and Food Safety 15, 646–667.

Ousegui, A., Le Bail, A., Havet, M., 2008. Numerical modeling of a high pressure thawing process of a biomaterial. AIChE Journal 54 (2), 544–553.

Palazoğlu, T.K., Miran, W., 2017. Experimental comparison of microwave and radio frequency tempering of frozen block of shrimp. Innovative Food Science and Emerging Technologies 41, 292–300.

Pham, Q.T., 2014. Food Freezing and Thawing Calculations. Springer Briefs in Food, Health, and Nutrition.

Piyasena, P., Dussault, C., Koutchma, T., Ramaswamy, H.S., Awuah, G.B., 2007. Radio frequency heating of foods: principles, applications and related properties—a review. Critical Reviews in Food Science and Nutrition 43 (6), 587–606.

Rincon, A.M., Singh, R.K., Stelzleni, A.M., 2015. Effects of endpoint temperature and thickness on quality of whole muscle non-intact steaks cooked in a radio frequency oven. LWT - Food Science and Technology 64, 1323–1328.

Roiha, I.S., Jónsson, A., Backi, C.J., Lunestad, B.T., Karlsdóttir, M.G., 2018a. A comparative study of quality and safety of Atlantic cod (*Gadus morhua*) fillets during cold storage, as affected by different thawing methods of pre-rigor frozen headed and gutted fish. Journal of the Science of Food and Agriculture 98, 400–409.

Roiha, I.S., Tveit, G.M., Backi, C.J., Jónsson, Á., Karlsdóttir, M.G., Lunestad, B.T., 2018b. Effects of controlled thawing media temperatures on quality and safety of pre-rigor frozen Atlantic cod (*Gadus morhua*). LWT - Food Science and Technology 90, 138–144.

Rouillé, J., Le Bail, A., Ramaswamy, H.S., Leclerc, L., 2002. High pressure thawing of fish and shellfish. Journal of Food Engineering 53, 83–88.

Sakai, N., Hanzawa, T., 1994. Applications and advances in far-infrared heating in Japan. Trends in Food Science and Technology 5, 357–362.

Salvadori, V.O., Mascheroni, R.H., 2002. Analysis of impingement freezers performance. Journal of Food Engineering 54 (2), 133–140.

Sarkar, A., Singh, R.P., 2004. Air impingement technology for food processing: visualization studies. LWT - Food Science and Technology 37 (8), 873–879.

Schubring, R., Meyer, C., Schlüter, O., Boguslawski, S., Knorr, D., 2003. Impact of high pressure assisted thawing on the quality of fillets from various fish species. Innovative Food Science and Emerging Technologies 4, 257–267.

Schwartzberg, H.G., 1976. Effective heat capacities for the freezing and thawing of food. Journal of Food Science 41 (1), 152–156.

Sigurgísladóttir, S., Ingvarsdóttir, H., Torrissen, O.J., Cardinal, M., Hafsteinsson, H., 2000. Effects of freezing/thawing on the microstructure and the texture of smoked Atlantic salmon (*Salmo salar*). Food Research International 33, 857–865.

Skjervold, P.O., Røra, A.M., Fjæra, S.O., Vegusdal, A., Vorre, A., Einen, O., 2001. Effects of pre-, in-, or post-rigor filleting of live chilled Atlantic salmon. Aquaculture (Amsterdam, Netherlands) 194, 315–326.

Soto, V., Bórquez, R., 2001. Impingement jet freezing of biomaterials. Food Control 12 (8), 515–522.

Sriket, P., Benjakul, S., Visessanguan, W., Kijroongrojana, K., 2007. Comparative studies on the effect of the freeze–thawing process on the physicochemical properties and microstructures of black tiger shrimp (*Penaeus monodon*) and white shrimp (*Penaeus vannamei*) muscle. Food Chemistry 104, 113–121.

Srinivasan, S., Xiong, Y.L., Blanchard, S.P., 1997. Effects of freezing and thawing methods and storage time on thermal properties of freshwater prawns (*Macrobrachium rosenbergii*). Journal of the Science of Food and Agriculture 75, 37–44.

Srinivasan, S., Xiong, Y.L., Blanchard, S.P., Tidwell, J.H., 1998. Effects of freezing and thawing methods and storage time on physicochemical properties of freshwater prawns (*Macrobrachium rosenbergii*). Journal of Aquatic Food Product Technology 7 (2), 47–68.

Stroud, G.D., 1969. Rigor in Fish—The Effect on Quality, Torry Advisory Note No. 36. Ministry of Technology, Torry Research Station.

Taher, B.J., Farid, M.M., 2001. Cyclic microwave thawing of frozen meat: experimental and theoretical investigation. Chemical Engineering and Processing 40, 379–389.

Tironi, V., de Lamballerie, M., Le Bail, A., 2010. Quality changes during the frozen storage of sea bass (*Dicentrarchus labrax*) muscle after pressure shift freezing and pressure assisted thawing. Innovative Food Science and Emerging Technologies 11, 565–573.

Torres, J.A., Velazquez, G., 2005. Commercial opportunities and research challenges in the high pressure processing of foods. Journal of Food Engineering 67, 95–112.

Uyar, R., Bedane, T.F., Erdogdu, F., Palazoglu, T.K., Farag, K.W., Marra, F., 2015. Radio-frequency thawing of food products—a computational study. Journal of Food Engineering 146, 163–171.

Vyncke, W., 1983. Shelf life of thawed cod fillets kept in ice. Zeitschrift für Lebensmittel-Untersuchung und -Forschung 177, 19–21.

Wang, Y., Wig, T.D., Tang, J., Hallberg, L.M., 2003. Dielectric properties of foods relevant to RF and microwave pasteurization and sterilization. Journal of Food Engineering 57 (3), 257–268.

Xia, X., Kong, B., Liu, J., Diao, X., Liu, Q., 2012. Influence of different thawing methods on physicochemical changes and protein oxidation of porcine longissimus muscle. LWT - Food Science and Technology 46 (1), 280–286.

Zhao, Y., Flores, R.A., Olson, D.G., 1998. High hydrostatic pressure effects on rapid thawing of frozen beef. Journal of Food Science 63 (2), 272–275.

Zhao, Y., Flugstad, B., Kolbe, E., Park, J.W., Wells, J.H., 2000. Using capacitive (radio frequency) dielectric heating in food processing and preservation – a review. Journal of Food Process Engineering 23 (1), 25–55.

Zhu, S., Ramaswamy, H.S., Simpson, B.K., 2004. Effect of high-pressure versus conventional thawing on color, drip loss and texture of Atlantic salmon frozen by different methods. LWT - Food Science and Technology 37, 291–299.

Chapter | seven

Utilization of Fish Waste as a Value-Added Ingredient: Sources and Bioactive Properties of Fish Protein Hydrolysate

Ajay S. Desai[1,2,3], Margaret Brennan[1], S.S. Gangan[3] and Charles Brennan[1,2,4]

[1]Department of Wine, Food and Molecular Biosciences, Lincoln University, Christchurch, New Zealand, [2]Riddet Research Institute, Palmerston North, New Zealand, [3]Dr. Balasaheb Sawant Konkan Krishi Vidyapeeth, Dapoli, India, [4]School of Food Science and Engineering, South China University of Technology, Guangzhou, P.R. China

7.1 INTRODUCTION

The seafood industry plays a vital role in the economy of several countries worldwide. Currently, total world fish production is about 177 million metrics, out of this 40% is used for processing and human consumption (Delgado et al., 2020) while >60% of seafood are discarded as processing waste material (including, heads, skin, fins, frame, and viscera) without any recovery for human use (Halim et al., 2016; Mahanand et al., 2019). These waste products would create pollution and discharge problems in many countries. There is vast potential to utilize this fish processing waste and convert this waste into useful bioactive compounds. Currently, these wastes are used to produce fish meal, fertilizer, and fish silage because it contains a high amount of protein-rich material (Ahuja et al., 2020). The waste material may

Sustainable Fish Production and Processing. DOI: https://doi.org/10.1016/B978-0-12-824296-4.00004-9

also be processed into value-added products such as protein, enzymes, oil, hydrolysates, minerals, bioactive peptides, gelatin, amino acids, and gelatin (Ghaly, 2013). Presently, researchers have focused their study on utilizing fish waste for the production of fish protein hydrolysate (FPH) as value-added ingredients for the application in food products (Galla et al., 2012; Nasri et al., 2013). Several techniques have been used to refine potential elements from the fish waste product in order to enhance human health by protecting from several diseases (Ryan et al., 2011). One such method to obtain food products from fish waste material is the use of enzyme hydrolysates to break down the protein molecules into smaller peptides of different sizes. They are small fragments of peptides that have 2–20 amino acids. The development of protein hydrolysate from different sources is achieved with the addition of enzymes such as alcalase, trypsin, papain, pancreatin, pepsin, pronase, α-chymotrypsin, protamex, protease, bromelain, orientate, and validate (Chalamaiah et al., 2012). This protein hydrolysate is used as an emulsifier in salad dressing, spreads, and meat products like sausages (Khan et al., 2020). Although a successful method of protein extraction process often leads to a bitter taste in the product, the bitterness restricts the practical uses of these hydrolysates. Hydrolysis of whole fish and viscera may also cause bitterness in FPHs due to the presence of bile. These hydrolysates have numerous applications in pharmaceutical products, animal nutrition, and cosmetics (Halim et al., 2016). Many studies over the recent years concluded that FPH has functional and physicochemical applications (Galla et al., 2012). The best utilization of fish waste products could be achieved by converting these materials to FPH. The conversion of low-cost fish waste material to valuable products is of great concern all over the world. FPH is considered a rich source of bioactive peptides and protein. Kristinsson and Rasco (2000a) reviewed the production and functional properties of the FPH. The application and effect of FPH in human nutrition have been documented in the literature (Deraz, 2015; Jayathilakan et al., 2012). Furthermore, FPHs act as antioxidants against the free radical by providing hydrogen atoms (Zhang et al., 2019).

In recent years, different works have been carried out to produce valuable ingredients such as FPHs from the fish waste product. Also the presence of potential macro and micronutrients in FPHs makes it a useful product in industrial applications.

7.2 FISH PROTEIN HYDROLYSATE FROM FISH SKIN

FPH obtained from fish waste material includes skin, head, fins, frames, viscera. Generally, lean fish used for the FPH has been comparing to those of fatty fish, which consist of dark meat. Dark meat has a negative impact due to its ability to oxidate lipid (Chalamaiah et al., 2012). Whereas the use of FPH has been shown to improve the shelf life of products based on the increased amount of antioxidants in the food (Parvathy et al., 2018). Much of the work carried on the preparation of FPH from the fish skin waste product. Several studies reviewed the protein hydrolysate from the different species of fish skin. These include Grass carp (*Ctenopharyngodon idella*) (Wasswa et al., 2007), *Rachycentron canadum* (Yang et al., 2008), *Theragra chalcogramma*, Tilapia (Yang et al., 2009), sole (Giménez et al., 2009), *Priacanthus macracanthus* (Phanturat et al., 2010), North Atlantic lean fish (Picot et al., 2010), Channel catfish (*Ictalurus punctatus*) (Yin et al., 2010), *Oreochromis niloticus* (Ngo et al., 2010), *Aphanopus carbo* (Batista et al., 2010), Raja Kenojei (Lee et al., 2011), *Magalaspis cordyla* and *Otolithes ruber* (Sampath Kumar et al., 2011), sheela fish (*Sphyraena barracuda*) and ribbon fish (*Lepturacanthus savala*) (Nazeer and Deeptha, 2013), striped catfish (*Pangasianodon hypophthalmus*) (Tanuja et al., 2014), Black carp (*Mylopharyngodon piceus*), grass carp (*C. idella*), silver carp (*Hypophthalmichthys molitrix*) and bighead carp (*Hypophthalmichthys nobilis*) (Wu et al., 2015).

Giannetto et al. (2020) reviewed the molecular weight pattern of the protein hydrolysate obtained from *Engraulis encrasicolus* waste and reported that this protein hydrolysate help to protect against inflammation and decreases the protein expression of mediators. Yin et al. (2010) prepared the hydrolysate from the catfish skin and evaluate the functional (water holding capacity, emulsion capacity, and oil binding properties) and rheological characteristics of the protein hydrolysate. They concluded that proximate composition (protein, ash, and moisture), yield color characteristics (light yellow), and viscoelastic properties of protein hydrolysate of catfish skin were found to increase with the treatment of different enzymes. Sampath Kumar et al. (2011) reported skin of the horse mackerel (*M. cordyla*) and croaker (*O. ruber*) utilized for the production of protein hydrolysate in the presence of pepsin and trypsin enzymes. A peptide with high antioxidant activity was obtained from horse mackerel hydrolysate, and it reduced the oxidation of fat (PUFA) in food products compared to natural antioxidants (alpha-tocopherol). Nazeer and Deeptha (2013) investigated the enzymatic hydrolysis of protein hydrolysate from the skin of sheela fish (*S. barracuda*) and ribbon fish (*L. savala*). They described the amino acid profiling and antioxidative capacity of the hydrolysates. The protein hydrolysate exhibited high antioxidant activity using the trypsin enzyme extraction method, and its DPPH scavenging activity was reported to be 66%–60%. Amino acid sequences obtained from sheela and ribbon fish hydrolysate showed antioxidant properties and many essential amino acids. Tanuja et al. (2014) reported the functional and antioxidative properties of the protein hydrolysate obtained from the skin of striped catfish (*P. hypophthalmus*). Degree of hydrolysis of developed protein hydrolysate improved with the addition of alcalase enzyme and play an essential role in enhancing the antioxidant and functional properties (foaming capacity, emulsification capacity, oil binding capacity). Recently, Wu et al. (2015) prepared the protein hydrolysate from the skin of four freshwater fishes by using pepsin and described the antioxidant capacity of hydrolysates of different species. Protein hydrolysate developed from the skin of carp species showed a higher amount of hydrophobic amino acid residue and potent antioxidant activities that are capable of scavenging 61%–69% of hydroxyl radicals.

7.3 FISH PROTEIN HYDROLYSATE FROM FISH HEAD

The seafood industry produces a considerable amount of protein-rich fish heads as waste products. Fish heads are a potential source of FPH and prepared from different fish species includes salmon (*Salmo salar*) (Gbogouri et al., 2004), red salmon (*Oncorhynchus nerka*) (Sathivel et al., 2005), *Sardinella aurita* (Souissi et al., 2007), *A. carbo* (Batista et al., 2010), *S. aurita* (Bougatef et al., 2010a,b), bigeye tuna (*Thunnus obesus*) (Yang et al., 2011), yellowfin tuna (*Thunnus albacares*) (Nguyen et al., 2012), freshwater fish (*Catla catla* and *Labeo rohita*) (Ruthu et al., 2014), Bluefin leather jacket (*Navodon septentrionalis*) (Chi et al., 2015a,b), red scorpionfish (*Scorpaena notata*) (Aissaoui et al., 2015), and *sardine* (Martínez-Alvarez et al., 2016).

Sathivel et al. (2003) have described the protein hydrolysate from the head of the herring (*Clupea harengus*) with the use of alcalase as a proteolytic enzyme. Gbogouri et al. (2004) also reported the protein hydrolysates from the salmon head by alcalase enzyme and improve pH, temperature, and enzyme to substrate ratio of the protein hydrolysate using response surface methodology. Protein hydrolysate was obtained from the head of red salmon (*O. nerka*) by Sathivel et al. (2005) and observed the effect on nutritional and functional properties of protein hydrolysates with the use of neutrase, alcalase, flavourzyme, and palatase enzymes and time duration 25, 50, and 75 minutes. Souissi et al. (2007) investigated the effect of degree of hydrolysis of 6.62%, 9.31%, and 10.16% on the production of protein hydrolysate from the head waste of sardinelle (*S. aurita*) with alcalase. Bougatef et al. (2010a,b) demonstrated the sardinelle protein hydrolysate from the head waste product by treatment with various enzymes include alcalase and chymotrypsin. Yang et al. (2011) reported the protein hydrolysates obtained from the head of bigeye tuna (*T. obesus*) with alcalase enzyme and investigated the antioxidative capacity and molecular weight distribution pattern of protein hydrolysate. Ruthu et al. (2014) prepared protein hydrolysate from freshwater fish (*C. catla*) head and reported the antimicrobial and antioxidative characteristics of the protein hydrolysates. Chi et al. (2015a,b) have described the protein hydrolysate from head of Bluefin leather jacket (*N. septentrionalis*) using papain enzyme and reported three potential antioxidative peptides obtained from protein hydrolysate. Valcarcel et al. (2020) have prepared protein hydrolysate from the head waste product of seabass (*Dicentrarchus labrax*) by enzymatic hydrolysis using alcalase and investigated the foaming and gelling properties of protein hydrolysate.

7.4 FISH PROTEIN HYDROLYSATE FROM FISH VISCERA

Fish viscera as a waste material of the fish processing industry considered a potential source of protein and amino acid fraction. These waste materials can be used to produce protein-rich hydrolysate, which may have functional properties for industrial applications. A number of researchers have been made attempts for the development of protein hydrolysate from the fish viscera by the use of protease (Souissi et al., 2007; Ovissipour et al., 2012; Batista et al., 2010; Deraz, 2015).

Guerard et al. (2002) demonstrated the practical extraction of protein hydrolysate from the viscera of yellowfin tuna (*T. albacares*) by commercial enzymes. The canning industry waste was used for the production of hydrolysate through enzymatic hydrolysis (alcalase) purposes, and the degree of hydrolysis used was 22.5% and 0.1%–1.5% (enzyme–substrate ratio). Aspmo et al. (2005) investigated the protein hydrolysate from viscera of Atlantic cod (*Gadus morhua*) and reported that potential source of nitrogen from protein hydrolysate used to produce different bacterial strains of lactic acid bacteria. The developed FPH was compared with commercially available extract or peptones and found to be more effective as protein sources. Souissi et al. (2007) have reported the hydrolysis of sardinella (*S. aurita*) viscera waste proteins into protein hydrolysates using alcalase and investigated the lipid content solubility and antioxidant protein hydrolysate. Bhaskar and Mahendrakar (2008) demonstrated the protein hydrolysate from viscera waste protein of a catla (*C. catla*) and optimized the hydrolysis conditions like an enzyme to substrate ratio, pH, time, and temperature for preparing protein hydrolysate from fish viscera. Kechaou et al. (2009) have attempted to produce protein hydrolysate from the entrails of cuttlefish (*Sepia officinalis*) and sardine (*Sardina pilchardus*) using protease enzyme. Three different enzymes (protease, alcalase, and flavourzyme) were used to extract the protein hydrolysate from cuttlefish and sardine fish. Protein recovery rate (total amino acid residue) and lipid extractability increased phospholipid percentage through hydrolysis of FPH. Barkia et al. (2010) studied the hydrolysis parameters for the production of protein hydrolysate from sardinelle (*S. aurita*) viscera. Attempts have been made by Sampath Kumar et al. (2011) to prepare the protein hydrolysate from the entrails of horse mackerel (*M. cordyla*) and reported the antioxidative capacity and functional properties of the hydrolysate are increased significantly. Liu et al. (2014) described the protein hydrolysate from the surimi processing waste material contains fish viscera with the use of protamex and alcalase and presented structural and functional properties of the FPHs. Deraz (2015) investigated the protein hydrolysate from visceral waste proteins of Bolti fish (*Tilapia niloticus*) and the reported ratio of essential amino acid to nonessential amino acid (0.77–1.59) and essential amino acid to total amino acid (43.35–61.35) was found to be increased. The protein efficiency ratio and proximate composition of FPH were influenced by the degree of hydrolysis.

7.5 PROXIMATE COMPOSITION OF FISH PROTEIN HYDROLYSATE

From the nutritional point of view the chemical composition of food material plays a vital role in human health. The chemical composition of FPH obtained from the fish waste material includes fish head, fish skin, and fish viscera supply essential nutrients. Table 7.1 depicted the chemical composition of FPH obtained from various sources. Many of researchers have reported the proximate composition of FPH which are produced from waste of fish protein sources from Atlantic salmon muscle (*S. salar*) (Kristinsson and Rasco, 2000b), Icelandic scallop (*Chlamys islandica*) (Mukhin et al., 2001), black tilapia (*Oreochromis mossambicus*) (Abdul-Hamid et al., 2002), herring byproduct (*C. harengus*) (Sathivel et al., 2003), salmon head (*S. salar*) (Gbogouri et al., 2004), tuna byproduct (Nilsang et al., 2005), red salmon head (*O. nerka*) (Sathivel et al., 2005), grass carp skin (*C. idella*) (Wasswa et al., 2007), round scad (*Decapterus maruadsi*) (Thiansilakul et al., 2007), sardinelle byproduct (*S. aurita*) (Souissi et al., 2007), catla visceral waste (*C. catla*) (Bhaskar and Mahendrakar, 2008), pacific whiting muscle (*Merluccius productus*) (Mazorra-Manzano et al., 2010), croaker (*P. argentata*) (Yeung Joon Choi et al., 2009), channel catfish skin (*I. punctatus*) (Yin et al., 2010), yellowfin tuna (*T. albacares*) (Nguyen et al., 2012), Striped catfish (*P. hypophthalmus*) (Tanuja et al., 2014), rainbow trout (*Onchorhynchus mykiss*) (Taheri et al., 2013), Spanish mackerel (*Scomberomorus niphonius*) (Chi et al., 2015a,b), Sardinelle (*S. aurita*) (Jemil et al., 2014), Thornback rays (*R. clavata*) (Lassoued et al., 2015), and Smooth hound (*Mustelus mustelus*) (Sayari et al., 2016).

As Table 7.1 illustrates, several authors have determined the composition of FPH prepared from different fish waste products. The protein content of the FPH range between 55% and 90% of the total chemical composition based on dry weight (Lassoued et al., 2015; Sayari et al., 2016). The process of using hydrolyzing enzyme help to solubilize the protein in a fish waste product, which in turn affects their bioaccessibility and protein utilization (Nguyen et al., 2012). Many researchers reported the lipid content for the FPH was below 5% (Mukhin et al., 2001; Sayari et al., 2016). Only a few studies account for fat content above 5% in the FPHs (Sathivel et al., 2005; Silva et al., 2014; Chi et al., 2015a,b). Low lipid content in the FPH is because of the efficient separation of lipid from the insoluble protein by centrifugation (Taheri et al., 2013). The moisture content of the FPH was reported by many studies below 10% (Kristinsson and Rasco, 2000a; Mukhin et al., 2001; Abdul-Hamid et al., 2002; Sathivel et al., 2003; Gbogouri et al., 2004; Sathivel et al., 2005; Wasswa et al., 2007; Choi et al., 2009; Mazorra-Manzano et al., 2010; Taheri et al., 2013; Lassoued et al., 2015). The low moisture observed in the protein hydrolysate is due to higher temperatures employed during the spray drying process (Bueno-Solano et al., 2009). Many of the studies described that the ash content of the FPHs ranged between 0.45% and 25% of the total chemical composition. Below 12% ash content reported by Kristinsson and Rasco (2000a), Abdul-Hamid et al. (2002), Sathivel et al. (2003), Gbogouri et al.

TABLE 7.1 Fish Protein Hydrolysate Proximate Composition Obtained from Different Fish Waste Material

Fish	Waste Material	Protein (%)	Lipid (%)	Moisture (%)	Ash (%)	References
S. salar	Muscle	88.39 ± 0.55	0.23 ± 0.12	0.92 ± 0.22	8.96 ± 0.13	Kristinsson and Rasco (2000a)
C. islandica	Processing waste	81.3	0.56	5.55	12.6	Mukhin et al. (2001)
O. mossambicus	Type A	49.6 ± 0.19	2.80 ± 0.13	3.93 ± 0.15	8.65 ± 0.01	Abdul-Hamid et al. (2002)
	Type B	37.7 ± 0.05	2.56 ± 0.10	1.58 ± 0.12	8.56 ± 0.03	
C. harengus	Muscle	87 ± 0.2	0.4 ± 0.1	3.0 ± 0.1	10.1 ± 0.1	Sathivel et al. (2003)
	Head	85.2 ± 0.4	1.2 ± 0.2	3.5 ± 0.2	10.1 ± 0.3	
	Gonad	77 ± 0.4	1.5 ± 0.2	6.2 ± 0.3	15.3 ± 0.1	
S. salar	Head	82.3 ± 1.9	0.8 ± 0.02	5.3 ± 0.2	10.4 ± 1.1	Gbogouri et al. (2004)
O. nerka	Head	64.8 ± 4.7	22.5 ± 5.7	5.7 ± 0.5	6.9 ± 0.6	Sathivel et al. (2005)
D. maruadsi	Skin	69.0 ± 3.57	0.15 ± 0.03	8.75 ± 0.40	24.56 ± 2.56	Thiansilakul et al. (2007)
C. idella	Skin	92.0 ± 1.75	0.2 ± 0.10	2.87 ± 0.05	5.86 ± 0.60	Wasswa et al. (2007)
C. catla	Viscera	13.85 ± 1.55	1.94 ± 0.35	3.85 ± 0.65	0.45 ± 0.15	Bhaskar and Mahendrakar (2008)
Merluccius	Muscle	88.6 ± 0.3	0.1 ± 0.1	3.6 ± 1.9	11.9 ± 0.1	
Acipenser persicus	Viscera	65.82 ± 7.02	0.18 ± 0.4	4.45 ± 0.67	7.67 ± 1.24	Ovissipour et al. (2011)
Pennahia argentata	Muscle	85.9 ± 0.6	–	4.4 ± 0.7	4.7 ± 0.2	Choi et al. (2009)
I. punctatus	Skin	88.34 ± 0.21	1.83 ± 0.35	6.75 ± 0.11	3.07 ± 0.18	Yin et al. (2010)
Lutjanus vitta	Muscle	87.36 ± 0.12	0.61 ± 0.06	11.16 ± 0.39	1.76 ± 0.73	

Continued

TABLE 7.1 continued

Fish	Waste Material	Protein (%)	Lipid (%)	Moisture (%)	Ash (%)	References
Merluccius products	Muscle	85.6 ± 2.3	0.3 ± 0.1	2.5 ± 0.6	16.6 ± 0.3	Mazorra-Manzano et al. (2010)
Channa striatus	Roe	58.68 ± 0.82	–	10.28 ± 0.74	–	Galla et al. (2012)
L. rohita		58.68 ± 0.82	–	15.52 ± 0.93	–	
P. hypophthalmus	Muscle	92.12 ± 0.59	0.61 ± 0.03	5.97 ± 0.49	6.26 ± 0.11	Tanuja et al. (2014)
O. mykiss	Viscera	88.32 ± 0.07	0.8 ± 0.06	3.45 ± 0.02	1.14 ± 0.088	Taheri et al. (2013)
S. niphonius	Skin	81.05 ± 2.24	8.57 ± 0.37	0.75 ± 0.05	8.48 ± 0.38	
Sardinelle	Muscle	74.28 ± 0.53	0.63 ± 0.18	14.72 ± 0.38	10.21 ± 0.12	Jemil et al. (2014)
Goby		75.23 ± 0.33	0.04 ± 0.01	15.90 ± 0.19	8.49 ± 0.19	
Ray		81.03 ± 0.32	–	12.91 ± 0.15	6.00 ± 0.14	
O. niloticus	Viscera	58.4	37.47	–	2.6	Silva et al. (2014)
Raja clavata	Skin	89.10 ± 0.77	0.03 ± 0.02	4.75 ± 0.02	4.87 ± 0.02	Lassoued et al. (2015)
M. mustelus	Head and Viscera	80.52 ± 1.87	2.89 0.83	–	15.38 1.23	Sayari et al. (2016)

(2004), Sathivel et al. (2005), Wasswa et al. (2007), Choi et al. (2009), Taheri et al. (2013), Silva et al. (2014), and Lassoued et al. (2015). High ash content in the FPH was reported by Thiansilakul et al. (2007), Mazorra-Manzano et al. (2010), and Sayari et al. (2016). The high ash content in the hydrolysate during hydrolysis added acid and alkali used for adjustment of pH medium.

7.6 APPLICATION OF FISH PROTEIN HYDROLYSATE FROM FISH WASTE MATERIAL

Due to the unique properties of FPH obtained from a fish waste product, such as protein solubility, water holding capacity, oil absorption ability, gelling capacity, emulsifying capacity, and foaming ability, it can be used as a potential value-added ingredient in food products (Taheri et al., 2013; Jemil et al., 2014). Protein hydrolysate produced from fish waste materials is a high nutritional profile as bioactive compounds can be easily digestible and utilized for the different metabolic processes (Nesse et al., 2011). Chalamaiah et al. (2012) reported that FPH has been used for adding to food products such as cereal, meat products, noodles, and crackers. Bjorndal et al. (2013) studied the effect of FPH diets on mitochondrial fatty acid metabolisms and inflammation in transgene mice. In the study, mice were fed high-fat diets (23%), including 15% FPH and 5% casein for 2 weeks. It is found that mice fed with FPH showed a higher n-3/n-6 fatty acid ratio than the control. FPH is also used as cryoprotective agents. A study conducted by Kittiphattanabawon et al. (2012) reported the cryoprotective ability of gelatin hydrolysate prepared from blacktip shark (*Carcharhinus limbatus*) skin with sucrose and sorbitol. They investigated blacktip shark hydrolysate as a potential source to the 3:1 blend of sucrose-sorbitol, which is regularly used in frozen fish meat storage, and the result described a 10% degree of hydrolysis hydrolysates prevent the denaturation of myosin heavy chain. FPH can be used as a potential source of nitrogen to grow different beneficial microorganisms. Klompong et al. (2010) obtained FPH from Yellow Stripe Trevally (*Selaroides leptolepis*) with alcalase and flavourzyme enzyme and reported optimal results obtained with FPH 25% degree of hydrolysis prepared with flavourzyme enzyme. In another study by Safari et al. (2011) the peptone obtained from Silver Carp (*H. molitrix*) head waste with alcalase was shown to be a more effective as a growing medium for *Vibrio anguillarum* compared to commercially used peptone. FPHs have been used in fish aquaculture to enhance the growth and survival of culture-based fisheries. Masuda et al. (2013) prepared five diets for feeding sea bass larvae with five different FPHs and reported that a high proportion of small peptides in the protein hydrolysates affected the growth performance of seabass larvae. In another study, three different hydrolysates from shrimp, tilapia, and krill fish were used to evaluate the supplemental effect on growth performance, innate immune parameters, and disease resistance of red sea bream juvenile and concluded that red sea bream juvenile fed with enriched shrimp hydrolysates stimulated the production of lysozyme and antiprotease activities and increased the feed conversation ratio and digestibility

(Khosravi et al., 2015). FPH has been tested for its potential application as antioxidant agents. In the food industry, oxidation of lipid is the main problem in the food product. Due to this, unpleasant flavor and odor developed in the processed product, affecting the quality of the final product (Domínguez et al., 2019).

To avoid the deterioration of the food and protect against diseases created by lipid oxidation, it is essential to stop the food product's lipid oxidation. Antioxidants are used to inhibit the oxidation process by retarding the discoloration and deterioration. In muscle tissue, two types of antioxidants are present (glutathione and carnosine). These antioxidants inhibit lipid oxidation through the scavenging of hydroxyl radicals and oxygen (Samaranayaka and Li-Chan, 2011). Hydrolysate purified from the Hoki fish skin gelatin showed antioxidative characterizes. The antioxidative peptide, His-Gly-Pro-Leu-Gly-Pro-Leu, is effective in lipid oxidation (Zou et al., 2016). Nikoo et al. (2014) studied the antioxidant activity of gelatin extracted from Amur sturgeon's skin using DPPH radical scavenging activity, ABTS, and hydroxyl radical. The amino acid sequence of the peptide was Pro-Ala-Gly-Tyr, and the use of 25 ppm peptide inhibits the lipid oxidation in minced fish. Dekkers et al. (2011) reported the effect of tilapia hydrolysate on the storage stability of the mahi-mahi red muscle. They illustrated that mahi-mahi red muscle dipped in the tilapia hydrolysate for 2–4 minutes showed a reduced rate of the formation of lipid hydroperoxides and thiobarbituric acid. Saito et al. (2009) studied the effect of collagen hydrolysate extracted from the salmon and trout skin on the total lipid in the plasma. They reported salmon and trout hydrolysate reduced the full lipid level and triglycerides level in the blood. In a recent study, Ben Khaled et al. (2012) studied the protein hydrolysate from sardinelle (*S. aurita*) and its effect on the lipid level of rats. They found that rats fed with enriched sardinelle protein hydrolysate lower the malondialdehyde (MDA) level and increased the concentration of antioxidant.

7.7 BIOACTIVE PROPERTIES OF FISH PROTEIN HYDROLYSATE

Out of the total world fish production, 40% of the catch is used for processing and human consumption. A considerable amount of fish caught around 60% is discarded as processing waste material includes heads, skin, fins, and viscera (Dekkers et al., 2011). It is estimated at about 20 million tons of the total world fisheries (FAOSTAT, 2001). There is vast potential to utilize this fish processing waste and convert it into a useful bioactive compound. The bioactive compound extracted from protein-rich fish waste material would help to maintain human health by protecting from several diseases. The biologically active compound can be produced by solvent extraction, enzymatic hydrolysis, and microbial fermentation (Fig. 7.1). In the laboratory the solvent extraction method used, which has various disadvantages, such as lower efficiency, selectivity, and pollution issue (Alasalvar et al., 2010). In microbial fermentation, lactic

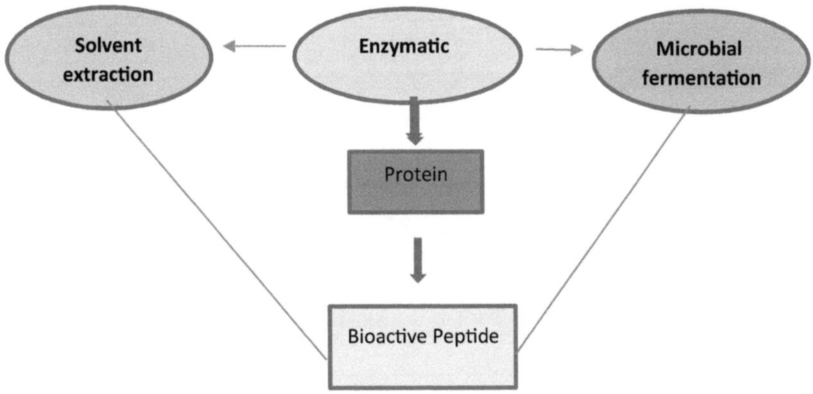

FIGURE 7.1 Exttraction of bioactive compound from fish waste.

acid bacteria are used for the hydrolysis of protein. However, the enzymatic hydrolysis method is used to obtain the bioactive peptides from the fish waste material in the industry. The rest of the ways can exhibit the toxic substances in the final product. In the protein sequence, bioactive peptides are inactive. They are obtained by the use of enzymatic hydrolysis (Vercruysse et al., 2005). Bioactive peptide obtained from fish processing material depends on structural properties and amino acid composition and sequences. These bioactive peptides may be involved in biological functions such as antioxidant activity (Ranathunga et al., 2006; Hung-Ngo et al., 2012).

7.8 ANTIOXIDANT ACTIVITY OF FISH PROTEIN HYDROLYSATE

Formation of free radical in the presence of free oxygen is an essential process in the human. The free radical in the form of reactive oxygen species includes superoxide anion, hydroxyl radical, and other nonfree radicals such as hydrogen peroxide and free oxygen, which affect the regular use of oxygen for respiration. Oxidation of unsaturated fats takes place by free radicals. The free radicals combine with oxygen and form lipid radicals. These lipids radical combine with the hydrogen atom of unsaturated fatty acids and a hydrogen peroxide product is created. Due to this undesirable and nasty odor and flavors produced in the rancid food (Bernardini et al., 2011).

The production of free radicals has been involved in many diseases such as heart disease, cancer, diabetes, and brain disorder (Fig. 7.2). Free radicals are generally observed in the oxidation process of food compounds in the industry. However, free radical is involved in the oxidation of protein in the food product by reaction with lipid oxidation (Viljanen et al., 2004). Antioxidants can impede the oxidation process by providing a hydrogen atom to the free radical formed from the unsaturated lipids and stopping further reaction by scavenging the oxygen molecule. In the food industry, preservation of food products was done by the use of antioxidants and resulted in the increased shelf life of the

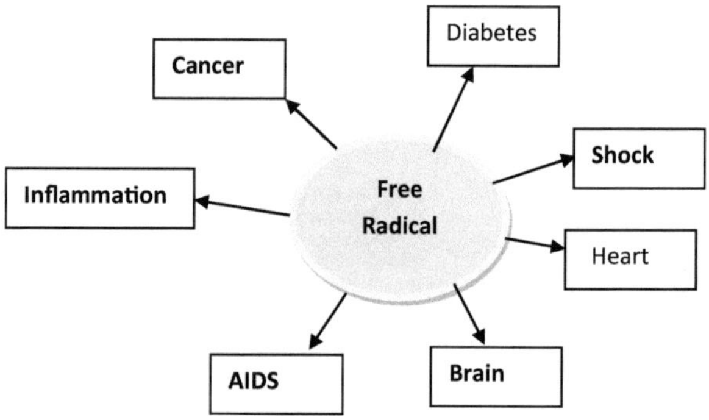

FIGURE 7.2 Free radical production and disease outbreaks.

product—antioxidant tetrad the progress of many chronic diseases and lipid peroxidation. The synthetic antioxidant used in the food industry includes butylated hydroxyl anisole, butylated hydroxyl toluene, and propyl gallate. However, the use of these antioxidants causes health issues and is restricted to use in the food product. Replacement of synthetic antioxidants with natural oxidants is an alternative to protect the food product and increase shelf life. Therefore there is a need to find out natural antioxidants derived from the fish waste material, which exerts no adverse effect on the food product (Bahareh and Ismail, 2010).

7.8.1 Production of Antioxidant Peptides by Enzymatic Hydrolysis

Production of antioxidant peptides by enzymatic hydrolysis:

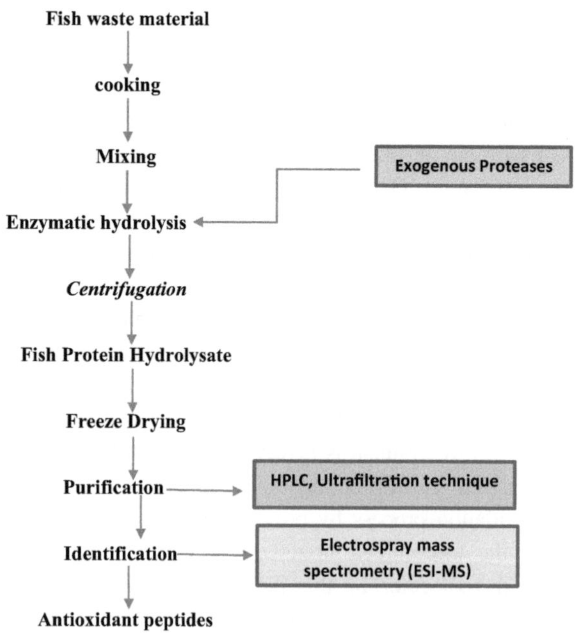

For the production of bioactive compounds from the fish waste material, enzymatic hydrolysis of the whole protein is the most common method. Time, temperature, and pH play an essential role in bioactive peptide production. Enzyme selection and utilization for the production of the bioactive peptide is most critical as it helps to break the sequence pattern of the peptide bonds (Shahidi and Zhong, 2008). Exogenous enzymes are generally used due to the short time required for the degree of hydrolysis and also obtain consistent molecular weight profiles and composition of bioactive peptides. For the production of bioactive compounds from fish protein, several enzymes were used for hydrolysis, which include alcalase, papain, pepsin, trypsin, α-chymotrypsin, pancreatin, pronase, protamex, bromelain, and cryptic (Hsu, 2010; Ngo et al., 2010). For the production of the antioxidant peptide from fish waste material, pepsin, chymotrypsin, and trypsin are used (Bougatef et al., 2009; Sampath Kumar et al., 2011). There is also a report about the use of alkaline protease for the production of bioactive peptides through microbial in tuna heads (Bougatef et al., 2012) and sardinelle (*S. aurita*) viscera and head (Bougatef et al., 2010a,b). They used *Bacillus licheniformis* and *Aspergillus clavatus* for the preparation of enzyme extract from sardine viscera to obtain antioxidative peptide. The result revealed that the use of hydrolysate obtained from visceral protease had DDPH radical scavenging activity 53.76% at 2 mg/mL, indicating that these proteases are very helpful for fish waste digestion and production of the bioactive antioxidant peptide. The degree of hydrolysis is a useful parameter for the detection of the antioxidant property of protein from waste material. Katari et al. (2013) worked on the production of hydrolysate from cuttlefish waste material. They found that antioxidant activity increased with the degree of hydrolysis. Measuring the antioxidative capacity of the peptide from the different fish waste materials is summarized in Table 7.2.

7.9 PURIFICATION AND IDENTIFICATION OF ANTIOXIDANT PEPTIDE OF FISH PROTEIN HYDROLYSATE

Protein hydrolysate obtained from fish waste material by different enzymes contains a mixture of active and inactive bioactive peptides with different molecular weights and amino acid sequences and exhibit potential activities. Chromatography gel filtration techniques have been used for the differentiation of protein and peptides from protein hydrolysate. This technique separates the peptides according to the amino acid composition and sequences (Byun et al., 2009; Rajapakse et al., 2005). Bougatef et al. (2010a,b) obtained protein hydrolysate from the sardine (*S. pilchardus*) with crude enzyme and different hydrolysis degrees. They found that hydrolysate showed 6% DH with the highest DDPH radical scavenging activity (87%). This hydrolysate was fractionated with the use of chromatography techniques with eight fractions. Fraction 4 exhibit 48% DDPH activity was again fractionated by

TABLE 7.2 Antioxidative Capacity of the Peptide from the Various Fish Waste Material

Fish Species	Fish Waste	Enzyme Used	Method Used	References
North Atlantic fish	Skin	Endopeptidase	2,2 Diphenyl 1 picrylhydrazyl (DDPH) radical scavenging activity	Picot et al. (2010)
Theragra chalcogramma	Skin	Protamex	DDPH radical scavenging activity	Jia et al. (2010)
Parastromateus niger	Viscera	Pepsin, trypsin, and α-chymotrypsin	DDPH radical scavenging activity	Jai ganesh et al. (2011)
Alaska Pollock	Skin	Trypsin	Iron chelating activity	Guo et al., 2013
Salmon	Fin	Pepsin	DDPH radical scavenging activity	Ahn et al. (2014)
Bluefin	Skin	Trypsin, alcalase, and pepsin	DDPH radical scavenging activity	Chi et al. (2015a)
O. niloticus	Skin	Trypsin	ABTS radical scavenging activity and ferric reducing antioxidant power assay	Choonpicharn et al. (2016)

reverse-phase high-performance liquid chromatography. The molecular weight and amino acid sequence of the purified peptides were determined using electrospray mass spectrometry (ESI-Ms). Ultrafiltration membrane system used to separate the peptides that have the desired molecular weights and functional properties (Jeon et al., 1999).

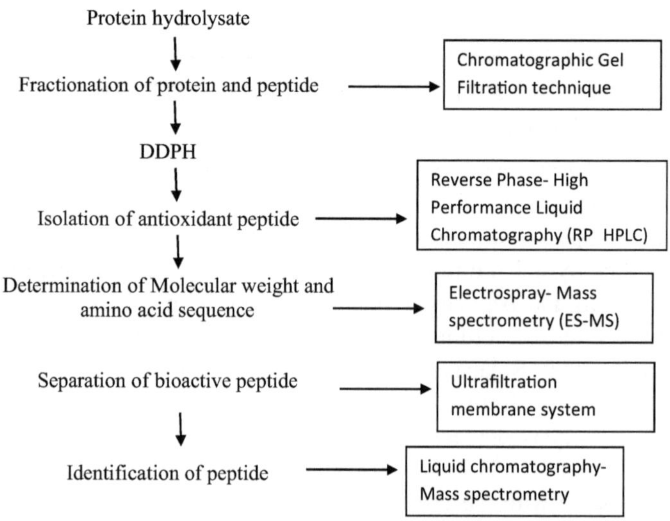

TABLE 7.3 Antioxidant Bioactive Peptides Purification Method and Amino Acid Sequence from Different Fish Waste Material

Fish Species and Waste	Purification and Identification Method	Molecular Weight and AA Sequence	References
O. niloticus (scale)	RP-HPLC, ESI-MS	*Asp-Pro-Ala-Leu-Ala-Thr-Glu-Pro-Asp-Pro-Met-Pro-Phe* (1382.57 Da)	Ngo et al. (2010)
Horse mackerel (viscera)	HPLC, ESI-MS	*Ala-Cys-Phe-Leu* (518.5 Da)	Sampath Kumar et al. (2011)
Alaska Pollock (skin)	RP-HPLC, ESI-MS	*Ser-Cus-His* (345 Da)	Guo et al. (2013)
Salmon (fin)	RP-HPLC, ESI-MS	*Phe-Leu-Asn-Glu-Phe-Leu-His-Val* (1018.48 Da)	Ahn et al. (2014)
Bluefin leather jacket (head)	RP-HPLC, ESI-MS	*Trp-Glu-Gly-ProLys*	Chi et al. (2015b)
O. niloticus (skin)	IE-Chromatography MALDI-TOF, MS	Glu-Asp-Cys-Ala_Lys-Thr-His-Asn-Arg-Tyr (1490.61 Da)	Choonpicharn et al. (2016)

Identification of antioxidant peptide from fish waste material is made by liquid chromatography followed by mass spectrometry (Perkins et al., 1999). Table 7.3 showed the antioxidant bioactive peptides purification method and amino acid sequence from different fish waste materials. Two to sixteen amino acid sequences were observed in the antioxidant peptides derived from FPH (Bougatef et al., 2010a,b). Ngo et al. (2010) identified an antioxidative peptide consisting of 13 amino acid sequences *Asp-Pro-Ala-Leu-Ala-Thr-Glu-Pro-Asp-Pro-Met-Pro-Phe* (1382.57 Da) from the scale of *O. niloticus*. Also, Sampath Kumar et al. (2011) isolated an antioxidant peptide *Ala-Cys-Phe-Leu* (518.5 Da) from the viscera of horse mackerel. Three antioxidant peptides were isolated from the skin of Alaska Pollock fish, *Ser-Cus-His* (345 Da) (Guo et al., 2013).

7.10 APPLICATION OF ANTIOXIDANT PEPTIDES FROM FISH WASTE MATERIAL IN FOOD PRODUCT

In the food industry, oxidation of lipid is the main problem in the food product. Due to this, unpleasant flavor and odor developed in the processed product, affecting the quality of the final product (Lin and Liang, 2002). To avoid the deterioration of the food and protect against diseases created by lipid oxidation, it is essential to stop the food product's lipid oxidation. Antioxidants are used to inhibit the oxidation process by retarding the discoloration and deterioration. In muscle tissue, two types of antioxidants were present includes glutathione and carnosine. They inhibit lipid oxidation through the scavenging of hydroxyl radicals and oxygen (Samaranayaka and Li-Chan, 2011). Hydrolysate purified from the Hoki fish skin gelatin showed antioxidative characterizes. The antioxidative peptide, His-Gly-Pro-Leu-Gly-Pro-Leu, was found to be effective in lipid oxidation (Mendis et al., 2005). Nikoo et al. (2014) studied the antioxidant activity of gelatin extracted from the skin of Amur sturgeon. The antioxidant activity was determined by DDPH radical scavenging activity, ABTS, and hydroxyl radical. The amino acid sequence of the peptide was Pro-Ala-Gly-Tyr. The use of 25 ppm peptide inhibits lipid oxidation in minced fish. Dekkers et al. (2011) worked on tilapia hydrolysate on the storage stability of the mahi-mahi red muscle. They observed mahi-mahi red muscle dipped in the tilapia hydrolysate for 2–4 minutes decreased the formation of lipid hydroperoxides and thiobarbituric acid for 120 minutes. storage. In a recent study, Khaled et al. (2012) studied the protein hydrolysate from sardinelle (*S. aurita*) and its effect on the lipid level of rats. They found that rats fed with enriched sardinelle protein hydrolysate lower the MDA level and increased the concentration of antioxidants. Saito et al. (2009) studied the effect of collagen hydrolysate extracted from the salmon and trout skin on the total lipid in the plasma. They reported salmon and trout hydrolysate reduced the whole lipid level and triglycerides level in the blood.

7.11 CONCLUSION

Presently, there is a tremendous demand for nutritional balanced, and rich food for human health. Decreasing the availability of fish raw materials and environmental pollution emphasizes the search for utilization of fish waste material to a useful product. The review comprises proximate composition and potential application of FPH obtained from fish waste materials. FPH can be used as a value-added ingredient that gives functional characteristics to various food products. The primary source of FPH from fish is the head, skin, and viscera. All over the world the market for FPH is increased enormously. Potential application has been found for FPH includes growth promoter, cryoprotectant, nitrogen source, and bioactive peptides. Fish waste material is the possible source of protein-rich compounds and peptides. The antioxidative property of bioactive peptides obtained from fish waste material depends on the molecular weight and amino acid composition. Bioactive peptides obtained from fish waste material such as skin, fin, viscera, and scale with antioxidative properties may be a potential alternative for the synthetic antioxidants, which exhibit the potential benefits in human nutrition.

REFERENCES

Abdul-Hamid, A., Bakar, J., Bee, G.H., 2002. Nutritional quality of spray dried protein hydrolysate from Black Tilapia (*Oreochromis mossambicus*). Food Chemistry 78 (1), 69–74. Available from: https://doi.org/10.1016/S0308-8146(01)00380-6.

Ahn, C.B., Kim, J.G., Je, J.Y., 2014. Purification and antioxidant properties of octapeptide from salmon byproduct protein hydrolysate by gastrointestinal digestion. Food Chemistry 147, 78–83.

Ahuja, I., Dauksas, E., Remme, J.F., et al., 2020. Fish and fish waste-based fertilizers in organic farming – with status in Norway: a review. Waste Management 115 (September), 95–112. Available from: https://doi.org/10.1016/j.wasman.2020.07.025.

Aissaoui, N., Abidi, F., Marzouki, M.N., 2015. ACE inhibitory and antioxidant activities of red scorpionfish (*Scorpaena notata*) protein hydrolysates. Journal of Food Science and Technology 52 (11), 7092–7102. Available from: https://doi.org/10.1007/s13197-015-1862-8.

Alasalvar, C., Shahidi, F., Miyashita, K., Wanasundara, U., 2010. Handbook of Seafood Quality, Safety and Health Applications. Wiley, New York, 518pp.

Aspmo, S.I., Horn, S.J., Eijsink, V.G.H., 2005. Use of hydrolysates from Atlantic cod (*Gadus morhua* L.) viscera as a complex nitrogen source for lactic acid bacteria. FEMS Microbiology Letters 248 (1), 65–68. Available from: https://doi.org/10.1016/j.femsle.2005.05.021.

Bahareh, H.S., Ismail, A., 2010. Antioxidative peptides from food proteins: a review. Peptides 31, 1949–1956.

Barkia, A., Bougatef, A., Khaled, H.B., et al., 2010. Antioxidant activities of sardinelle heads and/or viscera protein hydrolysates prepared by enzymatic treatment. Journal of Food Biochemistry 34 (1), 303–320. Available from: https://doi.org/10.1111/j.1745-4514.2009.00331.x.

Batista, I., Ramos, C., Coutinho, J., et al., 2010. Characterization of protein hydrolysates and lipids obtained from black scabbardfish (*Aphanopus carbo*) byproducts and antioxidative activity of the hydrolysates produced. Process Biochemistry 45 (1), 18–24. Available from: https://doi.org/10.1016/j.procbio.2009.07.019.

Ben Khaled, H., Ghlissi, Z., Chtourou, Y., et al., 2012. Effect of protein hydrolysates from sardinelle (*Sardinella aurita*) on the oxidative status and blood lipid profile of cholesterol-fed rats. Food Research International 45 (1), 60–68. Available from: https://doi.org/10.1016/j.foodres.2011.10.003.

Bernardini, R.D., Harnedy, P., Declan Bolton, D., et al., 2011. Antioxidant and antimicrobial peptide hydrolysates from muscle protein sources and byproducts: review. Food Chemistry 124, 1296–1307.

Bhaskar, N., Mahendrakar, N.S., 2008. Protein hydrolysate from visceral waste proteins of Catla (*Catla catla*): optimization of hydrolysis conditions for a commercial neutral protease. Bioresource Technology 99 (10), 4105–4111. Available from: https://doi.org/10.1016/j.biortech.2007.09.006.

Bjorndal, B., Berge, C., Ramsvik, M.S., Svardal, A., Bohov, P., Skorve, J., Berge, R.K., 2013. A fish protein hydrolysate alters fatty acid composition in liver and adipose tissue and increases plasma carnitine levels in a mouse model of chronic inflammation. Lipids in Health and Disease 12, 143–153.

Bougatef, A., Hajji, M., Balti, R., Lassoued, I., Triki-Ellouz, Y., Nasri, M., 2009. Antioxidant and free radical-scavenging activities of smooth hound (*Mustelus mustelus*) muscle protein hydrolysate obtained by gastrointestinal proteases. Food Chemistry 114, 1198–1205.

Bougatef, A., Nedjar-Arroume, N., Manni, L., et al., 2010a. Purification and identification of novel antioxidant peptides from enzymatic hydrolysates of sardinelle (*Sardinella aurita*) byproducts proteins. Food Chemistry 118 (3), 559–565. Available from: https://doi.org/10.1016/j.foodchem.2009.05.021.

Bougatef, A., Nedjar-Arroume, N., Manni, L., Ravallec, R., Barkia, A., Guillochon, D., et al., 2010b. Purification and identification of novel antioxidant peptides from enzymatic hydrolysate of sardinelle (*Sardinella aurita*) by products proteins. Food Chemistry 118, 559–565.

Bougatef, A., Balti, R., Haddar, A., Jellouli, K., Souissi, N., Nasri, M., 2012. Antioxidant and functional properties of protein hydrolysate of Bluefin tuna (*Thunnus thynnus*) head as influenced by the extent of enzymatic hydrolysis. Biotechnology and Bioprocess Engineering 17, 841–852.

Bueno-Solano, C., López-Cervantes, J., Campas-Baypoli, O.N., et al., 2009. Chemical and biological characteristics of protein hydrolysates from fermented shrimp byproducts.

Food Chemistry 112 (3), 671–675. Available from: https://doi.org/10.1016/j.foodchem.2008.06.029.

Byun, H.G., Lee, J.K., Park, H.G., Jeon, J.K., Kim, S.K., 2009. Antioxidant peptides isolated from the marine rotifer, *Brachionus rotundiformis*. Process Biochemistry 44, 842–846.

Chalamaiah, M., Dinesh Kumar, B., Hemalatha, R., et al., 2012. Fish protein hydrolysates: proximate composition, amino acid composition, antioxidant activities and applications: a review. Food Chemistry 135 (4), 3020–3038. Available from: https://doi.org/10.1016/j.foodchem.2012.06.100.

Chi, C.F., Wang, B., Wang, Y.M., et al., 2015a. Isolation and characterization of three antioxidant peptides from protein hydrolysate of Bluefin leatherjacket (*Navodon septentrionalis*) heads. Journal of Functional Foods 12, 1–10. Available from: https://doi.org/10.1016/j.jff.2014.10.027.

Chi, C.F., Wang, B., Hu, F.Y., et al., 2015b. Purification and identification of three novel antioxidant peptides from protein hydrolysate of Bluefin leatherjacket (*Navodon septentrionalis*) skin. Food Research International 73, 124–139.

Choi, Y.J., Hur, S., Choi, B.D., et al., 2009. Enzymatic hydrolysis of recovered protein from frozen small croaker and functional properties of its hydrolysates. Journal of Food Science 74 (1), C17–C24. Available from: https://doi.org/10.1111/j.1750-3841.2008.00988.x.

Choonpicharn, S., Tateing, S., Jaturasitha, S., et al., 2016. Identification of bioactive peptide from *Oreochromis niloticus* skin gelatin. Journal of Food Science and Technology 53, 1222–1229.

Dekkers, E., Raghavan, S., Kristinsson, H.G., et al., 2011. Oxidative stability of mahi mahi red muscle dipped in tilapia protein hydrolysates. Food Chemistry 124 (2), 640–645. Available from: https://doi.org/10.1016/j.foodchem.2010.06.088.

Delgado, C.L., Wada, N., Rosegrant, M.W., et al., 2020. Fish to 2020. International Food Policy Research Institute (IFPRI).

Deraz, S.F., 2015. Protein hydrolysate from visceral waste proteins of Bolti Fish (*Tilapia nilotica*): chemical and nutritional variations as affected by processing pHs and time of hydrolysis. Journal of Aquatic Food Product Technology 24 (6), 614–631. Available from: https://doi.org/10.1080/10498850.2013.797534.

Domínguez, R., Pateiro, M., Gagaoua, M., et al., 2019. A comprehensive review on lipid oxidation in meat and meat products. Antioxidants 8 (10), 1–31. Available from: https://doi.org/10.3390/antiox8100429.

FAOSTAT, 2001. FAO Statistical Databases, Fisheries Data. Food and Agriculture Organization of the United Nations, Rome.

Galla, N.R., Pamidighantam, P.R., Akula, S., et al., 2012. Functional properties and in vitro antioxidant activity of roe protein hydrolysates of *Channa striatus* and *Labeo rohita*. Food Chemistry 135 (3), 1479–1484. Available from: https://doi.org/10.1016/j.foodchem.2012.05.098.

Gbogouri, G.A., Linder, M., Fanni, J., et al., 2004. Influence of hydrolysis degree on the functional properties of salmon byproducts hydrolysates. Journal of Food Science 69 (8), C615–C622. Available from: https://doi.org/10.1111/j.1365-2621.2004.tb09909.x.

Ghaly, A., 2013. Microbial & biochemical technology fish processing wastes as a potential source of proteins, amino acids and oils: a critical review. Journal of Microbial & Biochemical Technology 54172 (5), 107–129. Available from: https://doi.org/10.4172/1948-5948.1000110.

Giannetto, A., Esposito, E., Lanza, M., et al., 2020. Protein hydrolysates from anchovy (*Engraulis encrasicolus*) waste: in vitro and in vivo biological activities. Marine Drugs 18 (2). Available from: https://doi.org/10.3390/md18020086.

Giménez, B., Alemán, A., Montero, P., et al., 2009. Antioxidant and functional properties of gelatin hydrolysates obtained from skin of sole and squid. Food Chemistry 114 (3), 976–983. Available from: https://doi.org/10.1016/j.foodchem.2008.10.050.

Guerard, F., Guimas, L., Binet, A., 2002. Production of tuna waste hydrolysates by a commercial neutral protease preparation. Journal of Molecular Catalysis B: Enzymatic 19 (20), 489–498. Available from: https://doi.org/10.1016/S1381-1177(02)00203-5.

Guo, l, Hou, H., Li, B., Zhang, Z., Wang, S., Zhao, X., 2013. Preparation, isolation and identification of iron-chelating peptides derived from Alaska Pollock skin. Process Biochemistry 48, 988–993.

Halim, N.R.A., Yusof, H.M., Sarbon, N.M., 2016. Functional and bioactive properties of fish protein hydolysates and peptides: a comprehensive review. Trends in Food Science & Technology 51, 24–33. Available from: https://doi.org/10.1016/j.tifs.2016.02.007.

Hsu, K.C., 2010. Purification of antioxidative peptides prepared from enzymatic hydrolysates of tuna dark muscle byproduct. Food Chemistry 122, 42–48.

Hung-Ngo, D., Sang Vo, T., Nghiep, N.D., et al., 2012. Biological activities and potential health benefits of bioactive peptides derived from marine organisms. International Journal of Biological Macromolecules 51, 378–383.

Jai ganesh, R., Nazeer, R.A., Sampath Kumar, N.S., 2011. Purification and identification of antioxidant peptide from black pomfret, *Parastromateus niger* (Bloch, 1975) viscera protein hydrolysate. Food Science and Biotechnology 20 (4), 1087–1094. Available from: https://doi.org/10.1007/s10068-011-0147-x.

Jayathilakan, K., Sultana, K., Radhakrishna, K., et al., 2012. Utilization of byproducts and waste materials from meat, poultry and fish processing industries: a review. Journal of Food Science and Technology 49 (3), 278–293. Available from: https://doi.org/10.1007/s13197-011-0290-7.

Jemil, I., Jridi, M., Nasri, R., et al., 2014. Functional, antioxidant and antibacterial properties of protein hydrolysates prepared from fish meat fermented by *Bacillus subtilis* A26. Process Biochemistry 49 (6), 963–972. Available from: https://doi.org/10.1016/j.procbio.2014.03.004.

Jeon, Y.J., Byun, H.G., Kim, S.K., 1999. Improvement of functional properties of cod frame protein hydrolysates using ultrafiltration membranes. Process Biochemistry 35, 471–478.

Jia, J., Zhou, Y., Lu, J., Chen, A., Li, Y., Zheng, G., 2010. Enzymatic hydrolysis of Alaska Pollock (*Theragra Chalcogramma*) skin and antioxidant activity of the resulting hydrolysate. Journal of Science of Food and Agriculture 90, 635–640.

Kechaou, E.S., Dumay, J., Donnay-Moreno, C., et al., 2009. Enzymatic hydrolysis of cuttlefish (*Sepia officinalis*) and sardine (*Sardina pilchardus*) viscera using commercial proteases: effects on lipid distribution and amino acid composition. Journal of Bioscience and Bioengineering 107 (2), 158–164. Available from: https://doi.org/10.1016/j.jbiosc.2008.10.018.

Khaled, H.B., Ghlissi, Z., Chtourou, Y., et al., 2012. Effect of protein hydrolysates from sardinelle (*Sardinella aurita*) on the oxidative status and blood lipid profile of cholesterol-fed rats. Food Research International 45, 60–68.

Khan, S., Rehman, A., Shah, H., et al., 2020. Fish protein and its derivatives: the novel applications, bioactivities, and their functional significance in food products. Food Reviews International 1–28. Available from: https://doi.org/10.1080/87559129.2020.1828452.

Khosravi, S., Rahimnejad, S., Herault, M., et al., 2015. Effects of protein hydrolysates supplementation in low fish meal diets on growth performance, innate immunity and disease resistance of red sea bream *Pagrus major*. Fish and Shellfish Immunology 45 (2), 858–868. Available from: https://doi.org/10.1016/j.fsi.2015.05.039.

Kittiphattanabawon, P., Benjakul, S., Visessanguan, W., et al., 2012. Cryoprotective effect of gelatin hydrolysate from blacktip shark skin on surimi subjected to different freeze–thaw cycles. LWT - Food Science and Technology 47 (2), 437–442. Available from: https://doi.org/10.1016/j.lwt.2012.02.003.

Klompong, V., Benjakul, S., Kantachote, D., et al., 2010. Use of protein hydrolysate from yellow stripe trevally (*Selaroides leptolepis*) as microbial media. Food and Bioprocess Technology 5 (4), 1317–1327. Available from: https://doi.org/10.1007/s11947-010-0402-9.

Kristinsson, H.G., Rasco, B.A., 2000a. Fish protein hydrolysates: production, biochemical, and functional properties. Critical Reviews in Food Science and Nutrition 40 (1), 43–81. Available from: https://doi.org/10.1080/10408690091189266.

Kristinsson, H.G., Rasco, B.A., 2000b. Biochemical and functional properties of Atlantic salmon (*Salmo salar*) muscle proteins hydrolyzed with various alkaline proteases. Journal of Agricultural and Food Chemistry 48 (3), 657–666. Available from: https://doi.org/10.1021/jf990447v.

Lassoued, I., Mora, L., Nasri, R., et al., 2015. Characterization, antioxidative and ACE inhibitory properties of hydrolysates obtained from thornback ray (*Raja clavata*) muscle. Journal of Proteomics 128, 458–468. Available from: https://doi.org/10.1016/j.jprot.2015.05.007.

Lee, J.K., Jeon, J.K., Byun, H.G., 2011. Effect of angiotensin I converting enzyme inhibitory peptide purified from skate skin hydrolysate. Food Chemistry 125 (2), 495–499. Available from: https://doi.org/10.1016/j.foodchem.2010.09.039.

Lin, C.C., Liang, J.H., 2002. Effect of antioxidants on the oxidative stability of chicken breast meat in a dispersion system. Journal of Food Science 67, 530–533.

Liu, Y., Li, X., Chen, Z., et al., 2014. Characterization of structural and functional properties of fish protein hydrolysates from surimi processing byproducts. Food Chemistry 151, 459–465. Available from: https://doi.org/10.1016/j.foodchem.2013.11.089.

Mahanand, S.S., Sharma, S., Wangkheirakpam, R., 2019. Fish waste utilization with reference to fish protein hydrolysate – a review. Fishery Technology 1 (56), 169–178.

Martínez-Alvarez, O., Batista, I., Ramos, C., et al., 2016. Enhancement of ACE and prolyl oligopeptidase inhibitory potency of protein hydrolysates from sardine and tuna byproducts by simulated gastrointestinal digestion. Food and Function 7 (4), 2066–2073. Available from: https://doi.org/10.1039/c5fo01603g.

Masuda, Y., Jinbo, T., Imaizumi, H., et al., 2013. A step forward in development of fish protein hydrolysate-based diets for larvae of Japanese eel Anguilla japonica. Fisheries Science 79 (4), 681–688. Available from: https://doi.org/10.1007/s12562-013-0637-2.

Mazorra-Manzano, M.A., Pacheco-Aguilar, R., Ramírez-Suárez, J.C., et al., 2010. Endogenous proteases in Pacific whiting (*Merluccius productus*) muscle as a processing aid in functional fish protein hydrolysate production. Food and Bioprocess Technology 5 (1), 130–137. Available from: https://doi.org/10.1007/s11947-010-0374-9.

Mendis, E., Rajapakse, N., Kim, S.K., 2005. Antioxidant properties of a radical-scavenging peptide purified from enzymatically prepared fish skin gelatin hydrolysate. Journal of Agricultural and Food Chemistry 53, 581–587.

Mukhin, V.A., Novikov, V.Y., Ryzhikova, L.S., 2001. A protein hydrolysate enzymatically produced from the industrial waste of processing Icelandic scallop *Chlamys islandica*. Applied Biochemistry and Microbiology 37 (3), 292–296. Available from: https://doi.org/10.1023/A:1010289403850.

Nasri, R., Younes, I., Jridi, M., et al., 2013. ACE inhibitory and antioxidative activities of Goby (*Zosterissessor ophiocephalus*) fish protein hydrolysates: effect on meat lipid oxidation. Food Research International 54 (1), 552–561. Available from: https://doi.org/10.1016/j.foodres.2013.07.001.

Nazeer, R.A., Deeptha, R., 2013. Antioxidant activity and amino acid profiling of protein hydrolysates from the skin of *Sphyraena barracuda* and *Lepturacanthus savala*. International Journal of Food Properties 16 (3), 500–511. Available from: https://doi.org/10.1080/10942912.2011.553757.

Nesse, K.O., Nagalakshmi, A.P., Marimuthu, P., et al., 2011. Efficacy of a fish protein hydrolysate in malnourished children. Indian Journal of Clinical Biochemistry 26 (4), 360–365. Available from: https://doi.org/10.1007/s12291-011-0145-z.

Ngo, D.H., Qian, Z.J., Ryu, B.M., et al., 2010. In vitro antioxidant activity of a peptide isolated from Nile tilapia (*Oreochromis niloticus*) scale gelatin in free radical-mediated oxidative systems. Journal of Functional Foods 2 (2), 107–117. Available from: https://doi.org/10.1016/j.jff.2010.02.001.

Nguyen, H.T.M., Pérez-Gálvez, R., Bergé, J.P., 2012. Effect of diets containing tuna head hydrolysates on the survival and growth of shrimp *Penaeus vannamei*. Aquaculture 324–325, 127–134. Available from: https://doi.org/10.1016/j.aquaculture.2011.11.014.

Nikoo, M., Benjakul, S., Ehsani, A., et al., 2014. Antioxidant and cryoprotective effects of a tetrapeptide isolated from Amur sturgeon skin gelatin. Journal of Functional Foods 7 (1), 609–620. Available from: https://doi.org/10.1016/j.jff.2013.12.024.

Nilsang, S., Lertsiri, S., Suphantharika, M., et al., 2005. Optimization of enzymatic hydrolysis of fish soluble concentrate by commercial proteases. Journal of Food Engineering 70 (4), 571–578. Available from: https://doi.org/10.1016/j.jfoodeng.2004.10.011.

Ovissipour, M., Kenari, A.A., Motamedzadegan, A., et al., 2011. Optimization of protein recovery during hydrolysis of yellowfin tuna (*Thunnus albacares*) visceral proteins. Journal of Aquatic Food Product Technology 20 (2), 148–159. Available from: https://doi.org/10.1080/10498850.2010.548910.

Ovissipour, M., Abedian Kenari, A., Motamedzadegan, A., et al., 2012. Optimization of enzymatic hydrolysis of visceral waste proteins of yellowfin tuna (*Thunnus albacares*). Food and Bioprocess Technology 5 (2), 696–705. Available from: https://doi.org/10.1007/s11947-010-0357-x.

Parvathy, U., Nizam, K.M., Zynudheen, A.A., et al., 2018. Characterization of fish protein hydrolysate from red meat of *Euthynnus affinis* and its application as an antioxidant in iced sardine. Journal of Scientific and Industrial Research 77 (2), 111–119.

Perkins, D.N., Pappin, D.J., Creasy, D.M., Cottrell, J.S., 1999. Probability-based protein identification by searching sequences databases using mass spectrometry data. Electrophoresis 20, 3551–3567.

Phanturat, P., Benjakul, S., Visessanguan, W., et al., 2010. Use of pyloric caeca extract from bigeye snapper (*Priacanthus macracanthus*) for the production of gelatin hydrolysate with antioxidative activity. LWT - Food Science and Technology 43 (1), 86–97. Available from: https://doi.org/10.1016/j.lwt.2009.06.010.

Picot, L., Ravallec, R., Martine, F.P., et al., 2010. Impact of ultrafiltration and nanofiltration of an industrial fish protein hydrolysate on its bioactive properties. Journal of the Science of Food and Agriculture 90 (11), 1819–1826. Available from: https://doi.org/10.1002/jsfa.4020.

Rajapakse, N., Mendis, E., Jung, W.K., et al., 2005. Purification of a radical scavenging peptide from fermented mussel sauce and its antioxidant properties. Food Research International 38, 175–182.

Ranathunga, S., Rajapakse, N., Kim, S.K., 2006. Purification and characterization of antioxidative peptide derived from muscle of conger eel (*Conger myriaster*). European Food Research and Technology 222, 310–315.

Ruthu, Murthy, P.S., Rai, A.K., et al., 2014. Fermentative recovery of lipids and proteins from freshwater fish head waste with reference to antimicrobial and antioxidant properties of protein hydrolysate. Journal of Food Science and Technology 51 (9), 1884–1892. Available from: https://doi.org/10.1007/s13197-012-0730-z.

Ryan, J.T., Ross, R.P., Bolton, D., et al., 2011. Bioactive peptides from muscle sources: meat and fish. Nutrients 3 (9), 765–791. Available from: https://doi.org/10.3390/nu3090765.

Safari, R., Nasrollahzadeh Saravi, H., Pourgholam, R., et al., 2011. Use of hydrolysates from silver carp (*Hypophthalmichthys molitrix*) head as peptone for *Vibrio anguillarum* and optimization using response surface method (RSM). Journal of Aquatic Food Product Technology 20 (2), 247–257. Available from: https://doi.org/10.1080/10498850.2011.562064.

Saito, M., Kiyose, C., Higuchi, T., et al., 2009. Effect of collagen hydrolysates from salmon and trout skins on the lipid profile in rats. Journal of Agricultural and Food Chemistry 57 (21), 10477–10482. Available from: https://doi.org/10.1021/jf902355m.

Samaranayaka, A.G.P., Li-Chan, E.C.Y., 2011. Food-derived peptidic antioxidants: a review of their production, assessment, and potential applications. Journal of Functional Foods 3 (4), 229–254. Available from: https://doi.org/10.1016/j.jff.2011.05.006.

Sampath Kumar, N.S., Nazeer, R.A., Jaiganesh, R., 2011. Purification and biochemical characterization of antioxidant peptide from horse mackerel (*Magalaspis cordyla*) viscera protein. Peptides 32 (7), 1496–1501. Available from: https://doi.org/10.1016/j.peptides.2011.05.020.

Sathivel, S.S., Echtel, P.J.B., Abbitt, J.B., et al., 2003. Biochemical and functional properties of herring (*Clupea harengus*) byproduct hydrolysates. Journal of Food Science 68 (7), 2196–2200.

Sathivel, S., Smiley, S., Prinyawiwatkul, W., et al., 2005. Functional and nutritional properties of red salmon (*Oncorhynchus nerka*). Food Chemistry and Toxicology 70 (6), 401–406.

Sayari, N., Sila, A., Haddar, A., et al., 2016. Valorisation of smooth hound (*Mustelus mustelus*) waste biomass through recovery of functional, antioxidative and antihypertensive bioactive peptides. Environmental Science and Pollution Research 23 (1), 366–376. Available from: https://doi.org/10.1007/s11356-015-5244-6.

Shahidi, F., Zhong, Y., 2008. Bioactive peptides. Journal of AOAC International 9, 914–931.

Silva, J.F.X., Ribeiro, K., Silva, J.F., et al., 2014. Utilization of tilapia processing waste for the production of fish protein hydrolysate. Animal Feed Science and Technology 196, 96–106. Available from: https://doi.org/10.1016/j.anifeedsci.2014.06.010.

Souissi, N., Bougatef, A., Triki-Ellouz, Y., et al., 2007. Biochemical and functional properties of sardinella (*Sardinella aurita*) byproduct hydrolysates. Food Technology and Biotechnology 45 (2), 187–194.

Taheri, A., Anvar, S.A.A., Ahari, H., et al., 2013. Comparison the functional properties of protein hydrolysates from poultry byproducts and rainbow trout (*Onchorhynchus mykiss*) viscera. Iranian Journal of Fisheries Sciences 12 (1), 154–169.

Tanuja, S., Haridas, A., Zynudheen, A.A., et al., 2014. Functional and antioxidative properties of fish protein hydrolysate (FPH) produced from the frame meat of striped catfish *Pangasianodon hypophthalmus* (Sauvage, 1878) using alkaline protease alcalase. Indian Journal of Fisheries 61 (2), 82–89.

Thiansilakul, Y., Benjakul, S., Shahidi, F., 2007. Compositions, functional properties and antioxidative activity of protein hydrolysates prepared from round scad (*Decapterus maruadsi*). Food Chemistry 103 (4), 1385–1394. Available from: https://doi.org/10.1016/j.foodchem.2006.10.055.

Valcarcel, J., Sanz, N., Vázquez, J.A., 2020. Optimization of the enzymatic protein hydrolysis of by-products from seabream (Sparus aurata) and seabass (Dicentrarchus labrax), chemical and functional characterization. Foods 9, 1503. Available from: https://doi.org/10.3390/foods9101503.

Vercruysse, L., Van Camp, J., Smagghe, G., 2005. ACE inhibitory peptides derived from enzymatic hydrolysates of animal muscle protein: a review. Journal of Agricultural and Food Chemistry 53, 8106–8115.

Viljanen, K., Kivikari, R., Heinonen, M., 2004. Protein–lipid interaction during liposome oxidation with added anthocyanin and other phenolic compounds. Journal of Agriculture Food Chemistry 52, 1104–1111.

Wasswa, J., Tang, J., Gu, X., et al., 2007. Influence of the extent of enzymatic hydrolysis on the functional properties of protein hydrolysate from grass carp (*Ctenopharyngodon idella*) skin. Food Chemistry 104 (4), 1698–1704. Available from: https://doi.org/10.1016/j.foodchem.2007.03.044.

Wu, X., Cai, L., Zhang, Y., et al., 2015. Compositions and antioxidant properties of protein hydrolysates from the skins of four carp species. International Journal of Food Science and Technology 50 (12), 2589–2597. Available from: https://doi.org/10.1111/ijfs.12927.

Yang, J.I., Ho, H.Y., Chu, Y.J., et al., 2008. Characteristic and antioxidant activity of retorted gelatin hydrolysates from cobia (*Rachycentron canadum*) skin. Food Chemistry 110 (1), 128–136. Available from: https://doi.org/10.1016/j.foodchem.2008.01.072.

Yang, J.I., Liang, W.S., Chow, C.J., et al., 2009. Process for the production of tilapia retorted skin gelatin hydrolysates with optimized antioxidative properties. Process Biochemistry 44 (10), 1152–1157. Available from: https://doi.org/10.1016/j.procbio.2009.06.013.

Yang, P., Ke, H., Hong, P., et al., 2011. Antioxidant activity of bigeye tuna (*Thunnus obesus*) head protein hydrolysate prepared with Alcalase. International Journal of Food Science and Technology 46 (12), 2460–2466. Available from: https://doi.org/10.1111/j.1365-2621.2011.02768.x.

Yeung Joon Choi, H.S., Choi, B.D., et al., 2009. Enzymatic hydrolysis of recovered protein from frozen small croaker and functional properties of its hydrolysates. Journal of Food Science 74 (1). Available from: https://doi.org/10.1111/j.1750-3841.2008.00988.x.

Yin, H., Pu, J., Wan, Y., et al., 2010. Rheological and functional properties of catfish skin protein hydrolysates. Journal of Food Science 75 (1), 11–17. Available from: https://doi.org/10.1111/j.1750-3841.2009.01385.x.

Zhang, L., Zhao, G.X., Zhao, Y.Q., et al., 2019. Identification and active evaluation of antioxidant peptides from protein hydrolysates of Skipjack tuna (*Katsuwonus pelamis*) head. Antioxidants 8 (8). Available from: https://doi.org/10.3390/antiox8080318.

Zou, T.B., He, T.P., Li, H.B., et al., 2016. The structure–activity relationship of the antioxidant peptides from natural proteins. Molecules 21 (1), 1–14. Available from: https://doi.org/10.3390/molecules21010072.

Chapter | eight

Ready-to-Eat Products Elaborated With Mechanically Separated Fish Meat From Waste Processing

Kelvin Adrah and Reza Tahergorabi

Food and Nutritional Sciences Program, North Carolina Agricultural and Technical State University, Greensboro, NC, United States

8.1 INTRODUCTION

The growing human population demands increased food production, including fish and fish products (Kratky and Zamazal, 2020). The worldwide fish market increased from 7.6 million to 29.2 million tons between 1995 and 2008 and is expected to hit 71 million tons in 2020. Fisheries contribute immensely to global food and nutritional security, and the Food and Agricultural Organization estimates that it provides livelihoods to over 10%−12% of global production (Prasad and Murugadas, 2018). However, fish production at such a large scale consequently generates vast amounts of fish waste up to 50% depending on the species and the method of processing (Tahergorabi et al., 2015). Fish frame, liver, head, viscera, skin, bones, blood, gonads, guts, and muscle tissue are primary discard materials that generate 32 million tons of waste (Kerton et al., 2013; Prasad and Murugadas, 2018).

Solid fish wastes derived from these fish parts are usually deposited at landfills, and their liquid counterparts are discharged to the marine environment. However, the absence of an adequate site for disposing of fish waste

Sustainable Fish Production and Processing. DOI: https://doi.org/10.1016/B978-0-12-824296-4.00006-2

produces environmental problems. Currently the food industry is looking for ways to reduce food-processing byproducts due to their adverse effects on the environment (Azahrani et al., 2019). These problems could be mitigated by converting fish wastes into consumable products (De et al., 2020).

Fish and other seafood products such as mollusks, crustaceans, and echinoderms, consumed directly or processed, offer a wide variety of products that provide nutritional benefits to humans. Seafood proteins are highly digestible and have a high biological value because of a dominant composition of sarcoplasmic and myofibrillar proteins in their muscles and lower amounts of collagen, a connective protein. Lipids are another vital nutrient that could be derived from fish. However, their quality and quantity differ according to the species, age, sex, and period of the year. For example, long-chain polyunsaturated fatty acids (PUFAs) range from about 0.2% to 3% weight in lean fish to fatty fish. Fish is also regarded as a rich source of vitamins and minerals. Abundant in seafood are iodine, selenium, calcium, zinc, thiamin, riboflavin, niacin, pyridoxine, cobalamin, etc. Vitamin A and D are also found in elevated concentrations in fish and fish products (Gil and Gil, 2015; Oehlenschläger, 2012).

In recent times, essential approaches have been implemented to reduce waste and recover usable materials from fish during processing. The introduction of mechanical separation of meat in fish provides the fish industry the opportunity to recover maximum meat from fish processing to produce marketable and desired food products in various forms (Prasad and Murugadas, 2018). Processing fish in this manner would not only increase the yield of edible flesh but could, if applied, improve the economy of our domestic industry. The total flesh of minced meat from various species of fish ranges from 37% to 60% based upon round weight. This recovery rate compares to a yield of 25%–30% intact flesh using conventional filleting methods on the same fish species (Crawford et al., 1972). Several technologies have been adopted to utilize recovered fish meat into producing functional ingredients from bones and skin, semiready or ready-to-eat (RTE) products obtained from meat adhered to bone and flesh. The latter is commonly referred to as mechanically separated meat (MSM) and is derived from removing meat from bones by applying pressure (Husein et al., 2019a,b).

Consequently, mechanically separated fish could produce RTE products such as fish sticks, fish burgers, fish nuggets, patties. Mechanical separation provides an alternative way for the diversification of new products derived from fish meat and also a solution for the utilization of fish residues from the fish industry. New products derived from the mechanical separation of meat from fish have gained interest from a part of consumers in several studies, thus, concluding that these types of products have the potential to find space in the market (Husein et al., 2019a,b).

8.2 FISH PROCESSING

As described in Table 8.1, fish processing employs the following steps: stunning of fish, grading, removal of slime, scaling, washing, beheading, gutting,

TABLE 8.1 Describes the Processes Involved in Fish Processing

Processing Method	Description
Stunning	Stunning is a vital step in fish processing because a prolonged death by asphyxia produces undesired substances in fish tissues. Oxygen deprivation in the blood and muscle causes the buildup of lactic acid and leads to paralysis of the neural system, and this will impact its quality and texture. Electric stunning is a recommended slaughtering method for killing salmon and grass carp because it causes the early onset of rigor mortis.
Grading	Grading involves fish grading by species and size. This could be done by a preferred automated system or manually. The mechanical grading offers more precision for fish before and after rigor mortis than for fish in a state of rigor mortis. Also the automated system is cost-effective and increases the quality of fish products at the end of the processing chain.
Slime removal	Slime is secreted on the surface of fish as a protective mechanism against harmful conditions. However, slime secretion ceases before rigor mortis, providing a perfect medium for the growth of *Pseudomonas* spp. Anaerobic bacteria present during fish processing can also take sulfur compounds from the fish surface to produce hydrogen sulfur. Slime removal should be carried out by continuous washing. Slime present in eel, trout, and other freshwater species should be soaked in a solution of 2% baking soda and then washed in a cylindrical rotating washer.
Scaling	Scaling is a labor-intensive process that could be performed manually with a stiff brush or scaling blades. Scaling eliminates all bacterial pathogens that are harbored in scales of fish, keeping fish fresh during refrigeration. Scales of pike-perch, carp, and bream are first blanched in boiling water for 3–6 seconds and then scaled using mechanized handheld scalers.
Washing	Washing is done to get rid of accumulated bacteria on the fish. Water quality and fish: water ratios are parameters that determine the effectiveness of washing during processing. The recommended fish: water ratio is 1:1. This process is performed using vertical drum, horizontal drum, and conveyor belt washers. The fish product does not decline in quality during washing.
Beheading	This process can be executed manually or mechanically. Manual beheading is, however, recommended because it yields minimal flesh loss. A cut around the operculum is called a round-cut, and it results in a low meat loss. Mechanical beheading is mainly used for larger fish undergoing round or contour cuts. Similarly, manually operated machines are suitable for larger fish undergoing straight cuts.
Gutting	Gutting in fish processing demands removing the internal organs, which constitutes between 5% and 8% of the fish weight, and optionally cleaning the body cavity of the peritoneum, kidney tissue, and blood. Mechanical gutting machines are used for trout, eel, and other fishes. However, the use of these machines inflates the cost of production.
Cutting of fins	This process entails cutting of fins manually by the use of a knife or by a mechanized rotating disc knife. Fins are excised when passed manually through mechanical knives with slits.
Steaks and fillets	Fillets are meat pieces comprising the dorsal and abdominal muscles. Manual processing of fillets is carried out in small freshwater fish industries. The mechanical filleting is used for processing marine fish. Steaks are prepared by slicing beheaded whole fish perpendicular to the backbone.
Meat bone separation	In this process, meat is removed from skin, scales, and bones by actions of automated separators. Meat is squeezed through holes as fish is transported along a conveyor belt and the bones are scraped away.

Source*: Summarized from Ghaly, A.E., Dave, D., Budge, S., Brooks, M., 2010. Fish spoilage mechanisms and preservation techniques. American Journal of Applied Sciences 7 (7), 859.*

cutting of fins, slicing into steaks, filleting, meat-bone separation, labeling, and distribution (Ghaly et al., 2013).

8.3 NUTRITIONAL QUALITY AND CHEMICAL PROPERTIES OF FISH

Fish has several benefits that make it an essential part of a healthy diet, such as high levels of omega-3 fatty acids, which have crucial roles in the body and are associated with a lower risk of all-cause mortality (Thilsted et al., 2016; Zhao et al., 2016; Zheng et al., 2012). The dietary recommendations for increased fish consumption are predominantly attributed to marine PUFA (Jensen et al., 2013). Fish and seafood products also have a high nutritional value regarding beneficial amounts of protein and essential micronutrients required in human diets (Khalili Tilami and Sampels, 2018; Nettleton and Exler, 1992). Husein et al. (2019a,b) explain that the nutritional value of the low commercial fish is similar to that of the valuable ones. The chemical composition of discarded fish species presented in trawling discards demonstrates how the protein content of minor species is comparable to that of the more valuable ones and with the same nutritional value. Seafood is considered to be a reasonable but not a rich source of vitamins. This section elaborates on the macro- and relevant micronutrient compositions of seafood.

8.3.1 Protein Content

Proteins are composed of amino acids linked by peptide bonds. In the lumen of the gastrointestinal tract, proteins are hydrolyzed by the actions of proteases and peptidases to generate amino acids, dipeptides, and tripeptides. Amino acids are essential for cell-specific production of low-molecular-weight metabolites with enormous physiological importance. Therefore inadequate intake of proteins results in stunting, anemia, physical weakness, edema, vascular dysfunction, and impaired immunity (Wu, 2016). Fish has been recognized as a valuable source of high-quality protein in the human diet (Puwastien et al., 1999). Aquatic animal foods have a higher protein composition than other terrestrial meats. Furthermore, marine protein is highly digestible and rich in several peptides and essential amino acids that are short in terrestrial meat proteins, for example, methionine and lysine (Tacon and Metian, 2013).

The protein content of fish is relatively constant in a given species, and its composition only changes slightly with fish size. The amino acid content of fish muscle tissue is excellent in terms of matching the dietary needs of humans (Shearer, 1994). Dietary protein recommendations are traditionally expressed daily (Murphy et al., 2016). Based on short-term nitrogen balance studies the recommended daily allowance for protein intake for men and women with minimal physical activity is 0.8 g protein per kg body weight per day (Wu, 2016). Puwastien et al. (1999) reported high protein content (17–22 g/100 g) measured in commonly consumed freshwater and marine fishes. Jensen et al. (2013) examined protein content (15.7% ± 0.7% and 18.6% ± 1.4%) of wild cod and farmed cod, respectively. Likewise, Atka mackerel, Pacific sand lance, and Pacific sandfish had a median protein content (15.4%) (Payne et al., 1999).

8.3.2 Fatty Acid Profile/Lipid Content

Fatty acids are organic compounds formed by a hydrocarbonated chain and a carboxylic group typically bounded with glycerol creating acyl glycerides (mono-, di- or triglycerides). Fatty acids are classified as either saturated or unsaturated, depending on the nature of the hydrocarbonated chain, which can be monounsaturated or polyunsaturated (Rubio-Rodríguez et al., 2010). Otherwise, naturally occurring fatty acids can be classified into three categories based on the number of double bonds present in sidechains: saturated fatty acids (no double bonds), monounsaturated fatty acids (MUFA) (single-double bond), and PUFA (≥2 double bonds) (Saini and Keum, 2018). Fatty acids are characterized by an even number of carbon atoms in the range of 16–26. Fatty acids with only single bonds between adjacent carbon atoms are referred to as "saturated," whereas fatty acid with at least a C=C double bond is "unsaturated." PUFAs are named based on the position of the double bonds and the total chain length (Ruxton et al., 2004). MUFAs differ from PUFA based on having a single, double bond (Kris-Etherton, 1999). The geometric configuration of a MUFA is determined by the position of the hydrogen atoms around the double bond, resulting in either a *cis* or *trans* isomer. Humans can synthesize many of the fatty acids, but omega-3 and omega-6 fatty acids cannot be produced in the human body, so are thereby considered essential fatty acids (Rubio-Rodríguez et al., 2010). Animals located in the marine ecosystem are recognized as rich sources of PUFA, especially omega-3 fatty acids eicosapentaenoic acid (EPA) and docosahexaenoic acid (DHA) (Huynh and Kitts, 2009). That being said, the fatty acid composition of the lipid extracted from fish waste of the *Sardinella lemuru* was composed of saturated fatty acid, MUFA, and PUFA (Khoddami et al., 2009).

Seafood serves as a source of white meat for human consumption due to its health benefits. The fat composition of fish, which is mostly responsible for these health-associated benefits, converts fatty dienoic acid into tetraenoic, pentanoic, and hexanoic acids. Similar to regular fats and oils, fish oils are composed of several lipid fractions such as triacylglycerol, diacylglycerol, monoacylglycerol, phospholipids, steryl esters, sterols, and free fatty acids. However, these fractions vary between different species and various tissues and organs (Saify et al., 2003). A report by Mataix et al. (2003) showed the varying concentrations of EPA and DHA among different species of fish.

Furthermore, fatty acid parameters were compared to seed oils (Table 8.2). Fish oils recorded substantially higher levels of omega-3 fatty acids compared to seed oils. The total concentration of PUFA was, however, higher than fish oils. This suggested that omega-6 fatty acids were found in

TABLE 8.2 Lipids (g/100 g oil) in Different Kinds of Seed and Fish Oils

Oil	SFA	MUFA	PUFA	Omega-3	Omega-6
Sunflower	12.0	20.5	67.5	0.10	63.2
Corn	14.5	29.9	55.6	0.90	50.4
Soya	15.6	21.2	63.2	7.30	51.5
Olive	14.3	73.0	12.7	0.70	7.8
Cod liver	22.6	20.7	56.8	19.8	0.9
Herring	21.3	56.6	22.1	11.9	12
Salmon	19.9	17.0	63.1	35.3	1.06
Sardine	30.4	14.5	55.1	28.1	2.2

Source: *Adapted from Rubio-Rodríguez, N., Beltrán, S., Jaime, I., de Diego, S.M., Sanz, M.T., Carballido, J.R., 2010. Production of omega-3 polyunsaturated fatty acid concentrates: a review. Innovative Food Science & Emerging Technologies 11 (1), 1–12. https://doi.org/10.1016/j.ifset.2009.10.006.*

higher amounts in seed oils as against fish oils. Saturated fats were also in higher amounts in fish oils than in seed oils.

8.3.2.1 SATURATED FATTY ACIDS

Longstanding and current epidemiological findings highly recommend limiting the consumption of saturated fatty acids. Dietary saturated fat intake has been shown to increase low-density lipoprotein cholesterol and has been associated with an increased risk of cardiovascular disease (Siri-Tarino et al., 2010). The frequent consumption of saturated fatty acids is also linked to nonalcoholic fatty liver disease (Leamy et al., 2013) and lung cancer in men (Knekt et al., 1991). Abd Aziz et al. (2013) reported saturated fatty acids of red snapper 557.9 mg/100 mg, yellowstripe scad 869.6 mg/100 g, gray eel catfish 1340.8 mg/100 g, Indian mackerel 587.8 mg/100 g, Spanish mackerel 322.8 mg/100 g, and hardtail scad 708.8 mg/100 g. Unsurprisingly, saturated fatty acid measured in longtail shad, a high-fat fish ($>8\%$ fat), measured 17,620.2 mg/100 g. Similarly, Moonfish recorded saturated fatty acid composition of 2572.4 mg/100 g.

8.3.2.2 MONOUNSATURATED FATTY ACIDS

The most common MUFAs, palmitoleic acid, and oleic acid are both *cis* isomers of MUFA. However, oleic acid is the predominant MUFA in the diet, representing approximately 92% of *cis* MUFA (Gillingham et al., 2011). Fish lipids in saury, pollock, herring, capelin, sprats, and other marine mammals are enriched in long-chain MUFAs that originate from their food source, zooplankton (Yang et al., 2016b). Among the widely consumed fish, saury is mostly highly enriched fish in long-chain MUFA. Salmon and cod also contain some significant amounts of these fatty acids. Thus a considerable amount of PUFA is found in regularly consumed fish.

Following extensive research concerning the total elimination of trans-fatty acids and limiting saturated fatty acids to about 10% of total energy intake, scientific interest has been directed toward establishing evidence to support the cardioprotective effects of MUFAs (Hammad et al., 2016). Several studies have reported the impact of MUFA on risk factors of cardiovascular diseases. A western diet supplemented with 5% (w/w) low chain MUFAs and fed to diet-induced obese mice improved insulin resistance. It lowered blood lipids compared to western diet control (Yang et al., 2011). Arachidonic acid, through one of its various metabolites, competes with omega-3 PUFAs for several physiological processes and is reported to be associated with metabolic or cardiovascular disorders (Sugimoto et al., 2016). Studies have revealed intake of long-chain MUFAs decreased arachidonic acid levels and increased plasma/hepatic EPA and DHA levels (Yang et al., 2011, 2013, 2016a). Furthermore the consumption of MUFA and soluble fiber promotes benefits on glucose metabolism, independently of adiposity, during a practical lifestyle intervention in at-risk individuals. Total MUFAs measured in Hardtail scad 121.2 mg/100 g, golden snapper 94.0 mg/100 g, Indian mackerel 294.8 mg/100 g, yellow stripe scad 286.8 mg/100 g, moonfish 504.2 mg/100 g, and longtail shad 3059 mg/100 g (Abd Aziz et al., 2013).

8.3.2.3 POLYUNSATURATED FATTY ACIDS

PUFAs, such as DHA, EPA, and arachidonic acid, are essential fatty acids for humans digested from dietary sources such as fishes and fish oils (Hayashi et al., 2020). Essential fatty acids, omega-3, and omega-6 PUFAs have been shown to possess curative and preventive properties on cardiovascular diseases, neurodevelopmental diseases in infants, cancers, and fat glycemic control. However, amounts of these PUFAs vary considerably among fish (Osman et al., 2001). The effects of omega-3 fatty acids on cardiovascular health were evaluated from randomized controlled trials and clinical trials. The study revealed that the consumption of omega-3, either as food or supplement for 6 months, reduces cardiovascular events by 10%, cardiovascular death by 9%, and coronary events by 18%. Also the mortality rate is lowered. The results reported in this study were based on the evaluation of other studies that included persons with high cardiovascular risk and in studies that are heterogenic in the dose administered (Delgado-Lista et al., 2012). Consuming ≥ 3 g of omega-3 PUFAs per day as fish oil or bluefish decreases blood pressure. A high intake of these PUFAs can be attained by consuming salmon, mackerel, herring, tuna, and sardines twice or three times a week to achieve an amount of 500 mg of EPA or DHA a day (Cabo et al., 2012). Studies and reviews on cardiovascular effects have also been published (Bork et al., 2020; Oppedisano et al., 2020; Yamagata, 2020).

A common source of omega-3 fatty acids in the human diet is fish. Oily fishes such as Clupeidae and Salmonidae families are composed of very high quantities of DHA and EPA (Table 8.3). Fishes, therefore, are valuable products that are not only acknowledged for their high omega-3 concentrates but also as raw material that could be processed into functional foods.

TABLE 8.3 Shows EPA and DHA Content of Some Species of Fish

Species	Common Name	g/100g of Foodstuff Portion	
		C20:5n-3 (EPA)	C22:6n-3 (DHA)
Scomber scrombrus	Mackerel	1.10	2.56
Sardina pilchardus	Sardine	0.62	1.12
Salmo salar	Salmon	0.50	1.00
Engraulis encrasicolus	Fresh anchovy	0.14	0.80
Gadus morhua	Cod	0.23	0.47
Merluccius merluccius	Hake	0.10	0.54
Galeorhinus galeus	Dogfish	0.04	0.30

Source: *Adapted from Rubio-Rodríguez, N., Beltrán, S., Jaime, I., de Diego, S.M., Sanz, M.T., Carballido, J.R., 2010. Production of omega-3 polyunsaturated fatty acid concentrates: a review. Innovative Food Science & Emerging Technologies 11(1), 1–12. https://doi.org/10.1016/j.ifset.2009.10.006.*

8.3.3 Vitamins and Minerals

Micronutrient deficiency is a global problem that affects a third of the world population, affecting mental and physical abilities, thus lowering productivity and quality of life. Food-based strategies to enhance the intake of micronutrients are recognized as the long-term sustainable solution to micronutrient deficiencies in developing countries (Demment et al., 2003). Whole foods such as fish have been considered as a good source of minerals and vitamins in addition to that as an excellent source of protein and omega-3 fatty acids (Özyurt et al., 2009). The composition of potassium, sodium, chlorine, magnesium, phosphorus, and calcium is up to 1 mg/100 g, whereas iron, zinc, copper, and iodine are less than 1 mg/100 mg (Martínez-Valverde et al., 2000; Navarro, 1991). Yet, due to several factors such as species, sex, biological cycle, and the portion size of fish analyzed, establishing and generalizing mean mineral values is challenging. Also Bogard et al. (2015) argue that the high content of several vital minerals in small fish species may be naturally attributed to the inclusion and exclusion of various fish (bones, skin, head, viscera, etc.) in analyzed samples.

Iron deficiency causes retarded growth and impaired cognitive development (Moy, 2006) and, in women, has a severe impact on health and morbidity. Calculations based on the iron content and high bioavailability of heme iron revealed that a traditional fish meal, sour soup, prepared with *Esomus longimanus*, covered 45% of the daily iron requirement for women (Roos et al., 2007b). Also sous vide salmon fish increases the bioavailability of iron (Navas-Carretero et al., 2008). Comparatively, fish has a low total iron concentration to meat, for example, sous-vide salmon contains only 0.46 mg Fe/100 g edible portion, with even minimal levels of heme iron compared with grilled beef steak, which has 3.6 mg Fe/100 g (Navas-Carretero et al., 2008). However, there is a lack of understanding regarding how meat enhances iron absorption.

Conversely a study by Harvey et al. (2005) observed that the frequent consumption of poultry/fish-rich diet resulted in higher iron stores, measured as serum ferritin, in premenopausal women compared to red meat consumers. Meats obtained via mechanical separation are composed of a higher content of iron, mainly if higher pressures are applied. Iron content is twice as high as of the manually deboned meat due to bone marrow incorporation. This increase is attributed to the presence of hemoglobin, which varies considerably depending on the bone–meat ratio, setting of the mechanical separator, the skin content, and the age of the animal. Also separator steel components also contain metal particles that could find their way with separated meat (Husein et al., 2019a,b).

Zinc is an essential trace element for all living organisms. It ensures the stability of biological molecules such as DNA and biological structures such as membranes and ribosomes. The concentration of zinc (μg/g wet weight) in muscle tissue of 28 fish species ranged from 1.02 to 8.47 μg/g (wet weight) (Jones et al., 2000). Haase et al. (2020) report the recommended intake of zinc for adults, pregnant and lactating women range from 7 to 16 mg/day, depending on sex and dietary phytate intake. The average human contains between 2 and 3 g of zinc, located in muscles and the skeleton (Maares and Haase, 2020). Zinc is always present as a divalent cation (Zn^{2+}); a significant proportion is bound to proteins (Krężel and Maret, 2016). Also the human body excretes about 0.1% of the zinc pool daily; therefore, regular zinc intake is required to prevent deficiency (Maret and Sandstead, 2014). Fish could provide the right amount of zinc to consumers.

Iodine is an essential micronutrient element necessary for the synthesis of thyroid hormones, thyroxine, and triiodothyronine. Frequent consumption of an iodine-deficient diet causes a broad spectrum of illnesses, including goiter and mental retardation (Haldimann et al., 2005). Seawater fish and other marine foods are considered rich in dietary iodine (Eckhoff et al., 1997). The iodine concentrations in wild fish species varied between 18 μg/100 g (Atlantic halibut) and 1210 μg/100 g (pollack) (Nerhus et al., 2018). According to García-Casal et al. (2013) the recommended dietary allowance for iodine is 150 μg/day for adults.

Fish is also a significant source of vitamin D. Among the various illnesses associated with an inadequate intake of vitamin D, rickets, osteomalacia, and osteoporosis are the most common. Fan et al. (2013) established the relationship between higher fish intake a lower risk of hip fractures in elderly Chinese. A review article also reported that levels of 25-hydroxyvitamin D below 20 ng per milliliter are associated with a 30%–50% increased risk of incident colon, prostate, and breast cancer, along with higher mortality from these cancers (Holick, 2007). Nakamura et al. (2002) report the overall average vitamin D intake per day in a traditional Japanese diet was 7.10 μg, which is about 70% of the recommended dietary allowance of 10 μg in the Japanese diet. The amount of vitamin D from fish sources, however, measured 6.44 μg. Fish is the primary source of vitamin D in the Japanese diet. The most frequently consumed fish is salmon (10–32 μg of vitamin D/100 g). A common type of salmon consumed in Japan, Chum salmon, contains 32 μg of vitamin D per 100 g. Other types of regularly consumed fish are the sea bream (1–8 μg of vitamin D/100 g), baby sardines (46 μg of vitamin D/100 g), mackerel (11 μg) of vitamin D/100 g, and cod (1 μg of vitamin D/100 g).

The composition of vitamin A in fish is highly variable, depending on the species and the cleaning practice, which determines which parts of the fish are edible (Roos et al., 2002). Roos et al. (2007a) sampled common Cambodian fish species for vitamin A content screening. *Parachela siamensis* and *Rasbora tornieri* had very high vitamin A content (>1500 RAE/100 g). Six species (*Barbodes altus*, *Barbodes gonionatus*, *Dermogenys pusilla*, *Parchela siamensis*, *Punctioplites proctozysron*, and *Thynnichthys thynnoides*) had a high vitamin A content (500–1500 RAE/100 g).

8.4 MECHANICAL SEPARATION OF FISH MEAT

The seafood processing industry introduced mechanical flesh separating machines in the early 1970s to improve flesh recovery from fish. Compared to traditional hand filleting techniques, these machines increased edible meat from 34.3% to 101.9%. Thus food products can be manufactured or developed from the ample supply of the fish-flesh processed from underutilized fish species. Although the food industry has seen the substantial introduction of several fish products into the market the availability of these products is narrow and limited. To effectively utilize this technology, new uses and food products must be developed from minced fish. Additional product development research will further reveal the versatility of mechanically separated fish as a critical raw material in developing food products (Price et al., 1990). The mechanical separation process has, however, been demonstrated to be successfully applied in the fish processing industry (Secci et al., 2017). With the introduction of the meat-bone separator in the early 1970s to recover the maximum amount of fish flesh from whole fish and fish frames, a new fishery technology had begun (Mendelsohn, 1974). MSM, and more rarely mechanical deboned meat, is a technology used to obtain meat products by removing meat from bones through the application of high ($>10^4$ kPa) or low ($<10^4$ kPa) pressures. Borgogno et al. (2017) aimed to compare the impact of the mechanical separation process and manual minced technology applied in European sea bass (*Dicentrarchus labrax*), gilthead seabream (*Sparus aurata*), and rainbow trout (*Oncorhynchus mykiss*). The authors showed that the mechanical separation process yield was higher than the manual one when applied to sea bass and sea bream (42 g/100 g and 45 g/100 g, respectively against 39 g/100 g and 40 g/100 g). The lipid fraction of the three species remained unchanged, with no significant differences, was found in the fatty acid composition of the products. In effect the manufacturing of products by exploiting fish without altering the nutritional value of whole fish is a goal reached adopting the mechanical separation process.

This process removed the remaining meat from the skeleton and the carcasses of animals of various kinds. It was regarded as a useful and essential raw material of animal origin for the manufacture and preparation of food products (Husein et al., 2019a,b). The standard structure of the muscle fiber is altered during this process. However, quality meat has been obtained from fish byproducts (de Oliveira et al., 2015; Kirschnik et al., 2013). Amisy©, suppliers of food processing machinery, indicates fish meat separating machines are composed of a motor, feed-inlet, conveyor, and rolling meat picking barrel. When the machine starts operation, fishes are moved along the conveyor in the direction of the rolling barrel, where pure meat is extruded due to the combined pressure subjected by both conveyor and rolling barrel. The fish skin and bones are subsequently discharged from the outlet below the conveyor.

Even though mechanical meat separators have been significantly improved since their introduction, their mode of operation is still the basis of today's machinery (EFSA, 2013). In the beginning, primitive presses derived from other industries were used to separate meat from the bones, exerting high pressures of up to 200 bar. Mechanical separation at high pressures yields a fine-textured meat paste suitable for the manufacturing of only cooked sausages. After a time, advancement in technology resulted in the use of a much lower pressure (up to 20 bar) to press flesh-bearing bones. The application of low pressure produces a coarse texture of higher quality meat indistinguishable from traditional minced meat (EFSA, 2013).

Generally, there are two types of separators: press type and sieve type, with different names. However, the primary separators in the market are (1) drum and belt system, (2) endless screw system, and (3) hydraulically powered presses. These machines can operate at low or high pressure, but some machines are typically set to function at low pressure than other types. For example, drum and belt separators usually operate at low pressure, but an endless screw system can operate at both low and high pressures. Hydraulically powered presses are generally set to operate at high pressures but can be used at low pressures as well. Therefore MSM can be produced at both pressures by adjusting these mechanical separators (Husein et al., 2019a,b) (Fig. 8.1).

8.5 READY-TO-EAT PRODUCTS DERIVED FROM MECHANICALLY SEPARATED FISH MEAT

As mentioned earlier, fish products provide a rich source of high-quality protein, omega-3 fatty acids, and a range of micronutrients that have beneficial health effects on human health. These benefits include improving cardiovascular health, positive effects on brain development, antiinflammatory and anticarcinogenic effects. However, the amount of time required for its preparation discourages consumers, leading to the desire and preference for RTE fish products. On the other hand, RTE products bear high costs, which appears as a barrier to consumption and increased consumer demands.

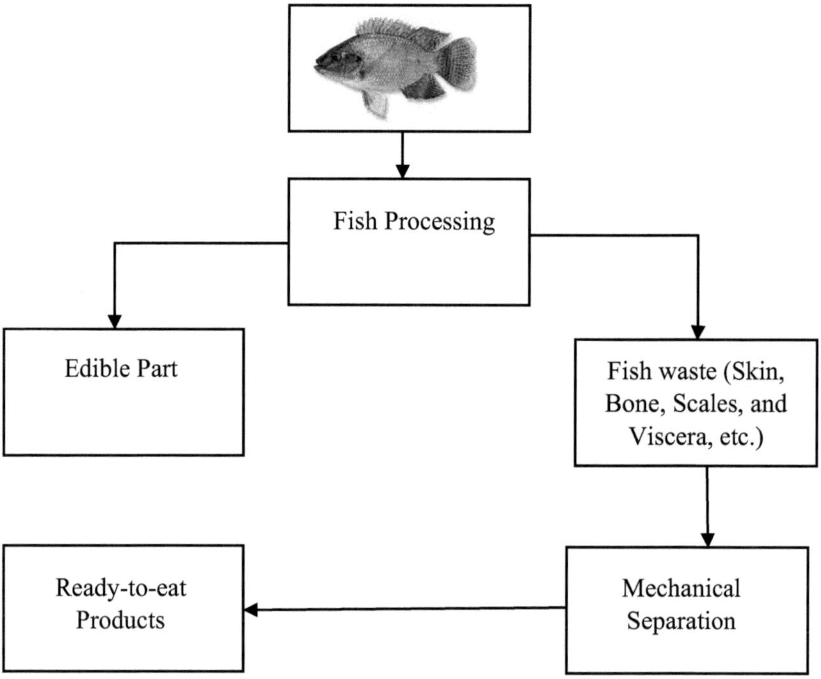

FIGURE 8.1 Flow diagram of the method to obtain MSM for production of ready-to-use products from fish processing.

However, research reveals consumers are likely to patronize RTE fish, mainly in sales channels (fast food and snack bars), suggesting an avenue for developing RTE fish products (Cosmina et al., 2012; Palmeira et al., 2016). RTE food products are not required to be fully cooked before consumption. They are defined as the status of the food that is ready for immediate consumption at the point of sale. It could be raw or cooked, hot or chilled, and consumed without further heat-treatment, including reheating (Chaudhury, 2010). They must also be appropriately handled during preparation and storage to inhibit bacterial growth resulting in foodborne illnesses. Common examples of RTE plates of seafood products are smoked fish, seafood salads, sushi or sashimi, cooked shrimps, crab or lobster meat, seafood sandwiches, and pickled seafood.

About 2–3 million metric tons of fish, representing 2%–3% of the total fishery, are used in surimi production. Briefly, surimi is a highly processed seafood from minced fish meat that has been leached by washing with water, mixed with sugar and other additives, and then frozen. After gutting, boning, mincing, and washing the fish meat is dehydrated using a screw press. The resulting product has a doughy texture, making it an excellent raw material for developing surimi-based foods. Traditionally, surimi is white-colored, bland-flavored onto which flavorings, colors, and other additives are applied during further processing. These, and by incorporating different shaping and cooking methods, various new products can be developed. Surimi can be used to produce foods such as imitation crab meat, kamaboko, chikuwa, fish ball, tempura, and satsuma-age. Surimi can now be developed into food for human needs (Satam et al., 2004). The United States stands a significant producer of surimi, with countries such as Japan and Thailand producing substantial quantities. China, Vietnam, Chile, France, and Malaysia are emerging producers of this fish product. Despite the decrease in surimi-based product consumption in Japan, which is attributed to the patronage of western foods, the production of surimi-based products has not been impacted. The sustainability of import–export balance results from the growing popularity of surimi products in the United States. Fish meat recovered from waste through mechanical separation provides enough raw material for surimi processing and product development (Hall and Ahmad, 1997).

A study has shown that a meat bone separator effectively increased the yield of muscles from various fish species. Further investigations explored developing these recovered muscles into useful products such as fish blocks. Fish blocks, intermediates from which fish sticks and portions are prepared, are made from fillets frozen under pressure into rectangular shapes. Owing to the short supply and high cost of producing conventional fillets the potential for utilizing recovered minced fish muscle appears to be promising (King and Carver, 1970; Miyauchi and Steinberg, 1970). Black rockfish, an underutilized marine fish, was successfully processed into a modified minced fish block prepared from machine-separated fish muscle (Teeny and Miyauchi, 1972). The modified fish block has a storage life of over 9 months at −18°C. The minced muscle was blended with several ingredients (e.g., antioxidant mixtures) to prevent fish blocks from rancidity and improve texture and flavor. Subsequently, raw breaded fish sticks were prepared from the modified fish blocks. The shelf-life of raw breaded fish sticks was limited to 2–6 months due to alteration of flavor and texture during storage.

In another study conducted by Cavenaghi-Altemio et al. (2013) frankfurter-type sausages were produced from mechanically separated tilapia surimi. Different formulations were developed with tilapia viscera surimi and two with mechanically separated tilapia meat surimi, all with pig lard addition. Nile tilapia, *Oreochromis niloticus*, a tropical fish with great potential for aquaculture, was used. Nile tilapia processing has frozen fillets as a priority; however, its low yield (30%–35%) results in large amounts of waste used for food product development.

Technological challenges encountered in the cooking of tilapia viscera surimi halted the further investigation. The sensory evaluation scores revealed that both frankfurters using mechanically separated tilapia meat surimi obtained the highest acceptability scores (50.0%). *Listeria monocytogenes* and *Salmonella* spp. were found to be absent. The authors concluded that frankfurter-type sausages from mechanically separated tilapia surimi produced in the study presented favorable characteristics for commercial purposes. Likewise, Oliveira Filho et al. (2010), developed sausages with varying concentrations of minced fish from Nile tilapia filleting waste. The study evaluated the sensory attributes, lipid oxidation, total volatile nitrogenous bases, pH, and microbiological condition of sausages prepared with varying concentrations of minced fish. The acceptability of the color attributes decreased with the increasing concentration of minced fish.

Sausages prepared with 40% and 60% minced fish had the best flavor, texture, and overall acceptability scores. Like Cavenaghi-Altemio et al. (2013), there was no presence of pathogenic microorganisms, but a decline in pH and proliferation of aerobic psychrotrophic bacteria were registered during storage. However, this did not compromise sensorial properties. Guimarães et al. (2019) developed Brazilian flathead (*Percophis brasiliensis*) sausages and evaluated their acceptability for human consumption. Fillets and filleting residue were derived from Brazilian flathead samples. The filleting residue, which comprises bone skeleton and skin, was processed in a fish MSM machine. The formulation employed for the sausage was 50% surimi, 35% fillet + 15% ingredients. The constituents were mixed until an emulsion was obtained. Following emulsification the mixture was removed from the cutter mixer and brought to manual filler using collagen strings. The centesimal composition of the Brazilian flathead sausage was as follows: humidity 73.78%, protein 16.17%, lipids 4.41%, carbohydrates 3.75%, and energetic value 119.37 kcal/100 g. According to legislation in Brazil, standard sausages made with beef, pork, or poultry must have the following chemical composition values: maximum humidity of 65%, the minimum protein content of 12%, and maximum lipid content of 30%. The sausages derived from Brazilian flathead, therefore, met the protein requirements but not the humidity values. The microbiological and toxicological results were within the legislation standards. Sensory evaluation showed acceptability of 87.30%. Overall the results demonstrate that Brazilian flathead sausage displays adequate hygienic-sanitary conditions, good nutritional quality, and a high index of sensorial acceptance. Thus, underutilized, and discarded fish could be a viable alternative for developing safe and nutritious consumer products.

Piotrowicz and Mellado (2015) developed sausages from Argentine anchovy surimi and evaluated their physical, chemical, and nutritional properties. Ingredients used in the formulation of the sausages included sausage mix, stabilizer, flavor enhancer, colorant carmine, curing salt, and smoke flavoring, antioxidants salt, garlic powder, soy protein isolate, and oil. MSM derived from Argentine anchovy was processed into surimi with two treatments (sodium bicarbonate—SB; phosphoric acid—PA). The sodium bicarbonate and phosphoric acid treatments were washed with 0.5% sodium bicarbonate and 0.05% phosphoric acid. Soybean protein (4.0%), sunflower oil (12.0%), and water (9.0%) were manually mixed to obtain an emulsion and subsequently processed into sausages. The proximate composition of MSM and surimi of Argentine anchovy obtained by different washes showed a higher protein content in MSM (89.2 ± 3.1 g/100 g) compared to surimi washed with SB (57.1 ± 2.3 g/100 g) and PA (63.6 ± 4.7 g/100 g). MSM also had a significantly higher fat content of 3.1 ± 0.2 g/100 g. These initial results can be explained by the lower moisture content reported in MSM. The high moisture content in surimi was attributed to the washing process (Ismail et al., 2010). The authors also compared the proximate composition of sausages produced with Argentine anchovy surimi to commercial sausages. Results revealed commercial sausages had a higher fat content (15.6% ± 0.4%) compared to the surimi sausages. They also noticed that sausages obtained with surimi SB and PA presented high moisture, and high protein composition (15.5% ± 0.4%), respectively. Textural analysis of sausages indicated that commercial sausages were firmer (0.72 kg) compared to Argentine anchovy surimi sausages (SB = 0.25 kg, PA = 0.27 kg). The high moisture content in surimi sausages could account for its low firmness value. Huda et al. (2012) explain that the quantity of starch used in the formulation may influence the texture of the product because higher starch composition increases hardness value. Generally, Argentine anchovy sausages washed with PA exhibited similar features to the commercial sausage. They were considered healthier due to their low sodium content and lower energy value, considering the high protein value and reduced fat content.

Among the residues produced in fish filleting, MSM of fish can easily be processed into flour. Tilapia flour produced from meat adhering to the skins and bones yields approximately 8% of the weight of the fish. It is considered a low-cost source of essential nutrients, constituting an alternative for nutritional supplementation and adding value to some products such as meatballs, burgers, pasta, bread, and biscuits (Godoy et al., 2010; Monteiro et al., 2014, 2018; Stevanato et al., 2010). Monteiro et al. (2014) showed that flours and instant soup could be developed from *Oreochromis niloticus* wastes with excellent nutritious and technological potential. The developed products displayed high protein and ash contents, whereas low amounts of carbohydrates were compared with similar products already commercialized in the food industry. The products also showed encouraging technological properties such as water absorption index, particle size average, water activity, and pH, which are important parameters for elaborating innovative food formulation. Conventional flours can be replaced by this developed flour to produce a beneficial impact on consumers. Shaviklo et al. (2011) fortified ice creams with 50 g/kg and 30 g/kg fish protein (FP). Results indicated that the application of FP to ice cream did not alter the sensory attributes of the products after production. Fish-flavor and off-odor were increased after 2 months of storage. The protein content of fortified products was higher than in control. The authors also highlighted that sensory attributes of ice cream fortified with 5% FP were similar to ice cream fortified with 3% FP, which was already accepted by the panelists. This finding implies ice creams could be developed fortified with FP to enhance their nutritional and functional value. Similarly, seasoning of extruded corn snacks with functional fishery-derived ingredients could provide accepted corn snacks (Shaviklo et al., 2014). A study was designed to investigate the liking and sensory stability of seasoned extruded corn snacks with FP and omega-3 fish oil during 16 weeks of storage at ambient temperature. The results from the study indicated that seasoning corn snacks with fishery-derived products produced accepted nutritious corn snacks than the widely consumed regular corn snacks. Seasoning of extruded corn snacks with 18% FP powder 17% omega-3 fish oil did not influence the sensory attributes of the products after development and within 12 months of storage. The fortified prototypes were also liked by Iranian children between the ages of 7 and 12. Negative attributes indicative of oxidation such as off-odors and flavors were affected in the fortified prototypes after 12 weeks of storage; attributes suggesting quality (corn snack odor and flavor, and cheddar cheese odor and flavor) remained unchanged for all products during the study. The overall conclusion is that this fortification provides an avenue for corn snack diversification, which could be a healthy option to boost children's nutritional intake and to increase the consumption of FPs. Similarly, minced fish from Nile tilapia (*Oreochromis niloticus*) filleting with other ingredients were developed into a fish snack (Cortez Netto et al., 2020). Filleting waste is composed of the backbone, with no head, guts, and fins. The formulation of the snack was calculated by varying the minced fish/ starch ratios (60% cassava starch and 40% corn flour), as follows: treatment 1%–20% minced fish and 80% starch; treatment 2%–30% minced fish and 70% starch; and treatment 3%–30% minced fish and 70% starch. All treatments contained the same salt concentration of 2%. The water content was standardized as 60% in Treatment 1 and adjusted according to ingredients' moisture content (53.79% and 47.58% for treatments 2 and 3, respectively) for the other formulations. Raw corn flour was mixed with water from the formulation and cooked to obtain a firm dough. Consequently the cooked dough was mixed with other ingredients: minced fish, salt, and sour starch for further processing into the final product. Lipid oxidation was observed to increase in the snacks during storage, from 1.10 to 2.03 mg malondialdehyde/kg, with no significant difference among the different treatments. The high susceptibility of snacks to lipid oxidation is easily ascribed to high PUFA concentration. Furthermore the disruption of muscle membranes allows the interaction of oxidizing agents with PUFA, resulting in the propagation of oxidative reactions. A microbiological assessment revealed the absence of pathogenic bacteria, confirming good manufacturing and storage practices. *Salmonella* spp. was not detected, and total coliform bacteria counts were lower than the 3.0 CFU/g sample, without the presence of coliform bacteria. The coagulase-positive *Staphylococcus* counts were less than 10 CFU/g sample in all treatments. Sensory acceptance of the attributes flavor, color, texture, global acceptance, and purchase intent of the snacks decreased with increasing the inclusion of minced fish. Still, they remained with scores above 5, equivalent to "not liked nor disliked." Also the sensory properties of snacks were stable at 45 days during storage. From the study, it was evident snacks could be made with the inclusion of 20%–40% of minced fish of Nile tilapia filleting waste without altering the sensory acceptance, chemical, and microbiological stability of the product during 45 days of storage at 25°C ± 0.3°C.

Palmeira et al. (2014) evaluated the storage effect on the bacteriological and chemical quality of the restructured fish product, meatball-type, prepared with rainbow trout (*Oncorhynchus mykiss*) waste according to five different treatments using 1% transglutaminase, 4% textured soy protein, and partial replacement of sodium chloride by potassium chloride (75%/25%) and stored frozen at $-25°C \pm 2°C$ for 60 days. The treatments are elaborated below: starch addition (control); 1% transglutaminase; 4% soy protein addition; 4% soy protein addition and 1% transglutaminase; and 4% soy protein addition, 1% transglutaminase and partial replacement of salt sodium chloride by potassium chloride (75%/25%). The results showed no significant difference in microbiological quality and pH after storage. Correspondingly the pH value of treatments in the study was within the range of 6.0 and 6.19 over the 60-day storage, but no significant difference was observed. Samples with 4% soy protein addition, 1% transglutaminase, and partial replacement of salt sodium chloride by potassium chloride (75%/25%) presented the lowest total biogenic amine content (256.84 mg/kg), whereas samples with only 4% soy protein addition and 4% soy protein addition, 1% transglutaminase and partial replacement of salt sodium chloride by potassium chloride (75%/25%) had the highest value (791.36 and 707.19 mg/kg, respectively) in this parameter. The authors added that biogenic amines proved to be a potent quality indicator in the studied products, mainly in treatments with soy protein addition, and should be carefully evaluated in products prepared with fish waste and salt substitutes, because despite its health appeal the technological effect of this ingredient may be compromised. The study showed that the combined use of these ingredients could be applied as an alternative for elaborating new semiprepared products manufactured from trout wastes. Secci et al. (2017) conducted a study to evaluate the effect of the mechanical separation technique on the physicochemical properties of horse mackerel. Atlantic horse mackerel (*Trachurus trachurus*) were killed and minced by the belt-drum MSM machine in one-step separation. Meat from MSM was processed into fish burgers and analyzed during the storage period at $-20°C$ for quality parameters. Results showed mechanical separation inhibited water losses (2.67% against 4.57% and 5.57% in whole fillets and burgers from minced muscle, respectively). Color properties measured in MSM were duller and redder than samples from whole fillets and mined muscle. Due to the high PUFA content (52 g/100 g) of horse mackerel, all treatments were prone to oxidation. This observation was revealed by the high TBARS levels recorded in all samples (>8 mg MDA-eq/kg). Secci and Parisi (2016) indicated that increased lipid oxidation is attributed to the stress during fishing and the killing of fishes. Another explanation for the high TBARS value might be the presence of red muscle in horse mackerel, which is rich in myoglobin, a pro-oxidant (Maqsood et al., 2012; Tokur and Korkmaz, 2007). Despite the susceptibility to oxidative changes the authors concluded MSM could be a high-quality ingredient in burgers, nuggets, sticks, or even sauces, which would represent a way of improving the value of discard species.

Likewise, silver catfish (*Rhamdia quelen*, Heptapterdae) was formulated into fish burgers by replacing grounded fish fillets with pulp obtained from filleting wastes (PFW) (Bochi et al., 2008). The ventral muscles and backbone without fins were considered as the wastes of the filleting process. Filleting wastes were passed through a meat grinder fitted with a 2 mm plate to obtain fish pulp without visible fishbone. Formulation of the control fish burgers consisted of 88 g silver catfish grounded fillet and 12 g of textured soy protein. Three other treatment formulations were made by replacing the grounded fish fillets with fish pulp from filleting waste at different concentrations (20%, 50%, and 80%). Ingredients used in the formulation of all fish burgers were hydrogenated vegetable fat (5%), ice slurry (10%), wheat flour (5%), salt (1.25%), white pepper powder (0.04%), onion powder (0.1%), garlic powder (0.1%), coriander powder (0.2%), monosodium glutamate (0.5%), sodium ascorbate (0.3%), sodium tripolyphosphate (0.3%), and sorbitol (4%). Results revealed that fat content increased with increasing concentrations of PFW. Also the incorporation of PFW had no significant impact on fat and moisture retention after cooking. Similarly, texture and juiciness were not affected by PFW. Fish burgers formulated with 80% PFW, however, had lower overall acceptance than controls. Generally, fish burgers produced from 50% PFW had a higher nutritional value, better cooking characteristics, and the same acceptance of control fish burgers. The utilization of filleting waste from silver catfish can be a valuable alternative in developing food products with great acceptance and high nutritional value. Current consumer demand for healthy clean-label products inspired Husein et al. (2019b) to explore the need to valorize discarded fish. Four "clean label" formulations of fish burgers made with mechanically separated fish meat were characterized in sensory, textural, and chemical terms. MSM from European sea bass (*Dicentrarchus labrax*) and pigmented rainbow trout (*Oncorhynchus mykiss*) were used to prepare fish burgers. Other ingredients included in the formulated fish burger were rehydrated potato flakes, a starch providing texture, clear lemon juice, and sodium chloride. Operation parameters considered in the development of the recipes were the moisture content, fish-to-potato (dry matter ratio, DMR), and mechanically separated fish ratio between pigmented rainbow trout and sea bass. Generally, recipes with higher DMR were positively correlated with sandy, dry and crusty attributes, salty taste, and overall flavor. Fish burgers with a soft texture were perceived for recipes with lower DMR. As expected, lowering DMR decreased protein content while increasing ash and water composition. The results also revealed that 100 g of fish burger provided essential fatty acids than the recommended daily intake, irrespective of the formulation. The major changes detected seemed to be attributed to the fish-to-potato ratio. This study would optimize fish burger formulations to match the tastes and expectations of different consumer groups.

Freitas et al. (2012) used MSM from tilapia (*Oreochromis niloticus*) to formulate fish spread using either commercial salt or seasoned salt alongside other ingredients such as water, oil, and starch aimed at assessing consumer acceptability. A sensory evaluation based on a 9-point hedonic scale revealed lower scores for spreads formulated with common salt considering the overall impression, appearance, and flavor. Tilapia MSM-based spread formulated with seasoned salt was well accepted among consumers, mostly due to the improvement in its appearance and flavor.

Nevertheless, both products were well received. According to the technical regulations for microbiological standards, tilapia MSM presented sufficient quality as a raw material. Findings from this study are relevant to the fish industry in the production and marketing of new products obtained from Tilapia MSM.

Furthermore, de Oliveira et al. (2015) prepared fish flour from MSM from Brazilian catfish (*Brachyplatystoma vaillantii*). Residues of Brazilian catfish filleting comprising carcasses, pectoral fins, and scrap were deboned and the MSM, packaged and stored at approximately $-18°C$. Subsequently the MSM was processed into fish flour. Microbiological analysis of Brazilian catfish MSM and fish flour showed no *Salmonella* spp. or sulfite-reducing clostridia in the samples analyzed. The counts of coliforms at 45°C (<3.0 MPN/g) and coagulase-positive *staphylococcus* ($<1 \times 10^1$ CFU/g) were within the range set by the Brazilian legislation. The MSM analysis identified 17 fatty acids, of which the main ones were palmitic, oleic, stearic, palmitoleic, myristic, EPA, and DHA. Proximate analysis of MSM and Brazilian catfish flour showed that protein content ($9.52\ g \pm 0.07$ and $76.16\ g \pm 0.23$, respectively) and caloric value (137.44 and 374.80 kcal, respectively) per 100 g composition. The results suggest that byproducts from Brazilian catfish filleting to make fish flour resulted in a highly nutritious product. The addition of value to these byproducts has environmental, economic, scientific, and technological importance.

Pacu (*Piaractus mesopotamicus*) is a highly nutritious fish, but its availability to consumers does not meet the demand. Zuanazzi et al. (2017) prepared breaded portions of mechanically separated pacu, evaluating the physical, chemical, and microbiological parameters for a 120-day storage period at $-18°C$ and assessing its sensory properties after preparation of the breaded portions. The fish were beheaded, gutted, and the trunk was emptied and cleaned (without skin and fins). After immersion in chlorinated water the MSM was extracted and used in the formulation of breaded portions. Ingredients (MSM of pacu, starch, textured soy protein, dried parsley, monosodium glutamate, garlic powder, salt, water, fat, dehydrated onion, chili powder, sodium erythorbate, and oregano) were mixed, predusted, battered, and breaded. Breaded portions were stored for analysis after 30, 60, 90, and 120 days. Results showed higher moisture ($54.76\% \pm 2.49\%$) and lipid content ($18.24\% \pm 0.33\%$) in raw material (MSM) compared to breaded pacu MSM. This indicates moisture loss in raw pacu MSM after processing into breaded portions.

On the other hand, protein ($17.22\% \pm 0.39\%$) and mineral matter ($3.42\% \pm 0.23\%$) were found in higher amounts in breaded pacu MSM. The higher ash content is attributed to incorporating ingredients such as sodium chloride and monosodium glutamate into the breaded portions. The instrumental color values of breaded pacu MSM stored for 120 days ranged from 53.90 ± 0.68 to 56.11 ± 1.31, -7.34 ± 0.20 to -9.58 ± 0.55, and 21.41 ± 0.36 to 22.90 ± 0.76 for L^*, a^*, and b^*, respectively. Velho et al. (2009) observed that color considerably stimulates consumer interest in meat products. Thus it is of great essence to main the color parameters of the product without any significant changes, as it may be vital to its successful marketing. The values found for lipid oxidation ranged from 0.11 to 0.30 mg malonaldehyde/kg in breaded pacu stored for 120 days. The TBARS values were increased to 0.30 mg malonaldehyde/kg after 90 days and were reduced to 0.21 mg malonaldehyde/kg after 120 days. According to Osawa et al. (2005), TBARS can decrease over storage due to chemical interactions between malonaldehyde and proteins, making these proteins molecules insoluble components. Considering this event, secondary products originating from the oxidation of lipids cannot be detected by methods used, explaining the low values measured for TBARS. Throughout the 120-day period the breaded portions were suitable for consumption. Typical colonies were not observed on Baird-Parker agar plates. Sensory evaluation by tasters indicated that breaded MSM was well accepted. Additionally, they declared interest and intention to purchase. The results of the evaluated parameters for the prepared breaded pacu MSM were enough to make its consumption possible during frozen storage up to 120 days at $-18°C$.

8.6 QUALITY EVALUATION OF FISH PRODUCTS

Fish is a highly perishable commodity that starts deteriorating as soon as it dies. In tropical conditions, fish spoils quite rapidly, within a few hours of landing, if not adequately cooled (Lakshmanan, 2000). Rigor mortis accounts for all the quality changes a fish undergoes after its death. During this process, fish muscles become stiff a few hours after it dies. Digestive enzymes, surface bacteria that cause spoilage and partial and complete oxidation, are responsible for the degradation process (Prabhakar et al., 2020). Fish spoilage results from three primary mechanisms: enzymatic autolysis, oxidation, and microbial growth.

8.6.1 Enzyme Degradation

Enzyme degradation results in chemical and physical changes in fish species after they die. Autolytic enzymes such as ATPase and ADPase reduce the quality of the texture in the initial stages of deterioration but produce no off-odors and off-flavors. These endogenous autolytic enzymes act on fish muscle and cause substantial changes in the physicochemical properties in fish muscle during rigor mortis. This means autolytic deterioration occurs even with relatively low numbers of spoilage organisms. Most of the impact is on the textural quality with the production of hypoxanthine and formaldehyde as byproducts of autolysis. These result in tissue softening, cell wall rupturing, and blood with water draining out, which contain both protein and oil. These enzymes found in the muscle and viscera take responsibility for the postmortem degradation in fish muscle and fish products during storage and processing. Autolysis could also produce peptides and free amino acids, which would contribute to the spoilage of fish muscle as an outcome of microbial growth and production of biogenic amines (Benjakul et al., 2003; Fraser and Sumar, 1998; Ghaly et al., 2010; Prabhakar et al., 2020).

8.6.2 Oxidation

The high content of PUFAs in marine lipids makes it highly susceptible to rancidity. This is attributed to the degradation of PUFAs by autooxidation, and the processing of fish oils and fatty fish quickly leads to the formation of volatiles associated with rancidity. The rate and degree of the reaction are dependent on factors such as the fish species, the absence or presence of activators and inhibitors, and the treatment or storage. Rancidity produces off-flavors that are opposing factors for the production and commercialization of fish and foodstuffs containing fish oils (Pazos et al., 2005). Similarly, deterioration of flavor, texture, and the production of toxic compounds result from lipid oxidation (Kanner, 1994). Lipid oxidation is a three-stage free radical mechanism: initiation, propagation, and termination. Initiation involves forming lipid-free radicals through actions of catalysts such as heat, metal ions, and irradiation. These free radicals react with oxygen to form peroxyl radicals reacting with other lipids molecules to form hydroperoxides and a new free radical—the termination phase sets in when a buildup of these radicals interacts to create nonradical products.

Oxidation usually involves the reaction of oxygen with the double bonds of fatty acids, explaining why PUFAs are easily susceptible (Ghaly et al., 2010). During enzymatic oxidation, lipases cleave glycerides into free fatty acids, which produce off-flavor, frequently referred to as rancidity. The lipolytic enzymes could either be endogenous of the food product or derived from psychrotrophic microorganisms. The major lipase enzymes in fish lipid hydrolysis are triacyl lipase, phospholipase A2, and phospholipase B. These hydrolytic enzymes are present in the skin, blood, and tissues (Ghaly et al., 2010; Prabhakar et al., 2020). The nonenzymatic oxidation of lipids is ascribed to hematin compounds (hemoglobin, myoglobin, and cytochrome) and their catalysis generating hydroperoxides. The fatty acids produced during hydrolysis of fish lipids react with sarcoplasmic and myofibrillar proteins causing denaturation (Ghaly et al., 2010). Hemoglobin, a blood component, is reported to have the potential to promote or delay lipid oxidation. Pathways that implicate hemoglobin in promoting lipid oxidation have been described, including pseudolipoxygenase activity (Everse and Hsia, 1997). Hemoglobin can be a source of activated oxygen due to hemoglobin autooxidation, and heme or iron can be released from the protein to promote lipid oxidation. As mentioned earlier, metal particles could merge into MSM. Iron, acting similarly to other metal ions, could act as a catalyst for lipid oxidation, so the extent of its presence can compromise the quality of MSM during storage (Husein et al., 2019a,b).

8.6.3 Microbial Growth

The spoilage of fish initiates after the autolytic activity of endogenous enzymes. Microbial and lipid oxidation follow suit. Typical sources of bacteria flora that contributes to fish spoilage are the environment, skin, or digestive tract, or handling and processing. Thus storage conditions determine the activity rate of in situ bacteria present in the processed and packaged product. Large amounts of data generated have proposed three specific bacterial species to be the primary specific spoilage organisms (SSOs) in marine fish from temperate waters, namely *Shewanella putrefaciens*, *Pseudomonas* spp., *Photobacterium phosphoreum* (Reynisson et al., 2008). The composition of the microflora on newly caught fish depends on the microbial contents of the water in which the fish live. The microbial environment found typically in fishes includes Pseudomonas, Alcaligenes, Vibrio, Serratia, and Micrococcus. The growth and metabolic activities of these microorganisms result in fish spoilage and produce amines, biogenic amines such as putrescine, histamine and cadaverine, organic acids, sulfides, alcohols, aldehydes, and ketones with unpleasant and undesirable off-flavors (Ghaly et al., 2010). Spoilage in unpreserved fish is attributed to actions of gram-negative and fermentative bacteria (e.g., Vibrionaceae), whereas *Pseudomonas* spp. and *Shewanella* spp. are psychrotolerant gram-negative that grow on chilled fish. Carbon dioxide packing inhibits respiratory organisms and selects for *P. phosphoreum* and lactic acid bacteria. This points to the essence of clearly distinguishing between nonspoilage microflora from spoilage bacteria, as many of the bacteria present do no actually contribute to spoilage. Early studies of seafood microbiology acknowledged only a section of the spoilage microflora engaged in product deterioration; however, recent scientific discoveries of SSOs have contributed substantially to our understanding of seafood spoilage. The SSOs usually constitute a small fraction of the microflora on newly processed foods. The identification of SSOs is mostly dependent on their qualitative ability to produce off-odors (spoilage potential) and the quantitative ability to produce spoilage metabolites (spoilage activity). These sensory and chemical characteristics are critical in identifying SSOs (Gram and Dalgaard, 2002).

Trimethylamine (TMA), a volatile aliphatic molecule, is used generally to determine microbial deterioration leading to fish spoilage. Many fish species contain trimethylamine oxide (TMAO), used as an osmoregulation agent to avoid dehydration in marine environments and tissue waterlogging in freshwater. High quantities of TMAO are converted to TMA by bacteria such as *S. putrefaciens*, *Aeromonas* spp., psychrotolerant Enterobacteriaceae,

P. phosphoreum, and Vibrio spp. creating ammonia-like-off flavors and producing energy in the process.

In descending order of spoilage activity the spoilage compounds produced by specific spoilage bacteria are:

1. *S. putrefaciens*: TMA, hydrogen sulfide, methyl mercaptan, dimethylsulphide, hypoxanthine
2. *P. phosphoreum*: TMA, hypoxanthine
3. Pseudomonas spp.: Ketones, aldehydes, esters, nonhydrogen sulfides
4. Vibrionaceae: TMA, hydrogen sulfide
5. Aerobic spoilers: Ammonia, acetic, butyric, and propionic acid

As mentioned earlier, some spoilage metabolites can be used as quality parameters. The use of microbiological techniques in evaluating spoilage is slow compared to chemical analyses, which may be significantly faster. That being said, some compounds might not be produced in noticeable amounts in the early stages of food deterioration, as they are only generated in substantial quantities until close to spoilage. A classical single-compound quality index (SCQI) for seafood includes measurements of total volatile nitrogen (TVN), TMA, and hypoxanthine. Also the ratio between ATP degradation products (*K* value) and biogenic amines have even been used for some time as quality indices. Multiple-compound quality indices, in which statistical methods identify combinations of several metabolites, have recently been introduced and correlate better with sensory properties and shelf life in some products. This new combination of sophisticated chemical analyses, sensory assessment, and multivariate statistics will be a vital area in food spoilage research in the years to come (Gram and Dalgaard, 2002).

8.6.4 pH

pH variation is related to storage conditions and the procedures to which the fish is submitted immediately after its capture; therefore an isolated value is not an accurate representation of meat freshness or quality. According to Kyrana et al. (1997) the pH value for fresh fish meat is 6.0–6.5, increasing during storage because of the accumulation of metabolites. However, pH must be interpreted alongside other parameters. That being said, an increase in the pH value in food matrices is regarded as an indicator for loss of quality, indicating an accumulation of the metabolites mentioned above, including ammonia and other organic bases produced by the microbial and endogenous enzyme actions (Palmeira et al., 2014).

8.7 PRESERVATION TECHNIQUES

Fish is considered to be a problematic culinary object due to its ease of deterioration, susceptibility to oxidation, and the development of off-flavors. Methods of preserving the high nutritional quality of fish counts fundamentally on several parameters such as the diet of the fish, handling, and storage. Different types of preservation methods have been reported to improve the shelf-life of seafood and products. Ghaly et al. (2010) outline and describe the following fish preservation techniques: low-temperature storage, controlling water activity, controlling autolytic, enzymatic spoilage, gutting, the addition of sodium chloride, the addition of acids, use of phenolic antioxidants, using ethylenediaminetetraacetic acid, etc. Fish and fish product spoilage depend on some factors, as mentioned earlier. These factors need to be fully understood to develop effective and efficient handling and pretreatment methods and preservation techniques for food products.

8.8 OVERALL CONSUMER ACCEPTABILITY OF FISH PRODUCTS AND RTE FISH PRODUCTS

Currently, urban food consumers express more interest in food products available in RTE or ready-to-cook forms such as fillets, fingers, cutlets, patties, burgers, sausages, and fish balls (Kolekar and Pagarkar, 2013). The Food and Drug Administration directs that RTE foods should be in an edible form without an additional preparation step to achieve food safety. RTE products utilize byproducts of fish processing or unfavorite fish among consumers due to factors such as size, ugly shape, too much spiny body, and unfavorable taste. This, in turn, enhances their acceptability among consumers, their nutritious value, their sensory characteristics, and their shelf life and convenience (Husein et al., 2019a,b). Generally, food consumers lack time to prepare meals from scratch. The most common occasions among consumers for eating RTE meals are dinner, with the least likely being breakfast. Typically, we will find consumers prefer buying RTE products in shops they usually buy, which enhances their frequency of purchase. The quality of food such as flavor, appearance, texture, and odor; socioeconomic factors such as availability, price, and culture; biological factors such as energy and nutrient requirements; and psychological factors including behaviors, moods, and attitude toward eating influence food choice among consumers (Malhotra and Malhotra, 2014). The growing popularity of RTE foods against the traditional meal is pushed by factors such as reduced time spent in the kitchen and changes in family structures. The new status of women today, having advanced degrees and demanding jobs spend less time in culinary preparations and therefore are in high demand for RTE foods. Thus prepacked fixed weight portions with no waste and pre-prepared items such as fish portions with some culinary content are increasingly appreciated (Husein et al., 2019a,b).

The fish processing sectors are experiencing a drive that is forcing the promotion of RTE products. These products are derived from fish itself including beheaded, gutted, and filleted products, or from processed fish such as fish burgers, meatballs, nuggets, fish sticks, and different formulations of breaded fish. RTE products already available on the market are mainly codfish or Pollock. The perception of the consumer on seafood products would fall primarily on its healthful attributes. From this perspective, consumers consider fillets as the healthiest product, followed by fish burgers and fish nuggets. Fish burgers and fish nuggets are deemed to be processed products, and in fact, fish nuggets need to be fried before consuming, and fish burgers are consumed after grilling. The varying forms of presenting products impact the responses by consumers, and the health concerns associated with a product are expected to be a decisive factor for choosing a food product.

Furthermore the cost associated with buying RTE products could be a barrier to consumption. However, current consumers are willing to pay for these products. Thus the fish industry could exploit this demand from consumers (Cosmina et al., 2012).

8.9 CONCLUSIONS

There is an ever-increasing awareness about healthy foods, advancing knowledge of the prospects in fish processing, and acceptance of fish products. Product preference of consumers over the years has revealed the need for healthy foods, especially in the older population. MSM from fish waste would serve as the raw material for the development of RTE products like frozen surimi, frozen mince block, fish powders which would attend to the needs of growing consumers. The current food processing industry is well-equipped with the technology and technical know-how to drive RTE products into the market.

REFERENCES

Abd Aziz, N., Azlan, A., Ismail, A., Mohd Alinafiah, S., Razman, M.R., 2013. Quantitative determination of fatty acids in marine fish and shellfish from warm water of Straits of Malacca for nutraceutical purposes. BioMed Research International 2013.

Azahrani, M.H., Ananey-Obiri, D., Matthews, L., Tahergorabi, R., 2019. Development of low-fat fried fish using a two-prong strategy. CyTA – Journal of Food 17 (1), 882–891. Available from: https://doi.org/10.1080/19476337.2019.1661878.

Benjakul, S., Visessanguan, W., Tueksuban, J., 2003. Changes in physico-chemical properties and gel-forming ability of lizardfish (*Saurida tumbil*) during postmortem storage in ice. Food Chemistry 80 (4), 535–544.

Bochi, V.C., Weber, J., Ribeiro, C.P., Victório, A.D.M., Emanuelli, T., 2008. Fishburgers with silver catfish (Rhamdia quelen) filleting residue. Bioresource Technology 99 (18), 8844–8849.

Bogard, J.R., Thilsted, S.H., Marks, G.C., Wahab, M.A., Hossain, M.A., Jakobsen, J., et al., 2015. Nutrient composition of important fish species in Bangladesh and potential contribution to recommended nutrient intakes. Journal of Food Composition and Analysis 42, 120–133.

Borgogno, M., Husein, Y., Secci, G., Masi, S., Parisi, G., 2017. Technological and nutritional advantages of mechanical separation process applied to three European aquacultured species. LWT 84, 298–305. Available from: https://doi.org/10.1016/j.lwt.2017.05.068.

Bork, C.S., Venø, S.K., Lasota, A.N., Lundbye-Christensen, S., Schmidt, E.B., 2020. Marine and plant-based n-3 PUFA and atherosclerotic cardiovascular disease. Proceedings of the Nutrition Society 79 (1), 22–29.

Cabo, J., Alonso, R., Mata, P., 2012. Omega-3 fatty acids and blood pressure. British Journal of Nutrition 107 (S2), S195–S200.

Cavenaghi-Altemio, A.D., Alcade, L.B., Fonseca, G.G., 2013. Low-fat frankfurters from protein concentrates of tilapia viscera and mechanically separated tilapia meat. Food Science & Nutrition 1 (6), 445–451.

Chaudhury, R., 2010. Determinants of consumer behavior in buying RTE foods. Journal of Business and Retail Management Research 5 (1).

Cortez Netto, Oliveira Filho, Lapa-Guimarães, J., Viegas, E.M.M., 2020. Stability of snacks made with minced Nile tilapia stored at room temperature. Revista Ciência Agronômica 51 (1).

Cosmina, M., Demartini, E., Gaviglio, A., Mauracher, C., Prestamburgo, S., Trevisan, G., 2012. Italian consumers' attitudes towards small pelagic fish. New Medit 11 (1), 52–57.

Crawford, D.L., Law, D.K., Babbitt, J.K., McGill, L.S., 1972. Yield and acceptability of machine separated minced flesh from some marine food fish. Journal of Food Science 37 (4), 551–553. Available from: https://doi.org/10.1111/j.1365-2621.1972.tb02690.x.

De, D., Sandeep, K.P., Kumar, S., Raja, R.A., Mahalakshmi, P., Sivaramakrishnan, T., et al., 2020. Effect of fish waste hydrolysate on growth, survival, health of *Penaeus vannamei* and plankton diversity in culture systems. Aquaculture 524, 735240. Available from: https://doi.org/10.1016/j.aquaculture.2020.735240.

de Oliveira, I.S., Lourenço, L.D.F.H., Sousa, C.L., Joele, M.R.S.P., do Amaral Ribeiro, S.D.C., 2015. Composition of MSM from Brazilian catfish and technological properties of fish flour. Food Control 50, 38–44.

Delgado-Lista, J., Perez-Martinez, P., Lopez-Miranda, J., Perez-Jimenez, F., 2012. Long chain omega-3 fatty acids and cardiovascular disease: a systematic review. British Journal of Nutrition 107 (S2), S201–S213.

Demment, M.W., Young, M.M., Sensenig, R.L., 2003. Providing micronutrients through food-based solutions: a key to human and national development. The Journal of Nutrition 133 (11), 3879S–3885S.

Eckhoff, K.M., Maage, A., 1997. Iodine content in fish and other food products from East Africa analyzed by ICP-MS. Journal of Food Composition and Analysis 10 (3), 270–282.

EFSA, 2013. Scientific Opinion on the public health risks related to mechanically separated meat (MSM) derived from poultry and swine. EFSA Journal 11 (3), 3137. Available from: https://doi.org/10.2903/j.efsa.2013.3137.

Everse, J., Hsia, N., 1997. The toxicities of native and modified hemoglobins. Free Radical Biology and Medicine 22 (6), 1075–1099.

Fan, F., Xue, W.-Q., Wu, B.-H., He, M.-G., Xie, H.-L., Ouyang, W.-F., et al., 2013. Higher fish intake is associated with a lower risk of hip fractures in Chinese men and women: a matched case-control study. PLoS One 8 (2), e56849.

Fraser, O.P., Sumar, S., 1998. Compositional changes and spoilage in fish (part II)—microbiological induced deterioration. Nutrition & Food Science 98 (6), 325–329.

Freitas, D.D.G.C., Resende, A.L.D.S.S., Furtado, A.A.L., Tashima, L., Bechara, H.M., 2012. The sensory acceptability of a tilapia (*Oreochromis niloticus*) mechanically separated meat-based spread. Brazilian Journal of Food Technology 15 (2), 166–173.

García-Casal, M.N., Landaeta, M., Adrianza de Baptista, G., Murillo, C., Rincon, M., Bilbao, A., et al., 2013. Reference values of iron, iodine, zinc, selenium, copper, molybdenum, vitamin C, vitamin E, vitamin K, carotenoids and polyphenols for the Venezuelan population. Archivos Latinoamericanos de Nutricion 63 (4), 338–361.

Ghaly, A.E., Dave, D., Budge, S., Brooks, M., 2010. Fish spoilage mechanisms and preservation techniques. American Journal of Applied Sciences 7 (7), 859.

Ghaly, A.E., Ramakrishnan, V., Brooks, M., Budge, S., Dave, D., 2013. Fish processing wastes as a potential source of proteins, amino acids and oils: a critical review. Journal of Microbial & Biochemical Technology 5 (4), 107–129.

Gil, A., Gil, F., 2015. Fish, a Mediterranean source of n-3 PUFA: benefits do not justify limiting consumption. British Journal of Nutrition 113 (S2), S58–S67.

Gillingham, L.G., Harris-Janz, S., Jones, P.J.H., 2011. Dietary monounsaturated fatty acids are protective against metabolic syndrome and cardiovascular disease risk factors. Lipids 46 (3), 209–228. Available from: https://doi.org/10.1007/s11745-010-3524-y.

Godoy, L.C., Franco, M.L.R.S., Franco, N.P., Silva, A.F., Assis, M.F., Souza, N.E., Visentainer, J.V. 2010. Análise sensorial de caldos e canjas elaborados com farinha de carcaças de peixe defumadas: Aplicação na merenda escolar. Ciência E Tecnologia De Alimentos, 30, 86–89. Campinas, v. Available from: https://doi.org/10.1590/S0101-20612010000500014.

Gram, L., Dalgaard, P., 2002. Fish spoilage bacteria – problems and solutions. Current Opinion in Biotechnology 13 (3), 262–266. Available from: https://doi.org/10.1016/S0958-1669(02)00309-9.

Guimarães, J.L.B., Calixto, F.A.A., Keller, L.A.M., Silva, E., Furtado, A.A.L., & Mesquita, F.M. 2018. Development of a low commercial value fish-sausage from the fish trawling "mix" category. Food Science Technology, 39(Suppl. 1), 1–7.

Haase, H., Ellinger, S., Linseisen, J., Neuhäuser-Berthold, M., Richter, M., 2020. Revised D-A-CH-reference values for the intake of zinc. Journal of Trace Elements in Medicine and Biology 61, 126536. Available from: https://doi.org/10.1016/j.jtemb.2020.126536.

Haldimann, M., Alt, A., Blanc, A., Blondeau, K., 2005. Iodine content of food groups. Journal of food Composition and Analysis 18 (6), 461–471.

Hall, G.M., Ahmad, N., 1997. Surimi and fish-mince products. Fish Processing Technology. Springer, Berlin, pp. 74–92.

Hammad, S., Pu, S., Jones, P.J., 2016. Current evidence supporting the link between dietary fatty acids and cardiovascular disease. Lipids 51 (5), 507–517.

Harvey, L.J., Armah, C.N., Dainty, J.R., Foxall, R.J., Lewis, D.J., Langford, N.J., et al., 2005. Impact of menstrual blood loss and diet on iron deficiency among women in the UK. British Journal of Nutrition 94 (4), 557–564.

Hayashi, S., Satoh, Y., Ogasawara, Y., Dairi, T., 2020. Recent advances in functional analysis of polyunsaturated fatty acid synthases. Current Opinion in Chemical Biology 59, 30–36. Available from: https://doi.org/10.1016/j.cbpa.2020.04.015.

Holick, M.F., 2007. Vitamin D deficiency. New England Journal of Medicine 357 (3), 266–281.

Husein, Y., Parisi, G., Pietramellara, G., 2019a. Application of the Mechanical Separation Process in Different Fish Species for the Development of a New Product Based on Fish (PhD Dissertation). University of Florence, Italy. Available from: https://core.ac.uk/download/pdf/301577542.pdf.

Huda, N., Alistair, T.L.J., Lim, H.W., Nopianti, R., 2012. Some quality characteristics of Malaysian commercial fish sausage. Pakistan Journal of Nutrition 11 (8), 700.

Husein, Y., Secci, G., Dinnella, C., Parisi, G., Fusi, R., Monteleone, E., et al., 2019b. Enhanced utilisation of nonmarketable fish: physical, nutritional and sensory properties of 'clean label' fish burgers. International Journal of Food Science & Technology 54 (3), 593–601.

Huynh, M.D., Kitts, D.D., 2009. Evaluating nutritional quality of pacific fish species from fatty acid signatures. Food Chemistry 114 (3), 912–918. Available from: https://doi.org/10.1016/j.foodchem.2008.10.038.

Ismail, I., Huda, N., Ariffin, F., Ismail, N., 2010. Effects of washing on the functional properties of duck meat. International Journal of Poultry Science 9 (6), 556–561.

Jensen, I.-J., Larsen, R., Rustad, T., Eilertsen, K.-E., 2013. Nutritional content and bioactive properties of wild and farmed cod (*Gadus morhua* L.) subjected to food preparation. Journal of Food Composition and Analysis 31 (2), 212–216. Available from: https://doi.org/10.1016/j.jfca.2013.05.013.

Jones, G.B., Mercurio, P., Olivier, F., 2000. Zinc in fish, crabs, oysters, and mangrove flora and fauna from Cleveland Bay. Marine Pollution Bulletin 41 (7–12), 345–352.

Kanner, J., 1994. Oxidative processes in meat and meat products: quality implications. Meat Science 36 (1–2), 169–189.

Kerton, F.M., Liu, Y., Omari, K.W., Hawboldt, K., 2013. Green chemistry and the ocean-based biorefinery. Green Chemistry 15 (4), 860–871.

Khalili Tilami, S., Sampels, S., 2018. Nutritional value of fish: lipids, proteins, vitamins, and minerals. Reviews in Fisheries Science & Aquaculture 26 (2), 243–253.

Khoddami, A., Ariffin, A., Bakar, J., Ghazali, H., 2009. Fatty acid profile of the oil extracted from fish waste (head, intestine and liver) (*Sardinella lemuru*). World Applied Sciences Journal 7 (1), 127–131.

King, F.J., Carver, J.H., 1970. How to use nearly all the ocean's food. Commercial Fisheries Review 32 (12), 12.

Kirschnik, P.G., Trindade, M.A., Gomide, C.A., Moro, M., Viegas, E.M.M., 2013. Storage stability of Nile tilapia meat mechanically separated, washed, added with preservatives and frozen. Pesquisa Agropecuária Brasileira 48 (8), 935–942.

Knekt, P., Seppänen, R., Järvinen, R., Virtamo, J., Hyvönen, L., Pukkala, E., et al., 1991. Dietary cholesterol, fatty acids, and the risk of lung cancer among men. Nutrition and Cancer 16 (3–4), 267–275. Available from: https://doi.org/10.1080/01635589109514165.

Kolekar, A., Pagarkar, A., 2013. Quality evaluation of ready-to-eat fish ball in curry. SAARC Journal of Agriculture 11 (1), 35–43.

Kratky, L., Zamazal, P., 2020. Economic feasibility and sensitivity analysis of fish waste processing biorefinery. Journal of Cleaner Production 243, 118677. Available from: https://doi.org/10.1016/j.jclepro.2019.118677.

Krężel, A., Maret, W., 2016. The biological inorganic chemistry of zinc ions. Archives of Biochemistry and Biophysics 611, 3–19.

Kris-Etherton, P.M., 1999. Monounsaturated fatty acids and risk of cardiovascular disease. Circulation 100 (11), 1253–1258.

Kyrana, V.R., Lougovois, V.P., Valsamis, D.S., 1997. Assessment of shelf-life of maricultured gilthead sea bream (*Sparus aurata*) stored in ice. International Journal of Food Science & Technology 32 (4), 339–347.

Lakshmanan, P.T., 2000. Fish spoilage and quality assessment. In: Iyer, T.S. G., Kandoran, M.K., Thomas, M., Mathew, T. (Eds.), Quality Assurance in Seafood Processing. Society Fisher Technology, Cochin, pp. 26–40.

Leamy, A.K., Egnatchik, R.A., Young, J.D., 2013. Molecular mechanisms and the role of saturated fatty acids in the progression of non-alcoholic fatty liver disease. Progress in Lipid Research 52 (1), 165–174. Available from: https://doi.org/10.1016/j.plipres.2012.10.004.

Maares, M., Haase, H., 2020. A guide to human zinc absorption: general overview and recent advances of in vitro intestinal models. Nutrients 12 (3), 762.

Malhotra, G., Malhotra, A., 2014. An Empirical Investigation of Consumer's Preference in RTE Market, vol. 5. Centre for Business & Economic Research, London, pp. 1–14.

Maqsood, S., Benjakul, S., Kamal-Eldin, A., 2012. Haemoglobin-mediated lipid oxidation in the fish muscle: A review. Trends in Food Science & Technology 28 (1), 33–43.

Maret, W., Sandstead, H.H., 2014. Zinkbedarf und Risiko und Nutzen einer Zinksupplementierung. Perspectives in Medicine 2 (1–4), 3–18.

Martínez-Valverde, I., Jesús Periago, M., Santaella, M., Ros, G., 2000. The content and nutritional significance of minerals on fish flesh in the presence and absence of bone. Food Chemistry 71 (4), 503–509. Available from: https://doi.org/10.1016/S0308-8146(00)00197-7.

Mataix, J., García, L., Mañas, M., Martínez, E., Llopis, J., 2003. Tabla de Composición de Alimentos. Universidad de Granada, Granada.

Mendelsohn, J.M., 1974. Minced fish in a new form. Marine Fisheries Review 36 (8), 34–36.

Miyauchi, D., Steinberg, M., 1970. Machine separation of edible flesh from fish. Fishery Industrial Research 6 (4), 165–168.

Monteiro, M.L.G., Mársico, E.T., Lázaro, C.A., Ribeiro, R.O., Jesus, R.S., Conte-Júnior, C.A., 2014. Flours and instant soup from tilapia wastes as healthy alternatives to the food industry. Food Science and Technology Research 20 (3), 571–581.

Monteiro, M.L.G., Mársico, E.T., Soares Junior, Deliza, R., de Oliveira, Conte-Junior, C.A., 2018. Tilapia-waste flour as a natural nutritional replacer for bread: A consumer perspective. PloS one 13 (5)e0196665.

Moy, R., 2006. Prevalence, consequences and prevention of childhood nutritional iron deficiency: a child public health perspective. Clinical & Laboratory Haematology 28 (5), 291–298.

Murphy, C., Oikawa, S., Phillips, S., 2016. Dietary protein to maintain muscle mass in aging: a case for per-meal protein recommendations. The Journal of Frailty & Aging 5 (1), 49–58.

Nakamura, K., Nashimoto, M., Okuda, Y., Ota, T., Yamamoto, M., 2002. Fish as a major source of vitamin D in the Japanese diet. Nutrition 18 (5), 415–416.

Navarro, P., 1991. Valor nutritivo del pescado. I: Pescado fresco. Revista de Agroquímica y Tecnología de Alimentos 31 (3), 330–342.

Navas-Carretero, S., Pérez-Granados, A.M., Sarriá, B., Carbajal, A., Pedrosa, M.M., Roe, M.A., et al., 2008. Oily fish increases iron bioavailability of a phytate rich meal in young iron deficient women. Journal of the American College of Nutrition 27 (1), 96–101. Available from: https://doi.org/10.1080/07315724.2008.10719680.

Nerhus, I., Wik Markhus, M., Nilsen, B.M., Øyen, J., Maage, A., Ødegård, E.R., et al., 2018. Iodine content of six fish species, Norwegian dairy products and hen's egg. Food & Nutrition Research 62. Available from: https://doi.org/10.29219/fnr.v62.1291.

Nettleton, J.A., Exler, J., 1992. Nutrients in wild and farmed fish and shellfish. Journal of Food Science 57 (2), 257–260.

Oehlenschläger, J., 2012. Seafood: nutritional benefits and risk aspects. International Journal for Vitamin and Nutrition Research 82 (3), 168–176.

Oliveira Filho, P.R.C.D., Fávaro-Trindade, C.S., Trindade, M.A., Balieiro, J.C.D.C., Viegas, E.M.M., 2010. Quality of sausage elaborated using minced Nile tilapia submitted to cold storage. Scientia Agricola 67 (2), 183–190.

Oppedisano, F., Macrì, R., Gliozzi, M., Musolino, V., Carresi, C., Maiuolo, J., et al., 2020. The anti-inflammatory and antioxidant properties of n-3 PUFAs: their role in cardiovascular protection. Biomedicines 8 (9), 306.

Osawa, C.C., Felício, P.E.D., Gonçalves, L.A.G., 2005. Teste de TBA aplicado a carnes e derivados: métodos tradicionais, modificados e alternativos. Química Nova 28 (4), 655–663.

Osman, H., Suriah, A.R., Law, E.C., 2001. Fatty acid composition and cholesterol content of selected marine fish in Malaysian waters. Food Chemistry 73 (1), 55–60. Available from: https://doi.org/10.1016/S0308-8146(00)00277-6.

Özyurt, G., Polat, A., Loker, G.B., 2009. Vitamin and mineral content of pike perch (*Sander lucioperca*), common carp (*Cyprinus carpio*), and European catfish (*Silurus glanis*). Turkish Journal of Veterinary and Animal Sciences 33 (4), 351–356.

Palmeira, K.R., Mársico, E.T., Doro, L., Teixeira, C.E., Paschoalin, V.M.F., Monteiro, M.L.G., et al., 2014. Quality of semi-prepared products from rainbow trout waste (*Onchorynchus mykiss*) by using different technological strategics. Food and Nutrition Sciences 2014.

Palmeira, K.R., Mársico, E.T., Monteiro, M.L.G., Lemos, M., Conte Jr., C.A., 2016. Ready-to-eat products elaborated with mechanically separated fish meat from waste processing: challenges and chemical quality. CyTA – Journal of Food 14 (2), 227–238.

Payne, S.A., Johnson, B.A., Otto, R.S., 1999. Proximate composition of some north-eastern Pacific forage fish species. Fisheries Oceanography 8 (3), 159–177. Available from: https://doi.org/10.1046/j.1365-2419.1999.00097.x.

Pazos, M., Gallardo, J.M., Torres, J.L., Medina, I., 2005. Activity of grape polyphenols as inhibitors of the oxidation of fish lipids and frozen fish muscle. Food Chemistry 92 (3), 547–557. Available from: https://doi.org/10.1016/j.foodchem.2004.07.036.

Piotrowicz, I.B.B., Mellado, M.M.S., 2015. Chemical, technological and nutritional quality of sausage processed with surimi. International Food Research Journal 22 (5), 2103.

Prabhakar, P.K., Vatsa, S., Srivastav, P.P., Pathak, S.S., 2020. A comprehensive review on freshness of fish and assessment: analytical methods and recent innovations. Food Research International 133, 109157. Available from: https://doi.org/10.1016/j.foodres.2020.109157.

Prasad, M.M., Murugadas, V., 2018. Microbial quality and safety of fish and fishery waste. ICAR-Central Institute of Fisheries Technology.

Price, R., Souder, J., Brown, W., 1990. Preparation of fish pates utilizing mechanically separated fish flesh. Available from: https://www.aboutseafood.com/sites/all/files/S-024.pdf.

Puwastien, P., Judprasong, K., Kettwan, E., Vasanachitt, K., Nakngamanong, Y., Bhattacharjee, L., 1999. Proximate composition of raw and cooked Thai freshwater and marine fish. Journal of Food Composition and Analysis 12 (1), 9–16.

Reynisson, E., Lauzon, H.L., Magnusson, H., Hreggvidsson, G.Ó., Marteinsson, V.T., 2008. Rapid quantitative monitoring method for the fish spoilage bacteria *Pseudomonas*. Journal of Environmental Monitoring 10 (11), 1357–1362.

Roos, N., Leth, T., Jakobsen, J., Thilsted, S.H., 2002. High vitamin A content in some small indigenous fish species in Bangladesh: perspectives for food-based strategies to reduce vitamin A deficiency. International Journal of Food Sciences and Nutrition 53 (5), 425–437.

Roos, N., Chamnan, C., Loeung, D., Jakobsen, J., Thilsted, S.H., 2007a. Freshwater fish as a dietary source of vitamin A in Cambodia. Food Chemistry 103 (4), 1104–1111. Available from: https://doi.org/10.1016/j.foodchem.2006.10.007.

Roos, N., Thorseng, H., Chamnan, C., Larsen, T., Gondolf, U.H., Bukhave, K., et al., 2007b. Iron content in common Cambodian fish species: perspectives for dietary iron intake in poor, rural households. Food Chemistry 104 (3), 1226–1235. Available from: https://doi.org/10.1016/j.foodchem.2007.01.038.

Rubio-Rodríguez, N., Beltrán, S., Jaime, I., de Diego, S.M., Sanz, M.T., Carballido, J.R., 2010. Production of omega-3 polyunsaturated fatty acid concentrates: a review. Innovative Food Science & Emerging Technologies 11 (1), 1–12. Available from: https://doi.org/10.1016/j.ifset.2009.10.006.

Ruxton, C., Reed, S.C., Simpson, M., Millington, K., 2004. The health benefits of omega-3 polyunsaturated fatty acids: a review of the evidence. Journal of Human Nutrition and Dietetics 17 (5), 449–459.

Saify, Z.S., Akhtar, S., Khan, K.M., Pervreen, S., Ayattollahi, S.A.M., Hassan, S., et al., 2003. A study on the fatty acid composition of fish liver oil from two marine fish, *Eusphyra blochii* and *Carcharhinus bleekeri*. Turkish Journal of Chemistry 27 (2), 251–258.

Saini, R.K., Keum, Y.-S., 2018. Omega-3 and omega-6 polyunsaturated fatty acids: dietary sources, metabolism, and significance—a review. Life Sciences 203, 255–267. Available from: https://doi.org/10.1016/j.lfs.2018.04.049.

Satam, S., Sharangdhar, S., Sharangdhar, M., Sajid, I., Sonawane, U.D., 2004. Surimi: the "high-tech" raw materials from minced fish meat. Fishing Chimes 24 (8), 49–55.

Secci, G., Borgogno, M., Mancini, S., Paci, G., Parisi, G., 2017. Mechanical separation process for the value enhancement of Atlantic horse mackerel (*Trachurus trachurus*), a discard fish. Innovative Food Science & Emerging Technologies 39, 13–18. Available from: https://doi.org/10.1016/j.ifset.2016.10.018.

Secci, G., Parisi, G., 2016. From farm to fork: Lipid oxidation in fish products. A review. Italian Journal of Animal Science 15 (1), 124–136.

Shaviklo, G.R., Thorkelsson, G., Sveinsdottir, K., Rafipour, F., 2011. Chemical properties and sensory quality of ice cream fortified with fish protein. Journal of the Science of Food and Agriculture 91 (7), 1199–1204.

Shaviklo, A.R., Dehkordi, A.K., Zangeneh, P., 2014. Interactions and effects of the seasoning mixture containing fish protein powder/omega-3 fish oil on children's liking and stability of extruded corn snacks using a mixture design approach. Journal of Food Processing and Preservation 38 (3), 1097–1105.

Shearer, K.D., 1994. Factors affecting the proximate composition of cultured fishes with emphasis on salmonids. Aquaculture 119 (1), 63–88.

Siri-Tarino, P.W., Sun, Q., Hu, F.B., Krauss, R.M., 2010. Saturated fatty acids and risk of coronary heart disease: modulation by replacement nutrients. Current Atherosclerosis Reports 12 (6), 384–390.

Stevanato, F.B., Cottica, S.M., Petenuci, M.E., Matsushita, M., Desouza, N.E., Visentainer, J.V., 2010. Evaluation of processing, preservation and chemical and fatty acid composition of Nile tilapia waste. Journal of Food Processing and Preservation 34, 373–383.

Sugimoto, M.A., Sousa, L.P., Pinho, V., Perretti, M., Teixeira, M.M., 2016. Resolution of inflammation: what controls its onset? Frontiers in Immunology 7, 160.

Tacon, A.G., Metian, M., 2013. Fish matters: importance of aquatic foods in human nutrition and global food supply. Reviews in Fisheries Science 21 (1), 22–38.

Tahergorabi, R., Matak, K.E., Jaczynski, J., 2015. Fish protein isolate: development of functional foods with nutraceutical ingredients. Journal of Functional Foods 18, 746–756. Available from: https://doi.org/10.1016/j.jff.2014.05.006.

Teeny, F., Miyauchi, D., 1972. Preparation and utilization of frozen blocks of minced black rockfish muscle. Journal of Milk and Food Technology 35 (7), 414–417.

Thilsted, S.H., Thorne-Lyman, A., Webb, P., Bogard, J.R., Subasinghe, R., Phillips, M.J., et al., 2016. Sustaining healthy diets: the role of capture fisheries and aquaculture for improving nutrition in the post-2015 era. Food Policy 61, 126–131.

Tokur, B., Korkmaz, K., 2007. The effects of an iron-catalyzed oxidation system on lipids and proteins of dark muscle fish. Food Chemistry 104 (2), 754–760.

Velho, J.P., Barcellos, J.O.J., Lengler, L., Elias, S.A.-A., Oliveira, T.E.D., 2009. Disposição dos consumidores porto-alegrenses à compra de carne bovina com certificação. Revista Brasileira de Zootecnia 38 (2), 399–404.

Wu, G., 2016. Dietary protein intake and human health. Food & Function 7 (3), 1251–1265.

Yamagata, K., 2020. Effects of dietary n-3 polyunsaturated fatty acids on cardiovascular disease, Studies in Natural Products Chemistry, vol. 64. Elsevier, Amsterdam, pp. 217–240.

Yang, Z.H., Miyahara, H., Mori, T., Doisaki, N., Hatanaka, A., 2011. Beneficial effects of dietary fish-oil-derived monounsaturated fatty acids on metabolic syndrome risk factors and insulin resistance in mice. Journal of Agricultural and Food Chemistry 59 (13), 7482–7489.

Yang, Z.H., Miyahara, H., Iwasaki, Y., Takeo, J., Katayama, M., 2013. Dietary supplementation with long-chain monounsaturated fatty acids attenuates obesity-related metabolic dysfunction and increases expression of PPAR gamma in adipose tissue in type 2 diabetic KK-A y mice. Nutrition & Metabolism 10 (1), 16.

Yang, Z.H., Bando, M., Sakurai, T., Chen, Y., Emma-Okon, B., Wilhite, B., et al., 2016a. Long-chain monounsaturated fatty acid-rich fish oil attenuates the development of atherosclerosis in mouse models. Molecular Nutrition & Food Research 60 (10), 2208–2218.

Yang, Z.-H., Emma-Okon, B., Remaley, A.T., 2016b. Dietary marine-derived long-chain monounsaturated fatty acids and cardiovascular disease risk: a mini review. Lipids in Health and Disease 15 (1), 201. Available from: https://doi.org/10.1186/s12944-016-0366-5.

Zhao, L., Sun, J., Yang, Y., Ma, X., Wang, Y., Xiang, Y., 2016. Fish consumption and all-cause mortality: a *meta*-analysis of cohort studies. European Journal of Clinical Nutrition 70 (2), 155–161.

Zheng, J., Huang, T., Yu, Y., Hu, X., Yang, B., Li, D., 2012. Fish consumption and CHD mortality: an updated *meta*-analysis of seventeen cohort studies. Public Health Nutrition 15 (4), 725–737.

Zuanazzi, J.S.G., Marengoni, N.G., Botazzo, Á.C., & Delbem, J.A.F.D.L. 2017. Physical-chemical, microbiological, instrumental and sensory characterization of meat breaded (cms) of the pacu cultived in the pantanal.

Chapter | nine

Sustainable Fish and Seafood Production and Processing

Norhan E. Saleh, Elham A. Wassef and Heba H. Abdel-Mohsen

National Institute of Oceanography and Fisheries (NIOF), Fish Nutrition Laboratory, Aquaculture Division, Cairo, Egypt

9.1 INTRODUCTION

The UN identified that 850 million people worldwide need food. By 2050 the global population will increase to approximately nine billion, so no food must be lost from the food chain. The full exploitation of all-natural food resources, comprising those from the aquatic environment, will become increasingly obligatory as time passes (FAO, 2018).

In 2018 more than 88% (approximately 165 million tonnes) of the global fish production accounted for 179 million tonnes was consumed as food for humans, whereas 80% of the later amount (approximately 18 million tonnes) was transformed into fish meal (FM) and fish oil (FO). In contrast the rest (4 million tonnes) was primarily utilized for a variety of purposes such as: in aquaculture for feeding fry, fingerlings or on-growing, pet food products, food for livestock, zoo and fur animals, and also for the pharmaceutical industry (FAO, 2020) (Fig. 9.1).

According to the European Union (EU), Commission Council Directive 2008/98/E.C., "waste" is defined as "any substance, which is discarded or intends to discard or is required to discard. Harvested fish and other aquatic food sources are corruptible, shortly after catching, causing significant postharvest losses. Accordingly, they should be conserved or processed within a short time after harvest, and any delay may lead to loss of food source valid for human utilization. Moreover, some fish species and also parts of fish are often not adequate as food, so that will be wasted, adding to postharvest losses. Accordingly, seafood processors are looking to add value to the raw materials. Hence they applied several processing stages such as cutting, de- heading, de-skinning, gutting, and filleting.

Sustainable Fish Production and Processing. DOI: https://doi.org/10.1016/B978-0-12-824296-4.00002-5

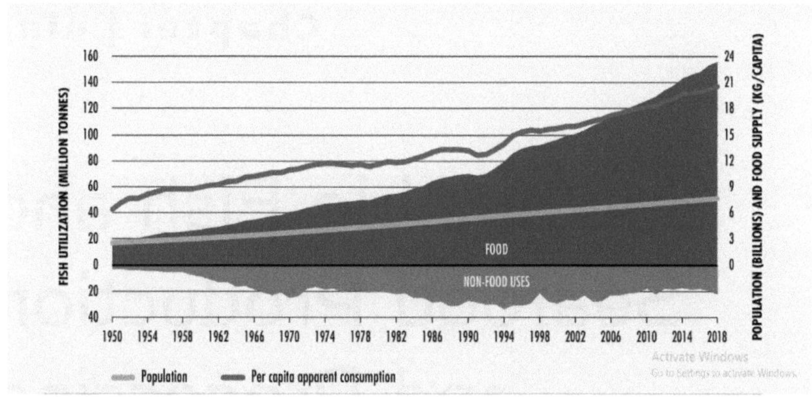

FIGURE 9.1 World fish utilization and apparent consumptions. FAO, 2020. The State of the World Fisheries Sustainability in Action. Food and Agricultural Organization, Rome.

The fish wastage-biomass usually consists of viscera (which represents about 12%–18% of total fish weight), heads (9%–12%), bones (9%–15%), skin (1%–3%), and scales (about 5%) (Dumay, 2006).

9.2 IMPACT OF THE FISH WASTES DISCHARGE ON THE ENVIRONMENT

Previously, fish by-products were often discarded as waste, however, recently, fishery wastes have become a great concern worldwide, and are influenced by various factors such as advancement in technology, biology, and economy (Kim and Mendis, 2006; Arvanitoyannis and Kassaveti, 2008).

Fish rendered products may cause environmental contamination, and even a large quantity of small-size fish can cause eutrophication in the aquatic environment. Accordingly, this massive amount of fish waste biomass may negatively influence the marine environment due to their massive microbial and endogenous enzyme load and their remarkable ability to degrade in a short time if not appropriately exploited, stored, or processed.

More recently, fish waste management has been one of the challenges facing the skate holders of environmental protection agencies or even an issue of public concern.

9.3 CLASSIFICATION OF THE FISH BY-PRODUCTS

As a consequence of the fish-processing industry's growth a remarkable amount of fish wastes biomass was resulted, representing upto 70% of processed fish and three-quarters of the total fish mass (Rustad et al., 2011). Fish by-products are susceptible to rapid decomposition unless appropriately processed or stored in favorable conditions, and can entail severe environmental and food-technical problems because of their high microbial and endogenous enzyme content (Arvanitoyannis and Kassaveti, 2008; Chalamaiah et al., 2012). Fish by-products can be categorized, according to Rustad et al. (2011), into two main types: the first is the wastes contained high enzymes level and are easily degradable; such as blood and viscera and the second comprises the more durable components (heads, bones, skin, and scales). In order to obtain high added-value outputs, fish wastes should be collected and treated within a short time after processing to ensure excellent quality to be used as raw materials (Rustad et al., 2011). In case that fish landing, processing, and production sites are geographically spread. Olsen et al. (2014) suggested that fish residues' processing should be conducted locally and immediately after being produced to create value from an otherwise waste product. This operation could make a considerable profit. However, equipped onboard fishing vessels should be available, and this is difficult to justify unless the markets for the new end-products are already present.

9.4 MANIPULATION OF WASTAGE AFTER FISH PROCESSING

Fish by-products can be utilized in different ways. For instance, directly or processed as food for human consumption, fillet cut-offs, heads, frames, and skin may be converted into soups, pâté, cakes, snacks, sausages, sauces, gelatin, and other food products. In some countries, tiny fish bones with little meat are salted and eaten as snacks. Additionally, fish processing residuals may be converted into low-quality commodities and less good value, such as FM, silage, or soil-fertilizer (Shahidi, 1994; Gómez-Guilén et al., 2002; Muyonga et al., 2004; Rustad et al., 2011). Nowadays, and due to the development of processing technologies, other applications for fish wastage have been initiated, and became a significant source of food and is expected to grow widely and adopted more efficiently (Al Khawli et al., 2019; Zhao et al., 2019). Due to the significant potential of using fish wastes biomass efficiently, several recent studies have been conducted on this topic (García-Moreno et al., 2017; Giannetto et al., 2020; Lopez-Pedrouso et al., 2020).

9.5 VALORIZATION OF SEAFOOD AND THEIR BY-PRODUCTS

There are several possibilities to valorize the marine by-products through processing to create more valuable ingredients or extract specific high-value compounds (FAO, 2002; Kelleher, 2005). Preserving marine fishery by-catch and by-products in the food chain can be adopted through the commercialization of discarded low-value components. By-products are utilized in the feed production (in the form of FM and FO, or converted into feed ingredients) and also as a supply of natural, healthy, and safe components that can be utilized in the pharmaceutical and nutraceutical industries (Kim et al., 2007; Ferraro et al., 2010; Shahidi and Ambigaipalan, 2015; Teves and Ragaza, 2016). This concept is a target toward a sustainable circular economy (or green approach). This creates the opportunity for efficient utilization of currently fish disposal, converting them into compounds that can be incorporated into aqua-or animal feeds or food and other valuable products.

By refining seafood discards, premium and bioactive ingredients for the manufacture of nutraceuticals can be acquired. The processing of fish proteins can produce bioactive peptides, amino acids, and other beneficial nitrogenous components (Harnedy and Fitz Gerald., 2012). Furthermore, the remains obtained from the FO refinery are excellent sources for the production of the essential highly unsaturated fatty acids concentrates, eicosapentaenoic acid (EPA) and docosahexaenoic acid (DHA), to be used as natural health food supplements (Kim and Mendis, 2006).

Prior studies reported that fish wastes biomass represents a significant source of biofunctional ingredients such as omega-3 oils, enzymes, polysaccharides, gelatin, and bioactive peptides (Messina et al., 2013; Pangestuti and Kim, 2017). Fish wastes are also a rich source of other valuable constituents such as dietetic compounds (chitosan), biofuel and biogas, cosmetics (collagen), natural pigments (after extraction), plastic alternates, and specific products for other manufactures (FAO, 2020). Furthermore, fish wastes are a rich source of enzymes and bioactive peptides that can be used as ingredients in animal feeds, fish silage, or fish sauce production. Fish proteolytic enzymes, generated from fish viscera, also have a growing market demand because of their variable applications, such as in food and pharmaceutical industries, in detergent, leather manufacture, as well as in the bioremediation operations (Mohanty et al., 2018). However, the biochemical composition of fish wastes biomass, that is, levels of protein, ash, and lipids, significantly differ between different source species (Le Gouic et al., 2018). Various valuable components extracted from fish wastes have been summarized in Table 9.1.

9.6 DEVELOPING TECHNOLOGIES FOR THE EXTRACTION OF BIOACTIVE COMPOUNDS FROM FISHERY BY-PRODUCTS

Seafood manufacturers want to see their fish and shrimps go to use, rather than literally going to waste. In order to properly utilize the marine fishery discards in a perfectly profitable manner, it is essential to apply the appropriate and safe techniques to extract the valuable nutrients and the bioactive compounds fully. Several methods are established to separate the bioactive compounds, hence valorizing fish by-products. The currently applied traditional techniques are among the downstream processing in the by-product refinery, which are widely used for the production of definite components, such as FM and FO (Schmidsdor, 1995), as well as oils rich in EPA and DHA (Breivik, 2007). However, the main disadvantages of these efficient methods are the potential thermal degradation of the target constituents due to the increased temperature during processing and the need for a high energy supply (Fig. 9.2). Another adopted technique for the extraction of bioactive components from disposing of fish is the use of organic solvents, which

TABLE 9.1 Valuable Components and the Utilization of Fish By-Products

By-Product	Valuable Components	Utilized as
Heads	Proteins, peptides, lipids, collagen, gelatine, minerals including calcium, flavor	Food, fish meal, FO, food-grade hydrolysates, animal grade hydrolysates, pet food, nutraceuticals, cosmetics
Frames (bones, flesh, fins)	Proteins, peptides, lipids, collagen, gelatine, minerals including calcium, flavor	Food, fish meal, FO, food-grade hydrolysates, animal grade hydrolysates, pet food, nutraceuticals, cosmetics
Trimmings	Proteins, peptides, lipids	Food, fish meal, FO, food-grade hydrolysates, animal grade hydrolysates, pet food
Viscera	Proteins, peptides, lipids, enzymes such as lipases	Food-grade hydrolysates, animal grade hydrolysates, fish meal, FO, fuel, fertilizers
Skin (with belly flap)	Collagen, gelatine, lipids, proteins, peptides, minerals, flavor	Fish meal, FO, cosmetics, food, fish meal, nutraceuticals, cosmetics, leather, fuel, fertilizers
Blood	Proteins, peptides, lipids, thrombin & fibrin	Fuel, fertilizer, therapeutants

Source: *Stevens, J.R., Newton, R.W., Tlusty, M., Little, D.C., 2018. The rise of aquaculture by-products: increasing food production, value, and sustainability through strategic utilisation. Marine Policy 90, 115–124.*

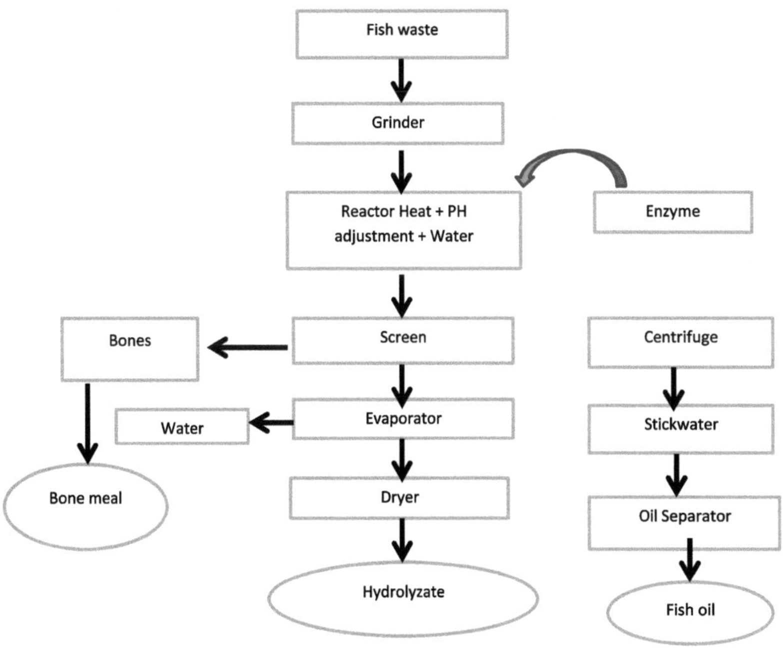

FIGURE 9.2 Flow diagram for the production of fish oil, bone meal, and hydrolysate from fish waste. Afreen and Ucak, 2020, Fish processing wastes used as feed ingredient for animal feed and aquaculture feed. Survey in Fisheries Sciences, 6(2), 55–64.

induce risks for both the environment and human health. These chemical extraction methods take prolonged extraction periods and may also result in undesirable compound degradation (Singh et al., 2015).

Conversely, enzymatic hydrolysis is a conventional method that is adopted for the extraction of natural and premium products from fishery discards. This method is applicable for solubilization and up the concentration of fish proteins, as mentioned by Aspevik et al. (2017). Soxhlet method with steam distillation and solvents is used for lipid extraction. Other adopted methods are characterized by their long extraction time and low yields, with high solvents and energy consumption, as well as potential health hazard aspects (Singh et al., 2015). These traditional methods involve extraction conditions (temperature, pH, concentration, extraction time, solvent type, etc.) anticipate changing the beneficial properties of the potentially valuable end-up products. Consequently, there is a great demand to search for more alternatives processing technologies that can conserve the aimed bioactive components (Galanakis, 2017; Ivanovs and Blumberga, 2017), evading as much as possible the use of the chemical solvents and operating at lower temperatures.

Recently, several technologies are available to convert fish biomass that is usually discarded, or resulted throughout the processing procedures as wastes, to valuable products for human consumption, for fish and animal feed, ingredients for the food industry, or pharmaceutical purposes. Nowadays, biotechnological practices such as fermentation and biocatalytic degradation have developed as a vital part of seafood manufacture. These treatment techniques serve not only as tools for recovering various valuable components but also as alternatives to the other costly physical, chemical, or mechanical methods used to benefit from seafood wastage (Suresh and Prabhu, 2012). The biotechnological approaches have been emerged to be a better way for fish wastes utilization than conventional methods that may lead to environmental pollution problems (Leduc et al., 2018). These latter processes are already established as eco-friendly methods that enable the recovery of more useful compounds from the raw materials aside from the target ones (Suresh and Prabhu, 2012).

The growing trend of green technology, supposing the utilization of more environmentally friend ways for manipulating seafood disposal, has further been highlighted in recent years (Ameer et al., 2017). Among the alternative extraction technologies that have been known as green extraction techniques for the extraction of high added-value components are supercritical fluid extraction, ultrasound-assisted extraction, microwave-assisted extraction, or pulsed electric fields (Poojary et al., 2016; Barba et al., 2020). These more recent technologies have many advantages, such as low chemical consumption rate (replacing with environmentally friend solvents), rapid extraction, unique constituents' recovery, and concomitant with higher selectivity.

Hou et al. (2017) reported that the addition of animal by-products, degraded through enzymatic, microbial, or chemical hydrolysis of proteins before feeding, is an appropriate way for obtaining high-quality functional peptides, which positively affect the nutrition and physiology of fish. These peptides of animal-source have shown antimicrobial, antioxidation, immune-modulatory and antihypertensive activities beyond their high nutritional value.

9.7 APPLICATIONS AND USES OF FISH WASTES

9.7.1 Uses of Marine Fishery By-products in the Animal Feed Industry

As mentioned before the formation of animal feed from unwanted products of seafood is beneficial for reducing environmental pollution and for providing cheap production of animals (Westendorf et al., 1998; Westendorf and Westendorf, 2000). Unwanted products of seafood that are used for animal feed composition are freely available because these are not used in human food and also a cheap source of feed element. Fish wastes biomass, which is usually discarded, can be considered as a promising novel source of protein in the production of aquafeeds.

In general, fish waste has proved to contain amino acids, fatty acids, vitamins, and minerals (Caruso, 2016). These are a rich source of protein and are used for the feed composition of aquaculture, swine, poultry, and other animal foods (Esteban et al., 2006). Seafood wastes obtained from small fish such as horse mackerel, chub mackerel, white croaker, flying fish, and sardine provide minerals, proteins, and fats sources (Rustad, 2003; Khan et al., 2003).

9.7.2 Fishery Wastes as a Source of Protein and Lipid

Fish meals and FOs can be produced from the fish as a whole, fish trimmings, or fish handling by-products. Small pelagic fish species, including *Peruvian anchoveta*, are caught in large volumes and mainly used in FM and FO production, in addition to a number of different other species (FAO, 2020). Therefore FM and fish oil production fluctuate according to the changes in yields of those species (FAO, 2002). Over time the application of appropriate management practices and accreditation schemes have decreased the number of unsustainable catches of species directed toward fish FM meal production. In 1994 the quantity of yield that is utilized for the production of FM and FO peaked at a level exceeded 30 million tonnes, then decreased in 2014 to be less than 14 million tonnes, and raised again to about 18 million tonnes in 2018. This progressive decline in supply has been coupled with the great demand for the fast-developing aquaculture industry, which has led to an increase in FM and FO prices (FAO, 2020). As a consequence, there is a growing share of FM and FO obtainable from fishery by-products. This fish waste biomass is currently estimated to produce approximately 25%–35% of the total bulk of FM and FO annual production, but with notable regional differences. For instance, in Europe, a relatively high proportion of fishery wastage, which accounted for about 54% of the total output, was exploited (Jackson and Newton, 2016). However, there is no expected increase in the raw material produced from whole wild fish (particularly, small pelagic). Therefore any elevation in FM production should be obtainable from fish discards, with varied nutritional values, being comparatively lower in protein but higher in minerals than FM manufactured from whole Fish (FAO, 2020).

FM and FO are still believed the utmost nutritious and highly digestible components in compound feed for aquaculture. FO is the primary source of EPA and DHA (EPA and DHA). However, the inclusion level of these latter fatty acids within aquafeeds has shown a clear decreasing trend, mainly as a result of supply and price variation, in addition to the continuously increasing demand from the aquaculture industry. More recently the incorporation of FM and FO in farmed fish grower diets is decreasing, however, they are selectively included at specific production phases, for instance, in hatcheries and for broodstock, and within finishing diets. The amount of FM and FO in diets for farmed grower Atlantic salmon, for example, is now often less than 10%. As demonstrated above a significant but declining proportion of fisheries production worldwide is processed into FM and FO (FAO, 2020).

9.7.2.1 FISH MEAL

Fish meal is a proteinaceous flour material. It is produced by cooking, pressing, drying, followed by grinding of complete fish or unwanted fish pieces to obtain Fish FM meal powder. FM is the primary protein source in aquafeeds, and concurrently it is a limiting factor in the aquaculture industry and conventionally used as a livestock feed supplement. For carnivorous fish species, aquafeeds have been mostly dependent on the FM as the principal protein source for several motives; for its superior essential amino acid profile, high protein content, better nutrients digestibility, and absence of antinutritional factors (ANFs). However, the inclusion of FM as the primary protein source in compounded aquafeeds is now at a hazard that threatens feed manufacturers to depend on it. Accordingly, feed formulators look for alternative feedstuffs that can replace an FM with no adverse effects on fish performance (Daniel, 2018). In the meantime, fish wastes meal is readily available, less expensive than FM, considered to be suitable, and has a sustainable supply for replacing FM in the commercial aquafeeds (Baeza-Ariño et al., 2016).

The efficacy of using trout-offal as a protein ingredient in gilthead bream *Sparus aurata* (L.) diets was researched by Kotzamanis et al. (2001). Trout-offal (intestine, frames, heads, skeletons, and fins) was finely minced and thoroughly mixed with the other dietary components to form a homogenous paste, then pelleted out. They concluded that trout-offal is a suitable, nonpolluting alternative for FM in seabream diets. Prior research with European sea bass, *Dicentrarchus labrax*, have evaluated different animal protein sources as dietary FM replacers such as fermented prawn waste liquor for juveniles at 30% of dietary FM (Nor et al., 2011), and a mixture of shrimp and tilapia protein hydrolysates associated with a combination of poultry by-products meal at 15% of FM (Robert, 2014). For other marine fish species, tuna-muscles by-product powder replaced 50% of dietary FM meanwhile, whole tuna by-products returned 30% of dietary FM without reducing the growth performance of olive flounder (*Paralichthys olivaceus*) (Uyan et al., 2006; Kim et al., 2014). Recently, Muttharasi et al. (2019) suggested that *Rastrelliger kanagurta*, *Sphyraena barracuda*, and *Fenneropenaeus* indicus disposal meals can be used as complete FM alternates to obtain low-cost feeds for farmed carp, *Cyprinus carpio*. In addition, Saleh et al. (2020) concluded that using a tilapia by-product can replace 30% of FM in the European sea bass (*D. labrax*) diet without compromising fish growth or health. Furthermore, fresh rendered fish can be utilized as a substitute ingredient for FM in swine feeds or other conventional protein sources (i.e., soybean meal and commercial FM) to partially fulfill their protein requirements (Esteban et al., 2006).

9.7.2.2 FISH OIL

FO is the second main ingredient in marine fish feeds, which can be extracted from fish and unwanted products of fish. FO is produced after the pressing of cooked fish, followed by subsequent centrifugation to obtain oil (FAO, 2020). The composition of FO is comprised mainly of triglycerides, phospholipids, wax esters, and glycerol ethers (Ruiter, 1995).

FO represents the primary source of long-chain polyunsaturated fatty acids (LC-PUFAs) goes into human consumption to carry out various essential roles linked to general health status and wellbeing. However, according to the estimation of the Marine Ingredients Organization, approximately 75% of the annual amount of FO globally produced is still directed to aquaculture feeds formulation (Auchterlonie, 2018). Due to FO output variability and associated changes in price, many researchers have evaluated other alternate PUFAs sources. Among them the sizable marine zooplankton stocks, such as the copepod *Calanus finmarchicus*, and Antarctic krill (*Euphausia superba*); however, several concerns remain over the effects on marine food webs. Nowadays, krill meal is directed to produce specific aquafeeds, and krill oil, mostly, is sold as a human nutrient supplement. However, there is a need to reduce krill raw materials fluoride content, as well as the cost of zooplankton production before their commercial inclusion as a standard used oil or protein ingredient in fish feeds (FAO, 2020), which may represent a practical challenge regarding its processing.

9.7.2.3 FISH SILAGE

A great deal of attention has been focused on exploiting fishery by-products, including fish filleting waste, by-catch, and low-grade industrial fish species. The ensiling technique may offer an alternative and inexpensive way for the direct utilization of these offal materials. Fish silage is a liquid product produced by mixing complete fish or some unwanted parts of fish with enzymes, acids, or required microorganisms. This process is carrying out with the action of fish enzymes (FAO, 2003). Mostly, fish offal remaining after fish handling will be an appropriate raw material for ensiling. It is essential to include fish viscera to provide enough enzymes for the hydrolysis process.

When the fish catch is available and cannot be utilized for direct human consumption for several reasons, it is used for the FM production, which has got ready market as an animal feed. However, if the catch is available at a place where there is no FM plant and no reasonable transport to the nearest plant, or the presence of severe restrictions on FM production because of fish odor, then it is mandatory to look for alternate processes. When animal farms are very near to landing centers, it is useful to go for silage production. The two applied techniques which have received the most attention are ensiling by either chemical acidification (conserved acid silage) or bacterial fermentation (fermented silage) and protein hydrolysis under controlled conditions employing selected exogenous enzymes (Tacon and Jackson, 1985). Both procedures rely on producing unfavorable conditions for the development of spoilage bacteria but advantageous conditions for protease enzymes (a low pH in the silage and a high temperature in the controlled hydrolysis).

Molasses is added to the fish or fish offal as a source of sugar. The final liquid product (which contains all water present in the original material) is stable and has excellent storage characteristics but with a malty odor. The ensilage process is simple and requires little capital equipment, mainly if nonoily fish are used. In the ensiling process of the fatty fish the temperature of the silage is raised to 70°C–90°C, then the oil is floated on the silage and then separated by decantation or centrifugation.

By using fish silage technology, small fish and fish by-products are not consumed as human food. Fish silage can be conserved and converted into a valued feed for use in aquaculture, chicken, livestock, and pork production. It also may be used in crop production as a natural fertilizer. Fish ensiling modifies fish wastes into a liquid blend of various hydrolyzed proteins, lipids, minerals, and also additional nutrients, which are simply digested and absorbed by aquatic and terrestrial animals. Silage manufactured by supplying an organic acid or adding a fermentable substrate can be stored for years and used on-demand (FAO, 2018). Fish silage has many advantages where it is an added source of animal feed, reduces the waste level making a positive impact on the environment, improves animal health by providing nutrients and bioactive compounds, and increases the economic output where wastes are transformed into a high-value product that can substitute expensive feed components in animal feeds (FAO, 2018).

The primary raw material to produce fish silage ought to be fresh and ensiled within a few hours after fish processing to the by-products. High-quality silage should be made from good-quality raw materials. If a low-quality raw material is used, the silage will not be suitable for animal feed production but can finally be utilized as fertilizer (FAO, 2018).

Fish silage is a mixture of hydrolyzed fish proteins and micronutrients. The enzymes are mostly liberated from the fish digestive system, and also to a less extent from the muscle and skin break down the proteins to amino acids and peptides. This cleavage makes the released nutrients content in silage easily digestible and highly absorptive for animals fed on diets supplied with silage.

In warm climates the oil will float on the surface of the silage, and if no supplementary antioxidant substance is used during the ensiling process, this oil will quickly be oxidized, reducing the silage quality. Generally, this oil ought to be separated from the produced silage. The separated oil is an excellent feed ingredient, particularly for aquaculture feed (FAO, 2002).

The high water content in the fish silage is one of its disadvantages, which increases the transport cost and the partial losses of tryptophan (Tacon and Jackson, 1985). Fat-free fish silage is composed of approximately 80% moisture, 15% protein, and less than 4% ash. Fish silage's nutritional properties are similar to FM but exceed the digestibility of the FM due to the presence of hydrolyzed proteins (FAO, 2018). The organic acids content in the silage shows the antibacterial property in the animal intestine and serves as a preservative for the silage itself. Due to the low acidity, fish silage can be fed directly without any previous treatment or mixing. Farmed fish should not be provided with processed protein from its species (FAO, 2018).

The use of fish silage in fish feeding has been extensively studied. It is characterized by its protein similarity with FM with lower production cost, so silage has high potential use in the aquaculture sector (Vidotti et al., 2003). Many authors assessed the nutritional value of diets containing microbial fish silage incompletely dehydrated by adding soybean meal, poultry by-products, or meat and bone powder. They recorded insignificant differences in the fish performance and protein utilization relative to diets based on FM as a primary protein source. The results showed that the experimental diets, particularly those diets containing silage and soy meal, possibly used in the North African catfish, *Clarias gariepinus* (carnivorous), and tilapia, *Oreochromis niloticus* (omnivorous) feeds, with no passive alterations in their performance, protein utilization and also carcass constituents (Fagbenro et al.,1994; Fagbenro and Jauncey,1998).

In salmon farming, seeking diets that induce fast growth and favor fish health to provide a high-quality product with low cost is mandatory. Many authors assess the use of fish silage in salmon (*Salmo salar*) diets and mentioned that although the fish silage did not stimulate fish growth, it did not induce a decline either, and obviously reduced the cost of the diet (Espe et al., 1992; Heras et al., 1994). Accepted results have been generally obtained with diets containing fish silage (Hardy et al., 1984; Arason, 1994). Several studies have evaluated the effects of using fish silage in fish diets on growth, feeding, and digestibility of many warm-water species such as tilapia, *O. niloticus* (Goddard et al., 2003; Wassef et al., 2001, 2003), Red tilapia, *Oreochromis mossambicus* × *O. niloticus* × *Oreochromis aureus* (Madage et al., 2015), the Indian carp, *Cirrhinus mrigala* (Ali et al., 1994), gilthead seabream, *S. aurata* (Wassef, 1991, 2005) and pacu, *Piriactus mesapotamicus* (Vidotti et al., 2002). The results indicated that fish silage is highly digestible and can replace upto

75% of FM in aquafeeds effectively. Silage is rich in free amino acids and peptides, owing to the highly hydrolyzed protein content, which has been induced better growth performance when incorporated in the fish diet, so it is recommended that fish silage partly substitutes FM in feeds. Also the use of fish silage for extruded feeds is well documented with accepted results. Incorporation of silage in meals has also revealed that the extruded pellets are more sturdy and higher resistant than ones without silage, thus decreasing feed losses due to dissolving or fractioning of the pellets before ingesting.

Fish silage is also used as cattle feed either as a whole mass, or the liquid portion can also be used. When a sustainable meal is desired, the silage is mixed with rice bran or other feed ingredients. Silage is a high-value feed ingredient with a unique potential that has been shown to improve feed quality for chicken and pork. As mentioned before the organic acids in fish silage have an antibacterial activity that enabling livestock to resist stress and diseases and thus inducing less animal mortality (FAO, 2018). This could help in eliminating the addition of nontherapeutic antibiotics in fish feeds. The presence of the free amino acids and peptides (predigested proteins) in the fish silage may result in enhanced growth. Fish feed formulators now substitute 5%–15% of FM with either fish silage or silage protein hydrolysates and also utilize silage oil in their produced feeds. Finally, making silage, wastes problems can be transformed to profit (FAO, 2018).

9.7.2.4 FISH PROTEIN HYDROLYSATES

Fish wastes are rich protein sources when converted into more nutritional value and bioavailable feed ingredients such as fish protein hydrolysate (FPH) (Benhabiles et al., 2012). FPH is a fish waste condensed high protein product which could be produced in two forms, either powder or liquid form, and contains a more considerable amount of small peptides. In unhydrolyzed product, these short-chain peptides are embedded within the parent proteins sequence, but they can be liberated by using the proteolytic enzymes such as trypsin, pepsin, alcalase, and protease (Sarmadi and Ismail, 2010; Chalamaiah et al., 2012) (Table 9.2) and furthermore, Wasswa et al. (2007) studied the effect of hydrolysis on functional properties of FPH. Idowu et al. (2020) reported that FPHs are valuable and rich sources of amino acids and peptides that are advantageous to human health. Generally, they have numerous bioactive potentials such as antimicrobial, antioxidant, anticoagulant, antitumor, calcium binder, and ACE inhibitor.

Fish dark muscles are highly exposed to oxidation and have off-flavor that result in low consumer demand and low marketing value of their artificial products (Chalamaiah et al., 2012). Consequently, some researchers have processed these dark fish muscles for the production of FPH, converting them into a high-value product (Naqash and Nazeer, 2013; Ghassem et al., 2014). Furthermore, many researchers investigated the possibility of extracting FBH from fish viscera and concluded that this raw material could be widely produced from marine and fresh fish viscera. (Bhaskar et al., 2008;

TABLE 9.2 The Chemical Composition of Dried Defatted Fish Protein Hydrolysate Produced from Whole Fish and/or Individual Parts of Fish

			Protein	Lipid	Ash	Moisture	
Persiansturgeon (*Acipenserpersicus*)	Viscera	Alcalase	65.82 ± 7.02	0.18 ± 0.4	7.67 ± 1.24	4.45 ± 0.67	Ovissipour et al. (2009)
Rainbowtrout (*Onchorhynchus mykiss*)	Viscera	Alcalase	88.32 ± 0.07	0.80 ± 0.60	1.14 ± 0.88	3.45 ± 0.02	Taheri et al. (2012)
Yellowfintuna (Thunnusalbacares)		Alcalase	72.34 ± 3.20	1.43 ± 0.57	2.82 ± 2.74	22.34 ± 1.38	Ovissipour et al. (2012)
Pacificwhiting, (*Merlucciusproductus*)	Muscle	Autolysis	85.6 ± 2.3	0.3 ± 0.1	16.6 ± 0.3	2.5 ± 0.6	Mazorra-Manzano et al. (2010)
Small-spottedcatshark, (*Scyliorhinuscanicular*)	Muscle	Alcalase Esperas and Protamex	89 ± 0.46	0.35 ± 0.06	1.11 ± 0.06	7.79 ± 0.70	Vazquez et al. (2017)
Tuna, anchovyandwild Fish	Tunaframe, anchovyand wildfishin aproportion of 5:4:1	Papainand Bromelain	69.94	1.77	17.5	3.03	Wu et al. (2018)
Atlantic salmon (*Salmo salar*)	Head	Alcalase2.4 L	82.3 ± 1.9	0.8 ± 0.02	10.4 ± 1.1	5.3 ± 0.2	Gbogouri et al. (2004)
Tuna	By-product	Flavourzyme	66.40 ± 0.27	2.37 ± 0.52	25.94 ± 0.04	7.25 ± 0.09	Nilsang et al. (2005)

Source: *Siddik, M.A., Howieson, J., Fotedar, R., Partridge, G.J., 2020. Enzymatic fish protein hydrolysates in finfish aquaculture: a review. Reviews in Aquaculture 1–25.*

Ovissipour et al., 2014; Villamil et al., 2017). Fish skin, as a by-product from fish processing, is rich in collagen and gelatin, and several studies supported the use of fish skin from different fish species to create hydrolysates (Ngo et al., 2010; Yin et al., 2010; Blanco et al., 2017). Fish heads are also a rich source of proteins, and some previous studies have tried to utilize the fish heads of different fish species and convert them into a hydrolysates source, such as for black parrotfish, *Chlorurus sordidus*, and skipjack tuna, *Katsuwonus pelamis* (Prihanto et al., 2019; Zhang et al., 2019). Several types of research evaluated the production of FPH from the frames and bones of different fish species, such as shortfin scad, *Decapterus macrosoma* (Šližytė et al., 2005), cod, *Gadus morhua (*Nazeer et al., 2011), and ribbonfish, *Lepturacanthus savala* (Kang et al., 2018). Additionally, some other studies utilize the entire fish body and by-products of various fish species for the production of FPH, such as round sardinella, *Sardinella aurita*, and Pacific hake, *Merluccius products* (Bougatef et al., 2010; Ho et al., 2014). The role of FPH in aquaculture production and the beneficial effects of its supplementation in aquafeeds on fish growth, feed intake, and utilization efficiency, and also biochemical composition are summarized in Siddik et al. (2020) review.

In aquaculture, dietary supplementation of FPH (rich in short-chain peptides), at proper levels, prompts fish growth performance, nutrient utilization, antioxidation status, and immune response (Tang et al., 2008; Zheng et al., 2013; Ospina-Salazar et al., 2016; Wei et al., 2016; Siddik et al., 2018). Siddik et al. (2020) reviewed most of the previous works studying the effects of using FBH on different fish species growth and immunity. In addition, peptides extracted from FPH have shown several fish physiological benefits comprising antioxidant, antimicrobial, antihypertensive, anticancer, and immunomodulatory activities when supplemented with food (Kang et al., 2019; Yaghoubzadeh et al., 2020).

9.7.2.5 FISH PROTEIN CONCENTRATE

Fish protein concentrate (FPC) is any stable proteinous product in which the protein is more concentrated relative to the original fish and manufactured for human consumption (http://www.fao.org/3/x5917e01.htm). FPC may be produced from any fish species or fishery wastes. FPC is prepared from fish by separating out the oil, screening, settling the bones out, and then drying. The rest product (FPC) is richer in protein (approximately 80%) and lesser in ash content relative to FM. FPC is comparable to fish hydrolysate except that the water and oil have been eliminated. The particle size of the FPC is less than FM and more homogenous in texture and color. FPC is more expensive than the FM due to the processing costs associated with its production, so it is only produced for human consumption and for some applications, primarily in milk replacers. Since the oil content of FPC is low so the unpleasant fishy taste is much less.

9.8 BIOACTIVE COMPOUNDS EXTRACTED FROM THE FISHERIES BY-PRODUCTS

(Fig. 9.3)

9.8.1 Minerals

Fish bones are excellent sources of calcium and several other minerals such as phosphorus, which can be utilized in preparing human food, animal feed, or even as food additives. Calcium phosphates, like hydroxyapatite, present in the fishbone can help in regenerating bones following surgery or significant trauma (FAO, 2020; Terzioğlu et al., 2018). Toppe et al. (2007) summarized the mineral composition of a wide range of fish species (Table 9.3).

9.8.2 Collagen and Gelatin

Fish skin and bones are primary sources of protein, including gelatin and collagen (Gómez-Guilén et al., 2002). Antifreeze proteins from polar fishes' skin tissue can be utilized in reducing the negative impacts that resulted from the frozen storage of meat (FAO, 2020).

9.8.2.1 COLLAGEN

Collagen is the largest high molecular weight protein in either invertebrate or vertebrate organisms (Silva et al., 2014). Collagen has been well studied and extracted from different marine sources, primarily from marine invertebrates.

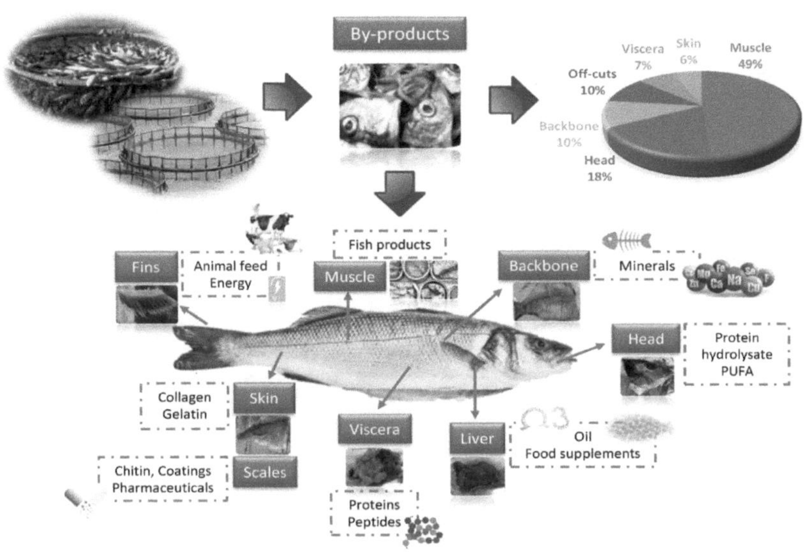

FIGURE 9.3 Fish processing by-product generation and end used opportunities. Al Khawli, F., Pateiro, M., Domínguez, R., José, M.L., Gullón, P., Kousoulaki, K., et al., 2019 Innovative green technologies of intensification for valorization of seafood and their by-products review. Marine Drugs 17(12), 689.

TABLE 9.3 Mineral Composition of Fish Bones Given as g/kg or mg/kg Lipid Free Dry Matter

	Cod	Saithe1	Blue Whiting	Salmon1	Trout	Herring (Small)	Herring (Large)	Mackerel	Horse Mackerel
Calcium (Ca)	g/kg 190	199	170	135	147	161	197	143	233
Phosphorous (P)	g/kg 113	108	87	81	87	94	95	86	111
Magnesium (Mg)	g/kg 3.0	3.0	3.2	2.2	2.4	2.6	2.9	2.6	3.6
Iron (Fe)	mg/kg 49	44	135	32	32	61	72	73	56
Zinc (Zn)	mg/kg 98	70	72	233	126	191	124	125	70
Copper (Cu)	mg/kg 1.0	1.2	3.0	0.9	0.9	2.6	0.8	2.2	0.5
Chromium (Cr)	mg/kg 10.8	9.8	16.9	5.5	6.7	3.6	2.4	3.9	5.5
Sodium (Na)	g/kg 7.7	7.1	4.6	5.7	5.8	3.3	7.8	6.5	7.1
Potassium (K)	mg/kg 5.2	4.9	2.6	8.2	7.7	5.0	7.7	6.7	4.4
Selenium (Se)	mg/kg n.d.	n.d.	n.d.	n.d.	n.d.	n.d.	0.3	n.d.	n.d.
Iodine (I)	mg/kg 3.7	2.6	1.4	2.7	2.5	1.2	3.6	2.2	2.1
Chlorine (Cl)	g/kg 4.8	4.1	1.9	4.4	4.2	1.4	4.5	4.1	4.0
Fluorine (F)	g/kg 0.19	0.17	0.07	0.10	0.09	0.03	n.d.	0.26	n.d.
Arsenic (As)	mg/kg n.d.	0.3	0.6	1.0	1.2	1.0	3.8	2.4	2.7
Cadmium (Cd)	mg/kg 0.01	0.02	0.13	n.d.	0.02	0.03	0.06	0.06	0.03
Mercury (Hg)	mg/kg 0.01	0.01	0.04	0.02	0.01	0.03	0.02	0.03	0.02
Lead (Pb)	mg/kg n.d.	0.07	0.12	n.d.	n.d.	0.26	0.09	0.24	0.11

Source: *Toppe, J., Albrektsen, S., Hope, B., Aksnes, A., 2007. Chemical composition, mineral content and amino acid and lipid profiles in bones from various fish species. Comparative Biochemistry and Physiology, Part B 146, 395–401.*

In marine vertebrate organisms, collagen extraction was instead focused on fish by-products including, skin (Ahmad and Benjakul, 2010; Hwang et al., 2007), scales (Chuaychan et al., 2015), bone (Nagai et al., 1999), cartilage (Jeevithan et al., 2014), and swim bladder (Liu et al., 2015). Additionally, large amounts of collagen are found in fish scales, skins (Ikoma et al., 2003), and also fins (Nagai and Suzuki, 2000).

Vertebrates contain high quantities of collagen in their bodies, which constitute about 30% of the total amount of protein (Birk and Bruckner, 2005). Actually, 27 kinds of collagen have been extracted and identified. Collagen type I is the most common one: it is recognized as fibrillar collagen and has a structural function as it is participating in the molecular shape, architecture, and even the mechanical characteristics of skin tissues (Birk and Bruckner, 2005). Owing to its excellent features (biocompatibility, biodegradability, safety, low allergenicity, antigenicity, and the capability to biofilm-forming); collagens are utilized in different industries such as pharmaceutical, medical, and cosmetics, and also as ingredients for food packaging (Cavallaro et al., 1994; Hood., 1987; Stainsby, 1987).

Firstly, collagen was mostly extracted from bovine and porcine sources. Later, people began to reticence toward these procedures owing to the religious background, besides the propagation of bovine spongiform encephalopathy (Wells et al., 1987), which represented a source of hazard contamination for the extracted protein (Jongjareonrak et al., 2005). Consequently, several researchers have been concentrated on the usage of marine collagen as an alternative because of the absence of dietary restriction and disease transmission (Jongjareonrak et al., 2005). For their extraction, collagens are frequently solubilized in organic acid, generally acetic acid, which extracts typically low collagen yield. Subsequently, researchers were concerned with collagen enzymatic extraction to enhance collagen solubilization and pepsin being amongst the most effective enzymes.

In the Mediterranean Sea the common coastal smooth-hound *Mustelus mustelus* is regularly caught all over the year either as a targeted species or a by-catch (Costantini et al., 2000; Bradai et al., 2006). The consumption of this species generates a significant amount of wastes that may be used as a source to extract collagen.

Slimane and Sadok (2018) extracted collagen from the skin of *Mustelus mustelus* and indicated the presence of collagen–chitosan interaction, which modified the secondary structure of collagen. Collagen–chitosan-based biofilm displayed prospective UV barrier properties and antioxidant activity, which can be used as green bioactive films preserving the nutraceutical products.

9.8.2.2 GELATIN

Protein is the major component of fish, and accordingly, fish by-products may represent the unconventional raw material for the production of high-protein ingredients, which are used for the manufacture of food-grade gelatin. Gelatin is attained by the collagen thermal denaturation, and it is widely used in the food industries to enhance the gelation procedure, water binding, foaming, and the emulsifying, viscosity, and elasticity properties of food (Jongjareonrak et al., 2006; Kittiphattanabawon et al., 2010). Furthermore, gelatin usage in edible film formation and encapsulation indicates its vital role in biomaterial-based packaging, pharmaceutical, and also the photographic industries (Gómez-Guilén et al., 2002; Jongjareonrak et al., 2006). Moreover, gelatin also has critical roles in maintaining joints, preventing osteoporosis, improving nail strength and growth, and promoting hair growth (Silvestrini, 1988). However, gelatin produced from cattle hides, pig skin, and cattle and pig bones has several restrictions in some religions such as Islam and Judaism, which prohibit the consumption of pork or slaughtered cow by-products (Zhou and Regenstein, 2005). Moreover, recently the outbreaks of diseases in the livestock accelerate the growing market need for alternative sources of gelatin (Muyonga et al., 2004). Although fish processing by-products are deemed a progressive source of gelatin, their utilization on a vast commercial scale is still restricted, mainly due to the low-quality properties of the obtaining gelatins relative to the mammalian source gelatin (Karim and Bhat, 2009).

9.8.3 Chitin and Chitosan

Finfish, crustaceans, and bivalves by-products are subjected to numerous applications, which increase the value of these by-products (FAO, 2020).

Chitin is a white hard inelastic nitrogenous polysaccharide compound extracted from crustacean shell waste, and it is the second most abundant organic compound on earth, next only to cellulose (Chawla et al., 2015). Squid, shrimp shells, and head waste constitute the single largest source of chitin (El Knidri et al., 2018). The crustacean exoskeletons contain 15%–20% chitin by dry weight.

Chitin and chitosan production from the food waste industries such as crustacean canning has become economically potential and environmentally attractive (Yadav et al., 2019), mostly when it comprises the retrieval of carotenoids. Substantial quantities of chitin are found in the by-products and are marketed as a fish feed supplement (Kumar, 2000). Shrimp head waste contained 20% ash, 64% crude protein, 18% lipids, chitin, and additional compounds; however, little amino acid degradation happened. The protein-rich materials may be utilized as supplements in monogastric animal feeds, and the remaining solids, which are rich in chitin material and calcium carbonate, can be used to extract chitin and chitosan. Chitin exhibits inherent exceptional characteristics, including biodegradability, biocompatibility, antimicrobial, antitumor, and also antioxidant activities (Casadidio et al., 2019; Morin-Crini et al., 2019).

Chitin is extensively used for several purposes. Chitin is a growth promoter for animals and birds and is used as an ingredient in their feed (Khempaka et al., 2011).

Chitosan, the acetylated chitin, is a natural flavoring material that can form ionic bonds and films. It is used in a wide range of applications, particularly in wastewater treatment, toiletries, cosmetics, beverages, food, agrochemicals, and furthermore in the pharmaceuticals industry (Morin-Crini et al., 2019).

Many types of research indicated that chitosan has several physiological functions in fish, including growth-promoting, immunostimulation, antioxidation, and antimicrobial effects. The previously published studies on these functions in various fish species are briefly reviewed by Abdel-Ghany and Salem (2020), and the significant results were represented in a comprehensive Table (Table 9.4).

9.8.4 Pigments

Several pigments such as astaxanthin and its esters, canthaxanthin, zeaxanthin, lutein, β-carotene, astacene are found in crustacean wastes. Some of these pigments have essential medical and biomedical uses owing to their high antioxidant activities and such as vitamin A precursors (FAO, 2020). Díaz-Jiménez et al. (2019) concluded that the supplementation of 1% of astaxanthin in the peppermint shrimp (*Lysmata wurdemanni)* diet improved growth, survival, and enhanced quality of the fertile eggs. However, the incorporation of other pigments such as β-carotene can induce beneficial effects when supplemented at levels higher than 1%. Also Rahman et al. (2016) indicated that using the dietary astaxanthin increasing the pigmentation and the antioxidant status in rainbow trout (*Oncorhynchus mykiss*), which consequently exerts many beneficial effects on human health. Saleh et al. (2018) reported that adding astaxanthin at 0.1 g/kg diet to European seabass (*D. labrax*) diets improving fish growth, survival, immunity, antioxidant competence, and stress tolerance against physical stressors.

9.8.5 Essential Amino Acids

The amino acid composition of the shrimp wastes and the powder attained after fermented shrimp wastes lyophilization is containing all the essential amino acids except tryptophan, which is lacking in both the original shrimp wastes and the fermented liquor (Bhaskar et al., 2008). The composition of fermented liquor is recommended for common carp (*C. carpio*) juveniles (NRC, 1993). It can be used as a growth promoter and act as an immunostimulant agent in Farmed Fish (Amar et al., 2000). Furthermore, shrimp waste protein hydrolysate is known to be nutritionally valuable as a feed ingredient due to its high content of the essential amino acids (Gildberg and Stenberg, 2001). During degradation the amino acid composition of the fermented liquor is also composed of essential amino acids that are necessary for marine shrimp, *Penaeus monodon* (Millamena et al., 1999). Later, Sachindra and Bhaskar (2008) have proved the in vitro antioxidant activities revealed by the fermented liquor powder obtained following lyophilization of the fermented shrimp by-products. These properties obviously pointed to the potential significance of this component as a nutritional and health-improving ingredient in the diets of aquaculture fish and penaeid shrimp. According to Fanimo et al. (2000), shrimp waste meal comprising heads, appendages, and exoskeleton are rich sources of lysine while the squid (*Loligo vulgaris*) is also an essential source of high-quality protein (Gokoglu et al., 2017). Red crab (*Pleuroncodes planipes*) is small in size, so it is not utilized as human food, but it is considered a valuable source for animal feeds due to its high nutritive

TABLE 9.4 Dietary Chitosan Concentrations Recommended for Farmed Fish and Shrimp

Species	Fish Size (g)	Tested Concentrations (g Chitosan kg/diet)	Recommended Concentration (g Chitosan kg/diet)	Effects	References
Rainbow trout (*Oncorhynchus mykiss*)	25.0	2.5, 5.0 and 10.0	2.5	Improved the hematological parameters and resistance against some environmental stresses.	Meshkini et al. (2012)
Sea bass (*Dicentrarchus labrax*)	0.21	0.5, 1.0, 2.0, 3.0 and 4.0	1.0 and 2.0	Decreased the total bacterial counts sharply.	Salem (2015)
	0.21	0.5, 1.0, 2.0, 3.0 and 4.0	1.0	Enhanced the growth performance and reduced mortality.	Zaki et al. (2015)
	0.05	5.0, 10.0 and 20.0	10.0	Enhanced the performance and reduced the fish mortality under stress conditions.	El-Sayed and Barakat (2016)
Asian seabass (*Lates calcarifer*)	2.0	5.0, 10.0 and 20.0	10.0	Chitosan-fed groups, challenged against *Vibrio anguillarum*, had higher hematological and innate immune parameters than the control group	Ranjan et al. (2012)
Gray Mullet (*Mugil cephalus*)	12.01	5.0, 10.0 and 15.0	10.0 and 15.0	Improved growth, hematology, and blood biochemical indices and innate immunity responses.	Akbary and Younesi (2017)
Juvenile loach (*Misgurnus anguillicaudatus*)	0.18 g	5.0, 10.0, 20.0 and 50.0	50.0	Enhanced growth, lipid metabolism, immune response and antioxidant-related gene expression.	Yan et al. (2017)
Common carp (*Cyprinus carpio*)	42.4	0.0025, 0.005 and 0.01	0.005 and 0.01	A positive effect on some immune parameters and resistance against bacterial infection.	Alishahi et al. (2014)
	45.0	10.0, 20.0 and 50.0	20.0	Promoted the nonspecific immunity, decreased the fish mortality and enhanced the growth under stress conditions.	Maqsood et al. (2010)
Juvenile carp (*Allogynogenetic crucian*)	8.0–11.0 g	1.0, 5.0, 7.5, 10.0 and 20.0	5.0 and 7.5	Some detected pathogenic bacteria either decreased or totally disappeared	Li et al. (2011)
Gibel carp (*Carassius auratus gibelio*)	4.80	1.8, 4.0, 7.5, 10.0 and 20.0	4.0 and 7.5	Improved fish growth performance, hematology, intestine morphology and immune response.	Chen et al. (2014)
Nile tilapia (*Oreochromis niloticus*)	5.66	0.0, 1.0, 3.0 and 5.0 (chitosan nanoparticles)	5.0	Decreased the total anaerobic and aerobic intestinal bacterial count. Improved the growth performance, and feed utilization. Enhanced the immune response.	Abd El-Naby et al. (2019)
Kelpgrouper (*Epinephelus bruneus*)	26.0	10.0 and 20.0	All concentrations	All concentrations of chitin and chitosan stimulated the hematological and immunological parameters.	Harikrishnan et al. (2012)
(Rutilusfrisiikutum, Kamenskii)	1.7	0.25, 0.5, 1.0 and 2.0	1.0 and 2.0	Enhanced the feed conversion ratio and survival rate. Increased Leukocyte count. Enhanced the intestinal villus length.	Najafabad et al. (2016)
Shrimp (*Litopenaeus vannamei*)	0.0012	0.5, 1.0, 2.0 and 4.0	2.13 and 2.67	Enhanced shrimp growth, improving the digestion and absorption of nutrients.	Niu et al. (2011)
Shrimp (*Penaeus monodon*)	1.49	4.0	4.0	Chitosanasan immunostimulant was superior to chitin orchitosan oligosaccharides.	Niu et al. (2013)

Source: *Abdel-Ghany and Salem (2020). Effects of dietary chitosan supplementation on farmed fish: a review. Reviews in Aquaculture, 12(1), 438–452.*

value. It contains a high amount of protein and minerals. Villarreal et al. (2004) reported good results when supplementing postlarvae and juvenile *Farfantepenaeus californiensis* feeds with red crab meal.

The head and shell of shrimp and also other crustaceans form the significant fishery waste when crustaceans are abundant. This waste contains a valuable percentage of protein, chitin, and also minerals. The proteins that are extracted to be converted into shrimp extract are combined with flavor-bearing compounds having potential use as a natural flavoring material (Suryawanshi et al., 2019).

9.8.6 Fatty Acids

During the FO refinery process the FO by-products can serve as necessary materials for the production of the essential highly unsaturated fatty acids concentrates, EPA, and DHA, to be used as food additives (Tocher et al., 2019). Extracted shrimp oil has a high content of health-promoting polyunsaturated fatty acids (PUFA), in particular omega-3 fatty acids (EPA) and DHA (Aneesh et al., 2020). Furthermore, muscles and gonads of Argentine red shrimp (*P. muelleri*) were documented to be rich in health beneficial long-chain omega-3-PUFA (HUFA) (Saini et al., 2020).

Adloo et al. (2012) results indicate that enrichment of *Artemia fransiscana* with HUFA increases survival in yellowfin seabream (*Acanthopagrus latus*) larvae at its first feeding.

9.8.7 Bioactive Peptides

The extracted proteins from the fish flesh contain peptides that possess many bioactivities such as antithrombotic, antihypertensive, antioxidant, and immune-modulatory (Kim et al., 2000; Kouhdasht and Moosavi-Nasab, 2018). Furthermore, these bioactive peptides have antiplatelet and anticoagulant properties, which are the principal reason for the ability of the peptides extracted from the fish to suppress clotting factors in the principle coagulation pathway (Je et al., 2004). The protein liberated from the fish muscle by the enzymatic hydrolysis possesses biologically active peptides that induce several nutritional and functional properties (Benkajul and Morrissey., 1997).

9.8.8 Enzymes

The different internal organs of the fish secrete several enzymes that show high catalytic activities at moderately low concentrations, such as trypsin, chymotrypsin, pepsin, and collagenase. These enzymes are separated from the fish viscera and available on a commercial wide scale (Kim et al., 2000; Zhou et al., 2011).

Monier (2020) studied the efficacy of dietary exogenous enzymes mixture (HX) supplementation on growth performance, digestive enzymes, and antioxidant activity of common carp (*C. carpio*) fry and concluded that the dietary HX at 1 g/kg diet supplementation level might be added to fish diets as a feed supplement to improve growth performance and health status of typical carp fry.

9.9 ROLE OF THE EXTRACTED BIOACTIVE MATERIALS IN HUMAN LIFE

Marine by-product-derived compounds have induced acute positive effects on human health attributed to their, for example, antidiabetic, anticancer, antihyperglycemic, antidepressant, antihypertensive, antimicrobial, antiinflammatory, antirheumatoid antioxidant, antiproliferative, and also immunomodulatory properties (Ishak and Sarbon, 2018; Létisse and Comeau, 2008; Zamora-Sillero et al., 2018). Furthermore, besides their biological activities, they are exploited by nutraceutical, cosmeceutical, and pharmaceutical industries (Atef and Ojagh, 2017).

Marine by-product compounds can also show critical technological properties when incorporated in some food products, increase solubility and water holding capacity, acting as emulsifying and foaming agents, and facilitating fat binding (Halim et al., 2016; He et al., 2013). Latest data indicate that it is the potential to alter the technical properties of fish burgers (juiciness, hardness, adhesiveness, cohesiveness) by adding little quantities of fish by-product protein powder or FPH (Hansen et al., 2019). There are numerous possibilities to valorize marine by-products through processing, as producing more valuable ingredients or either extracting definite high-value compounds.

The concept of the sustainable circular economy (green approach) may be achieved by preserving useless marine catch and by-products in the food chain. The commercialization of the by-products of low-value, or the production of foodstuffs and valuable bioactive molecules that can be utilized in the nutraceutical and pharmaceutical industries (Teves and Ragaza, 2016), satisfying this approach. The green process permits an efficient usage of fish by-products, converting them into ingredients that can be assimilated into food, animal feed, or additional valuable products (Fig. 9.4). Nowadays, using fish by-products as ingredients in animal feeds (flour and oil) is the most common practice (Aspevik et al., 2017). Finally, fish wastes may also be consumed in biofuel production or be consumed in other industrial and agronomic applications, such as fertilizers production (Kudre et al., 2017). The healthy compounds and their favorable properties associated with fish also exist in their by-products. Numerous bioactive compounds can be attained from fish by-products: protein, peptides, gelatin, collagen, chitin, PUFAs, minerals, enzymes, glycosaminoglycans, and vitamins. It should be considered that the long-chain omega-3 fatty acids, EPA and DHA, are amongst the most valuable compounds separated from fish by-products, reaching a high marketable value due to their positive effects on health (Harnedy and Fitz Gerald, 2012). Marine by-product ingredients can also provide desirable technical properties when incorporated in several food products, acting, for example, as emulsifying and foaming factors, facilitating lipid binding, solubility, and water carrying capacity (Halim et al., 2016).

9.10 CONCLUSION

Fish wastes management is an attractive topic, where it is the way to solve the severe impacts of fishery discards on the environment and prevent losing nutrients from the food chain. The use of fish by-products and fishery wastes allows the future decline of wastes that else will be dumped away in the aquatic environment, causing overload nutrients enrichment and finally water eutrophication. At the same time, they can be used as a source of aquafeeds for farmed fish, promoting aquaculture development and sustainability. Furthermore, fisheries waste and by-products are a fundamental source of complementary high-added valuable compounds, however possible hazards correlated to the presence of contaminants should be considered before their exploitation. Many techniques are used to treat fish wastes, such as mechanical, biotechnological techniques, ultrasound-assisted extraction, or microwave-assisted extraction technologies to be suitable for human and animal feeding.

REFERENCES

Abd El-Naby, F.S., Naiel, M.A.E., Al-Sagheer, A.A., Negm, S.S., 2019. Dietary chitosan nanoparticles enhance the growth, production performance, and immunity in *Oreochromis niloticus*. Aquaculture 501, 82–89.

Abdel-Ghany, H.M., Salem, M.E.S., 2020. Effects of dietary chitosan supplementation on farmed fish: a review. Reviews in Aquaculture 12 (1), 438–452.

Adloo, M.N., Matinfar, A., Sourinezhad, I., 2012. Effects of feeding enriched *Artemia Fransiscana* with HUFA, Vitamin C and E on growth performance, survival and stress resistance of yellowfin seabream larvae. Journal of Aquaculture Research and Development 3 (8), 2012.

Afreen, M., Ucak, I., 2020. Fish processing wastes used as feed ingredient for animal feed and aquaculture feed. Survey in Fisheries Sciences 6 (2), 55–64.

Ahmad, M., Benjakul, S., 2010. Extraction and characterisation of pepsin-solubilised collagen from the skin of unicorn leather jacket (*Aluterus monocerous*). Food Chemistry 120, 817–824.

Akbary, P., Younesi, A., 2017. Effect of dietary supplementation of chitosan on growth, hematology and innate immunity of grey mullet (*Mugil cephalus*). Products Biological & Researches 30, 194–203.

Al Khawli, F., Pateiro, M., Domínguez, R., José, M.L., Gullón, P., Kousoulaki, K., et al., 2019. Innovative green technologies of intensification for valorization of seafood and their by-products review. Marine Drugs 17 (12), 689.

Ali, M.Z., Gheyasuddin, S., Zaher, S., Hossain, M.A., Islam, M.N., 1994. Evaluation of fish silage from underutilized marine fishes as protein sources in the diet of major carp (*Ciirrhinus mrigald*). J. Aqua. Trop. 8, 247–254.

Amar, B., Philip, R., Singh, I.S.B., 2000. Efficacy of fermented prawn shell waste as a feed ingredient for Indian prawn *Fenneropenaeus indicus*. Aquacuture Nutrition 12, 433–442.

Ameer, K., Shahbaz, H.M., Kwon, J.-H., 2017. Green extraction methods for polyphenols from plant matrices and their by-products: a review. Comprehensive Reviews in Food Science and Food Safety 16, 295–315.

Aneesh, P.A., Anandan, R., Kumar, L.R.G., et al., 2020. A step to shell biorefinery—extraction of astaxanthin-rich oil, protein, chitin, and chitosan from shrimp processing waste. Biomass Conversion and Biorefinery. Available from: https://doi.org/10.1007/s13399-020-01074-5.

Arason, S., 1994. Production offish silage. In: Martin, A.M. (Ed.), Fisheries Processing: Biotechnological Applications. Chapman and Hall, London, pp. 244–272.

Arvanitoyannis, S.I., Kassaveti, A., 2008. Fish industry waste: treatments, environmental impacts, current and potential uses. International Journal of Food Science and Technology 43, 726–745.

Aspevik, T., Oterhals, Å., Rønning, S.B., Altintzoglou, T., Wubshet, S.G., Gildberg, A., et al., 2017. Valorization of proteins from co- and by-products from the fish and meat industry. Chemistry and Chemical Technologies in Waste Valorization. Springer, Cham, pp. 123–150.

Atef, M., Ojagh, S., 2017. Health benefits and food applications of bioactive compounds from fish by-products: a review. Journal of Functional Foods 35, 673–681.

Auchterlonie, N., 2018. The continuing importance of fishmeal and fish oil in aquafeeds. Presented at the Aquafarm Conference, 15–16 February Pordenone, Italy.

Baeza- Ariño, R., Martlnez-Llorens, Sm, Nogales-Merida, S., Jover-Cerda, M., Tomás-Vidal, A., 2016. Study of liver and gut alterations in seabream, *Sparus aurata* L., fed a mixture of vegetable protein concentrates. Aquaculture Research 47, 460–471.

Barba, F.J., Roselló-Soto, E., Brnčíc, M., Lorenzo, J.M., 2020. Green Extraction and Valorization of Byproducts from Food Processing from Food Processing. CRC Press, Boca Raton, FL.

Benhabiles, M.S., Abdi, N., Drouiche, N., Lounici, H., Pauss, A., Goosen, M.F.A., et al., 2012. Fish protein hydrolysate production from sardine solid waste by crude pepsin enzymatic hydrolysis in a bioreactor coupled to an ultrafiltration unit. Materials Science and Engineering: C 32 (4), 922–928.

Benkajul, S., Morrissey, M.T., 1997. Protein hydrolysates from Pacific whiting solid wastes. Journal of Agricultural and Food Chemistry 45, 3423–3430.

Bhaskar, N., Benila, T., Radha, C., Lalitha, R.G., 2008. Optimization of enzymatic hydrolysis of visceral waste proteins of Catla (*Catla catla*) for preparing protein hydrolysate using a commercial protease. Bioresource Technology 99, 335–343.

Birk, D.E., Bruckner, P., 2005. Collagen suprastructures. Collagen 247, 185–205.

Blanco, M., Vazquez, J.A., Perez-Martın, R.I., Sotelo, C.G., 2017. Hydrolysates of fish skin collagen: an opportunity for valorizing fish industry by-products. Marine Drugs 15, 131.

Bougatef, A., Nedjar-Arroume, N., Manni, L., Ravallec, R., Barkia, A., Guillochon, D., et al., 2010. Purification and identification of novel antioxidant peptides from enzymatic hydrolysates of sardinelle (*Sardinella aurita*) by-products proteins. Food Chemistry 118, 559–565.

Bradai, M.N., Saidi, B., Enajjar, S., Bouain, A., 2006. The gulf of Gabe's: a spot for the Mediterranean elasmobranches. In: Basusta, N., Keskin, C., Serena, F., Seret, B. (Eds.), The Proceedings of the Workshop on Mediterranean Cartilaginous Fish with Emphasis on Southern and Eastern Mediterranean. Turkish Marine Research Foundation, Istanbul, Turkey, pp. 107–117.

Breivik, H.L.-C., 2007. Omega-3 Specialty Oils. The Oily Press, Cambridge.

Caruso, G.F., 2016. Wastes and byproducts: a resource to be valorised. Journal of Fisheries Sciences 10 (1), 012–015.

Casadidio, C., Peregrina, D.V., Gigliobianco, M.R., Deng, S., Censi, R., Di, et al., 2019. Chitin and Chitosans: characteristics, eco-friendly processes, and applications in cosmetic science. Marine Drugs 17, 369.

Cavallaro, J.F., Kemp, P.D., Kraus, K.H., 1994. Collagen fabrics as biomaterials. Biotechnology and Bioengineering 43, 781–791.

Chalamaiah, M., Dinesh kumar, B., Hemalatha, R., Jyothirmayi, T., 2012. Fish protein hydrolysates: proximate composition, amino acid composition, antioxidant activities and applications: a review. Food Chemistry 135, 3020–3038.

Chawla, S.P., Kanatt, S.R., Sharma, A.K., 2015. Chitosan. In: Ramawat, K., Mérillon, J.M. (Eds.), Polysaccharides. Springer, Cham.

Chen, Y., Zhu, X., Yang, Y., Han, D., Jin, J., Xie, S., 2014. Effect of dietary chitosan on growth performance, haematology, immune response, intestine morphology, intestine microbiota and disease resistance in gibel carp (*Carassius auratus* gibelio). Aquaculture Nutrition 20, 532–546.

Chuaychan, S., Benjakul, S., Kishimura, H., 2015. Characteristics of acid- and pepsin-soluble collagens from scale of seabass (*Lates calcarifer*). LWT - Food Science and Technology 63 (1), 71–76.

Costantini, M., Bernardini, M., Cordone, P., Guilianini, P.G., Orel, G., 2000. Observations on fishery, feeding habits and reproductive biology of Mustelus mustelus (*Chondrichtye, Triakidae*). North. Adriat. Sea Biol. Mar. Mediterr 7, 427–432.

Daniel, N., 2018. A review on replacing fish meal in aqua feeds using plant protein sources. International Journal of Fisheries and Aquatic Studies 6 (2), 164–179.

Díaz-Jiménez, L., Hernández-Vergara, M.P., Pérez-Rostro, C.I., Ortega-Clemente, L.A., 2019. The effect of astaxanthin and β-carotene inclusion in diets for growth, reproduction and pigmentation of the peppermint shrimp *Lysmata wurdemanni*. Latin American Journal of Aquatic Research 47 (3), 559–567.

Dumay, J., 2006. Extraction de lipides en voie aqueuse par bioréacteur enzymatique combiné àl'ultrafiltration: application à la valorisation de co-produits de poisson (*Sardina pilchardus*). PhD Thesis.

El Knidri, H., Belaabed, R., Addaou, A., Laajeb, A., Lahsini, A., 2018. Review extraction, chemical modification and characterization of chitin and chitosan. International Journal of Biological Macromolecules 120 (Pt A), 1181–1189.

El-Sayed, H.S., Barakat, K.M., 2016. Effect of dietary chitosan on challenged *Dicentrarchus labrax* post larvae with Aeromonas hydrophila. Russian Journal of Marine Biology 42, 501–508.

Espe, M., Haaland, H., Njaa, L.R., 1992. Autolysed fish silage a feed ingredient for Atlantic Salmon (*Salmo salar*). Composition Biochemistry Physiology 103, 369–372.

Esteban, M.B., Garcia, A.J., Ramos, P., Marquez, M.C., 2006. Evaluation of fruit–vegetable and fish wastes as alternative feedstuffs in pig diets. Waste Management 27, 193–200.

Fagbenro, O., Jauncey, K., 1998. Physical and nutritional properties of moist fermented fish silage pellets as a protein supplement for tilapia (*Oreochromis niloticus*). Animal Feed Science and Technology 71, 11–18.

Fagbenro, O., Jauncey, K., Haylor, G., 1994. Nutritive value of diets containing dried lactic acid fermented fish silage and soybean meal for juvenile *Oreochromis niloticus* and *Clarias gariepinus*. Aquatic Living Resource 7, 79–85.

Fanimo, A.O., Oduguwa, O.O., Onifade, A.O., Olutunde, T.O., 2000. Protein quality of shrimp-waste meal. Bioresource Technology 72, 185–188.

FAO, 2002. Food and Agriculture Organization. Use of fishmeal and fish oil in auafeeds, Further thoughts on the fishmeal trap. FAO Fisheries Circular No. 975, Rome.

FAO, 2003. Animal Feed Resources Information System. Food and Agricultural Organization, Rome.

FAO, 2018. Production and Utilization of Fish Silage, A Manual on How to Turn Fish Waste Into Profit and a Valuable Feed Ingredient or Fertilizer. Food and Agricultural Organization, Rome.

FAO, 2020. The State of the World Fisheries Sustainability in Action. Food and Agricultural Organization, Rome.

Ferraro, V., Cruz, I.B., Jorge, R.F., Malcata, F.X., Pintado, M.E., Castro, P.M.L., 2010. Valorisation of natural extracts from marine source focused on marine by-products: a review. Food Research International 43, 2221–2233.

Galanakis, C.M., 2017. Nutraceutical and Functional Food Components: Effects of Innovative Processing Techniques. Academic Press, Cambridge, MA.

García-Moreno, P.J., Pérez-Gálvez, R., Espejo-Carpio, F.J., Ruiz-Quesada, C., Pérez-Morilla, A.I., Martínez-Agustín, O., et al., 2017. Functional, bioactive and antigenicity properties of blue whiting protein hydrolysates: effect of enzymatic treatment and degree of hydrolysis. Journal of the Science of Food and Agriculture 97 (1), 299–308.

Gbogouri, G.A., Linder, M., Fanni, J., Parmentier, M., 2004. Influence of hydrolysis degree on the functional properties of salmon by-products hydrolysates. Journal of Food Science 69, 615–623.

Ghassem, M., Fern, S.S., Said, M., Ali, Z.M., Ibrahim, S., Babji, A.S., 2014. Kinetic characterization of *Channa striatus* muscle sarcoplasmic and myofibrillar protein hydrolysates. Journal of Food Science and Technology 51, 467–475.

Giannetto, A., Esposito, E., Lanza, M., et al., 2020. Protein hydrolysates from anchovy (*Engraulis encrasicolus*) waste: in vitro and in vivo biological activities. Marine Drugs 18, 86.

Gildberg, A., Stenberg, E., 2001. A new process for advanced utilization of shrimp waste. Process Biochemistry 36, 809–812.

Goddard, J.S., McLean, E., Wille, K., 2003. Co-dried sardine silage as an ingredient in tilapia, *Oreochromis aureus*, diets. Journal of Aquaculture in the Tropics 18 (3), 257–264.

Gokoglu, N., Yerlikaya, P., Ucak, I., et al., 2017. Effect of bromelain and papain enzymes addition on physicochemical and textural properties of squid (*Loligo vulgaris*). Food Measure 11, 347–353.

Gómez-Guilén, M.C., Turnay, J., Fernández-Díaz, M.D., Ulmo, N., Lizarbe, M.A., Montero, P., 2002. Structural and physical properties of gelatin extracted from different marine species: a comparative study. Food Hydrocolloids 16, 25–34.

Halim, N.R.A., Yusof, H.M., Sarbon, N.M., 2016. Functional and bioactive properties of fish protein hydolysates and peptides: a comprehensive review. Trends in Food Science and Technology 51, 24–33.

Hansen, J.S., Rosnes, J.T., Skuland, A.V., Solstad, R.G., Kousoulaki, K., 2019. Technical and sensory properties of salmon burgers added soluble peptide-rich fish proteins. In: Proceedings of the WEFTA, Tórshavn, Faroe Islands.

Hardy, R.W., Shearer, K.D., Spinelli, J., 1984. The nutritional properties of co-dried fish silage in rainbow trout (*Salmo gairdnen*) dry diets. Aquaculture 38, 35–44.

Harikrishnan, R., Kim, J.S., Balasundaram, C., Heo, M.S., 2012. Immunomodulatory effects of chitin and chitosan enriched diets in *Epinephelus bruneus* against *Vibrio alginolyticus* infection. Aquaculture 326–329, 46–52.

Harnedy, P.A., Fitz Gerald, R.J., 2012. Bioactive peptides from marine processing waste and shell-fish: a review. Journal of Functional Foods 4, 6–24.

He, S., Franco, C., Zhang, W., 2013. Functions, applications and production of protein hydrolysates from fish processing co-products (FPCP). Food Research International 50, 289–297.

Heras, H., Mcleod, C.A., Ackman, R.G., 1994. Atlantic dogfish silage versus herring silage in diets for Atlantic salmon (*Salmo salar*): growth and sensory evaluation of fillets. Aquaculture 125, 93–103.

Ho, T.C., Li-Chan, E.C., Skura, B.J., Higgs, D.A., Dosanjh, B., 2014. Pacific hake (Merluccius productus Ayres, 1855) hydrolysates as feed attractants for juvenile Chinook salmon (Oncorhynchus tshawytscha Walbaum, 1792). Aquaculture Research 45 (7), 1140–1152.

Hood, L.L., 1987. Collagen in sausage casings. In: Pearson, A.M., Dutson, T.R., Bailey, A. J. (Eds.), Advances in Meat Research Volume 4: Collagen as Food, vol. 4. Van Nostrand Reinhold Company, New York, NY, pp. 109–129.

Hou, Y., Wu, Z., Dai, Z., Wang, G., Wu, G., 2017. Protein hydrolysates in animal nutrition: Industrial production, bioactive peptides, and functional significance. Journal of Animal Science and Biotechnology 8, 24.

Hwang, J.H., Mizuta, S., Yokoyama, Y., Yoshinaka, R., 2007. Purification and characterization of molecular species of collagen in the skin of skate (*Raja kenojei*). Food Chemistry 100, 921–925.

Idowu, A.T., Igiehon, O., Idowu, S., et al., 2020. Bioactivity potentials and general applications of fish protein hydrolysates. International Journal of Peptide Research and Therapeutics 27, 109–118.

Ikoma, T., Kobayashi, H., Tanaka, J., Walsh, D., Mann, S., 2003. Microstructure, mechanical, and biomimetic properties of fish scales from *Pagrus major*. Journal of Structural Biology 142, 327–333.

Ishak, N.H., Sarbon, N.M., 2018. A review of protein hydrolysates and bioactive peptides deriving from wastes generated by fish processing. Food and Bioprocess Technology 11, 2–16.

Ivanovs, K., Blumberga, D., 2017. Extraction of fish oil using green extraction methods: a short review. Energy Procedia 128, 477–483.

Jackson, A., Newton, R.W., 2016. Project to Model the Use of Fisheries By-Products in the Production of Marine Ingredients with Special Reference to Omega-3 Fatty Acids EPA and DHA. Institute of Aquaculture, University of Stirling and Marine Ingredients Organisation, London.

Je, J.Y., Park, P.J., Kwon, J.Y., Kim, S.K., 2004. A novel angiotensin I converting enzyme inhibitory peptide from Alaska pollack (*Theragra chalcogramma*) frame protein hydrolysate. Journal of Agricultural and Food Chemistry 52, 7842–7845.

Jeevithan, E., Bao, B., Bu, Y., Zhou, Y., Zhao, Q., Wu, W., 2014. Type II collagen and gelatin from silvertip shark (*Carcharhinus albimarginatus*) cartilage: isolation, purification, physicochemical and antioxidant properties. Marine Drugs 12, 3852.

Jongjareonrak, A., Benjakul, S., Visessanguan, W., Nagai, T., Tanaka, M., 2005. Isolation and characterisation of acid and pepsin-solubilised collagens from the skin of Brownstripe red snapper (*Lutjanus vitta*). Food Chemistry 93, 475–484.

Jongjareonrak, A., Benjakul, S., Visessanguan, W., Tanaka, M., 2006. Skin gelatin from bigeye snapper and brownstripe red snapper: chemical composition and effect of microbial transglutaminase on gel properties. Food Hydrocolloids 20, 1216–1222.

Kang, P.Y., Ishak, N.H., Sarbon, N.M., 2018. Optimization of enzymatic hydrolysis of shortfn scad (Decapterus macrosoma) myofibrillar protein with antioxidant effect using alcalase. International Food Research Journal 25 (5), 1808–1817.

Kang, H.K., Lee, H.H., Seo, C.H., Park, Y., 2019. Antimicrobial and immunomodulatory properties and applications of marine derived proteins and peptides. Marine Drugs 17, 350.

Karim, A.A., Bhat, R., 2009. Fish gelatin: properties, challenges, and prospects as an alternative to mammalian gelatins. Food Hydrocolloids 23, 563e576.

Kelleher, K., 2005. Discards in the World's Marine Fisheries. An Update. FAO Fisheries Technical Paper 470. FAO, Rome, p. 131p.

Khan, M.A.A., Hossain, M.A., Hara, K., Osatomi, K., Ishihara, T., Nozaki, Y., 2003. Effect of enzymatic fish-scrap protein hydrolysate on gel-forming ability and denaturation of lizard fish Saurida wanieso surimi during frozen storage. Fisheries Science 69 (6), 1271–1280.

Khempaka, S., Chitsatchapong, C., Molee, W., 2011. Effect of chitin and protein constituents in shrimp head meal on growth performance, nutrient digestibility, intestinal microbial populations, volatile fatty acids, and ammonia production in broilers. Journal of Applied Poultry Research 20 (1), 1–11.

Kim, S.-K., Mendis, E., 2006. Bioactive compounds from marine processing by-products—a review. Food Research International 39 (4), 383–393.

Kim, S.K., Choi, Y.R., Park, P.J., Choi, J.H., Moon, S.H., 2000. Screening of biofunctional peptides from cod processing wastes. Journal of the Korean Society of Agricultural Chemistry and Biotechnology 43, 225–227.

Kim, S.-K., Mendis, E., Shahidi, F., 2007. Marine fisheries by-products as potential nutraceuticals: an overview. In: Barrow, C.J., Shahidi, F. (Eds.), Marine Nutraceuticals and Functional Foods. CRC Press, Boca Raton, FL, pp. 1–23.

Kim, H.S., Jung, W.G., Myung, S.H., Cho, S.H., Kim, D.S., 2014. Substitution effects of fishmeal with tuna by-product meal in the diet on growth, body composition, plasma chemistry and amino acid profiles of juvenile olive flounder (*Paralichthy solivaceus*). Aquaculture 431, 92–98.

Kittiphattanabawon, P., Benjakul, S., Visessanguan, W., Kishimura, H., Shahidi, F., 2010. Isolation and Characterisation of collagen from the skin of brownbanded bamboo shark (*Chiloscyllium punctatum*). Food Chemistry 119, 1519–1526.

Kotzamanis, Y.P., Alexis, M.N., Andriopoulou, A., Castritsi-Cathariou, I., Fotis, G., 2001. Utilization of waste material resulting from trout processing in gilthead bream (*Sparus aurata* L.) diets. Aquaculture Research 32 (1), 288–295.

Kouhdasht, A.M., Moosavi-Nasab, M., 2018. Bioactive peptides derived from fish by-product collagen mini review. International Journal of Environmental Sciences & Natural Resources 13 (2), 47–50.

Kudre, T.G., Bhaskar, N., Sakhare, P.Z., 2017. Optimization and characterization of biodiesel production from rohu (*Labeo rohita*) processing waste. Renewable Energy 113, 1408–1418.

Kumar, M.N., 2000. A review of chitin and chitosan applications. Reactive and Functional Polymers 45, 1–27.

Le Gouic, A.V., Harnedy, P.A., FitzGerald, R.J., 2018. Bioactive peptides from fish protein byproducts. In: Mérillon, J.M., Ramawat, K. (Eds.), Bioactive Molecules in Food. Reference Series in Phytochemistry. Springer, Cham.

Leduc, A., Zatylny-Gaudin, C., Robert, M., Corre, E., Le Corguille, G., Caste, H., et al., 2018. Dietary aquaculture by-product hydrolysates: impact on the transcriptomic response of the intestinal mucosa of European sea bass (*Dicentrarchus labrax*) fed low fish meal diets. BMC Genomics 19, 396–416.

Létisse, M., Comeau, L., 2008. Enrichment of eicosapentaenoic acid and docosahexaenoic acid from sardine by-products by supercritical fluid fractionation. Journal of Separation Science 31, 1374–1380.

Li, X., Yu, Y., Xie, S., Yan, Q., 2011. Effect of chitosan on intestinal bacteria of Allogynogenetic Crucian Carp, *Carassius auratus* gibelio, as depicted by polymerase chain reaction denaturing gradient gel electrophoresis. Journal of World Aquaculture Society 42, 539–548.

Liu, D., Zhang, X., Li, T., Yang, H., Zhang, H., Regenstein, J.M., et al., 2015. Extraction and characterization of acid- and pepsin-soluble collagens from the scales, skins and swim-bladders of grass carp (*Ctenopharyngodon idella*). Food Bioscience 9, 68–74.

Lopez-Pedrouso, M., Lorenzo, J.M., Cantalapiedra, J., Zapata, C., Franco, J.M., Franco, D., 2020. Aquaculture and by-products: challenges and opportunities in the use of alternative protein sources and bioactive compounds. Advances in Food and Nutrition Research 92, 127–185.

Madage, S.S.K., Medis, W.U.D., Sultanbawa, Y., 2015. Fish silage as replacement of fishmeal in red tilapia feeds. Journal of Applied Aquaculture 27 (2), 95–106.

Maqsood, S., Singh, P., Samoon, M.H., Balange, A.K., 2010. Effect of dietary chitosan on non-specific immune response and growth of *Cyprinus carpio* challenged with *Aeromonas hydrophila*. International Aquatic Research 2, 77–85.

Mazorra-Manzano, M.A., Pacheco-Aguilar, R., RamırezSuarez, J.C., Garcia-Sanchez, G., Lugo-Sanchez, M.E., 2010. Endogenous proteases in Pacific whiting (*Merluccius productus*) muscle as a processing aid in functional fish protein hydrolysate production. Food and Bioprocess Technology 5, 130–137.

Meshkini, S., Tafy, A.A., Tukmechi, A., Farhang-Pajuh, F., 2012. Effects of chitosan on hematological parameters and stress resistance in rainbow trout (*Oncorhynchus mykiss*). Veterinary Research Forum 3, 49–54.

Messina, C.M., Renda, G., Barbera, L., Santulli, A., 2013. Byproducts of farmed European sea bass (*Dicentrarchus labrax* L.) as a potential source of n-3 PUFA. Biologia - Section Zoology. 68/2, 288–293.

Millamena, O.M., Bautista, M.N., Kanazawa, A., 1999. Quantitative dietary requirements of post larval tiger shrimp *Penaeus monodon* for histidine, isoleucine, leucine, phenylalanine and tryptophan. Aquaculture 179, 169–179.

Mohanty, B., Pattanaik, U., Panda, S., A. Jena, A., 2018. Future prospects and trends for effective utilization of fish processing wastes in India. Innovative Farming 3 (1), 1–5.

Monier, M.N., 2020. Efficacy of dietary exogenous enzyme supplementation on growth performance, antioxidant activity, and digestive enzymes of common carp (*Cyprinus carpio*) fry. Fish Physiology and Biochemistry 46, 713–723.

Morin-Crini, N., Lichtfouse, E., Torri, G., Crini, G., 2019. Applications of chitosan in food, pharmaceuticals, medicine, cosmetics, agriculture, textiles, pulp and paper, biotechnology, and environmental chemistry. Environmental Chemistry Letters 17 (4), 1667–1692.

Muttharasi, C., Muralisankar, T., Uthayakumar, V., Gayathri, V., Thangal, S.H., Anandhan, K., 2019. Utilization of marine fisheries wastes for the production of the freshwater fish *Cyprinus carpio*. Tropical Animal Health and Production 51 (8), 2305–2313.

Muyonga, J.H., Cole, C.G.B., Duodo, K.G., 2004. Extraction and physicochemical characterisation of Nile perch (*Lates niloticus*) skin and bone gelatin. Food Hydrocolloids 18, 581–592.

Nagai, T., Ogawa, T., Nakamura, T., Ito, T., Nakagawa, H., Fujiki, K., et al., 1999. Collagen of edible jelly fish exumbrella. Journal of the Science of Food and Agriculture 79, 855–858.

Nagai, T., Suzuki, N., 2000. Isolation of collagen from fish waste material – skin, bone and fin. Food Chemistry 68, 277–281.

Najafabad, K., M. Imanpoor, M.R., Taghizadeh, V., Alishahi, A., 2016. Effect of dietary chitosan on growth performance, hematological parameters, intestinal histology and stress resistance of Caspian kutum (*Rutilus frisii* kutum Kamenskii, 1901) fingerlings. Fish Physiology and Biochemistry 42, 1063–1071.

Naqash, S.Y., Nazeer, R.A., 2013. Antioxidant and functional properties of protein hydrolysates from pink perch (*Nemipterus japonicus*) muscle. Journal of Food Science and Technology 50, 972–978.

Nazeer, R.A., Deeptha, R., Jaiganesh, R., Sampathkumar, N.S., Naqash, S.Y., 2011. Radical scavenging activity of seela (*Sphyraena barracuda*) and ribbon fish (*Lepturacanthus savala*) backbone protein hydrolysates. International Journal of Peptide Research and Therapeutics 17, 209–216.

Ngo, D.-H., Qian, Z., Ryu, B., Park, J.W., Kim, S.-K., 2010. In vitro antioxidant activity of a peptide isolated from Nile tilapia (*Oreochromis niloticus*) scale gelatin in free radical-mediated oxidative systems. Journal of Functional Foods 2 (2), 107–117.

Nilsang, S., Lertsiri, S., Suphantharika, M., Assavanig, A., 2005. Optimization of enzymatic hydrolysis of fish soluble concentrate by commercial proteases. Journal of Food Engineering 70, 571–578.

Niu, J., Liu, Y.J., Lin, H.Z., Mai, K.S., Yang, H.J., Liang, G.Y., et al., 2011. Effect of dietary chitosan on growth and stress tolerance of postlarval (*Litopenaeus vannamei*). Aquaculture Nutrition 17, 406–412.

Niu, J., Lin, H.Z., Jiang, S.G., Chen, X., Wu, K.C., Liu, Y.J., et al., 2013. Comparison of effect of chitin, chitosan, chitosan oligosaccharide and N-acetyl-D-glucosamine on growth performance, antioxidant defenses and oxidative stress status of *Penaeus monodon*. Aquaculture 372–375, 1–8.

Nor, N.M., Zakaria, Z., Manaf, M.S., Salleh, M., 2011. The effect of partial replacement of dietary fishmeal with fermented prawn waste liquor on juvenile sea bass growth. Journal of Applied Aquaculture 23 (1), 51–57.

NRC, 1993. National Research Council—Nutrient Requirement of Fish. National Academy of Science, Washington, p124.

Olsen, R.L., Toppe, J., Karunasagar, I., 2014. Challenges and realistic opportunities in the use of by-products from processing of fish and shell-fish. Trends in Food Science & Technology 36, 144–151.

Ospina-Salazar, G.H., Rıos-Duran, M.G., Toledo-Cuevas, E.M., Marte effects of fish hydrolysate and snez-Palacios, C.A., 2016. The effects of fish hydrolysate and soy protein isolate on the growth performance, body composition and digestibility of juvenile pike silverside, *Chirostoma estor*. Animal Feed Science and Technology 220, 168–179.

Ovissipour, M., Abedian, A., Motamedzadegan, A., Rasco, B., Safari, R., Shahiri, H., 2009. The effect of enzymatic hydrolysis time and temperature on the properties of protein hydrolysates from Persian sturgeon (*Acipenser persicus*) viscera. Food Chemistry 115, 238–242.

Ovissipour, M., Kenari, A.A., Motamedzadegan, A., Nazari, R.M., 2012. Optimization of enzymatic hydrolysis of visceral waste proteins of yellowfin tuna (*Thunnus albacares*). Food and Bioprocess Technology 5, 696–705.

Ovissipour, M., Abedian, K.A., Nazari, R., Motamedzadegan, A., Rasco, B., 2014. Tuna viscera protein hydrolysate: nutritive and disease resistance properties for Persian sturgeon (*Acipenser persicus* L.) larvae. Aquaculture Research 45, 591–601.

Pangestuti, R., Kim, S.-K., 2017. Bioactive peptide of marine origin for the prevention and treatment of non-communicable diseases. Marine Drugs 15 (3), 67.

Poojary, M.M., Barba, F.J., Aliakbarian, B., Donsì, F., Pataro, G., Dias, D.A., et al., 2016. Innovative alternative technologies to extract carotenoids from microalgae and seaweeds. Marine Drugs 14 (11), 214.

Prihanto, A.A., Nurdiani, R., Bagus, A.D., 2019. Production and characteristics of fish protein hydrolysate from parrotfish (*Chlorurus sordidus*) head. PeerJ 7, e8297.

Rahman, M.M., Khosravi, S., Chang, K.H., Lee, S.M., 2016. Effects of dietary inclusion of astaxanthin on growth, muscle pigmentation and antioxidant capacity of juvenile rainbow trout (*Oncorhynchus mykiss*). Preventive Nutrition and Food Science 21 (3), 281–288.

Ranjan, R., Prasad, K.P., Vani, T., Kumar, R., 2012. Effect of dietary chitosan on haematology, innate immunity and disease resistance of Asian seabass *Lates calcarifer* (Bloch). Aquaculture Research 45, 983–993.

Robert, M., 2014. Development of hydrolysates for feeding aquaculture animals: molecular and functional characterization. Doctorate from the University of Caen Basse-Normandie.

Ruiter, A., 1995. Fish and Fishery Products: Composition, Nutritive Properties and Stability. Cab International, Wallingford.

Rustad, T., 2003. Utilisation of marine by-products. Electronic Journal of Environmental, Agricultural and Food Chemistry 2, 458–463.

Rustad, T., Storrø, I., Slizyte, R., 2011. Possibilities for the utilisation of marine by-products. International Journal of Food Science and Technology 46, 2001–2014.

Sachindra, N.M., Bhaskar, N., 2008. In vitro antioxidant activity of liquor from fermented shrimp biowaste. Bioresource Technology 99, 9013–9016.

Saini, R.K., Song, M.-H., Rengasamy, k R., Ko, E.-Y., Keum, Y.-S., 2020. Red shrimp are a rich source of nutritionally vital lipophilic compounds: a comparative study among edible flesh and processing waste. Foods 9, 1179.

Saleh, N.E., Wassef, E.A., Shalaby, S.M., 2018. The role of dietary astaxanthin in European sea bass (*Dicentrarchus labrax*) growth, immunity, antioxidant competence and stress tolerance. Egyptian Journal of Aquatic Biology & Fisheries 22 (5), 189–200.

Saleh, N.E., Wassef, E.A., El-bermawy, N.M., Barakat, K.M., Abdel-Mohsen, H.H., 2020. Bioconversion of fish-wastes biomass into a fish meal alternate for European seabass (*Dicentrarchus labrax*) diets. International Aquatic Research 12, 40–52.

Salem, M.E., 2015. Utilization of Some Feed Additives in Improving the Nutritive Value of Marine Fish Diet (PhD Dissertation). Alexandria University, Egypt.

Sarmadi, B.H., Ismail, A., 2010. Antioxidative peptides from food proteins: a review. Peptides 31, 1949–1956.

Schmidsdor, W., 1995. Fish meal and fish oil-not only by-products. In: Ruiter, A. (Ed.), Fish and Fishery Products, Composition, Nutritive Properties and Stability. CAB International, Wallingford, pp. 347–376.

Shahidi, F., 1994. Seafood processing by-products. In: Shahidi, F., Botta, J.R. (Eds.), Seafoods Chemistry. Blackie Academic & Professional, London.

Shahidi, F., Ambigaipalan, P., 2015. Novel functional food ingredients from marine sources. Current Opinion in Food Science 2, 123–129.

Siddik, M.A.B., Howieson, J., Partridge, G.J., Fotedar, R., Gholipourkanani, H., 2018. Dietary tuna hydrolysate modulates growth performance, immune response, intestinal morphology and resistance to *Streptococcus iniae* in juvenile barramundi, *Lates calcarifer*. Scientific Reports 8, 1–13.

Siddik, M.A., Howieson, J., Fotedar, R., Partridge, G.J., 2020. Enzymatic fish protein hydrolysates in finfish aquaculture: a review. Reviews in Aquaculture 1–25.

Silva, T., Moreira-Silva, J., Marques, A., Domingues, A., Bayon, Y., Reis, R., et al., 2014. Marine origin collagens and its potential applications. Marine Drugs 12, 5881–5901.

Silvestrini, B., 1988. United States Patent No. 4,749,684. United States Patent and Trademark Office, Washington, DC.

Singh, A., Ahmad, S., Ahmad, A. Green extraction methods and environmental applications of carotenoids – a review. RSC Advances, 5, 62358–62393, 2015.

Slimane, E.B., Sadok, S., 2018. Collagen from cartilaginous fish by-products for a potential application in bioactive film composite. Marine Drugs 16 (6), 211.

Šližytė, R., Dauksas, E., Falch, E., Storrøl, I., Rustad, T., 2005. Yield and composition of different fractions obtained after enzymatic hydrolysis of cod (*Gadus morhua*) by-products. Process Biochemistry 40, 1415–1424.

Stainsby, G., 1987. Gelatin gels. In: Pearson, A.M., Dutson, T.R., Bailey, A.J. (Eds.), Collagen as Food: Volume 4. Advances in Meat Research, vol. 4. Van Nostrand Reinhold Company, New York, NY.

Stevens, J.R., Newton, R.W., Tlusty, M., Little, D.C., 2018. The rise of aquaculture by-products: increasing food production, value, and sustainability through strategic utilisation. Marine Policy 90, 115–124.

Suresh, P.V., Prabhu, G.N., 2012. Seafood. In: Chandrasekaran, M. (Ed.), Valorization of Food Processing By-Products. CRC Press, Boca Raton, FL, pp. 685–736.

Suryawanshi, N., Jujjavarapu, S.E., Ayothiraman, S., 2019. Marine shell industrial wastes–an abundant source of chitin and its derivatives: constituents, pretreatment, fermentation, and pleiotropic applications-a revisit. International Journal of Environmental Science and Technology 16 (3), 3877–3898.

Tacon, A.G., Jackson, A.J., 1985. Utilization of conventional and unconventional protein sources in practical fish feeds. Nutrition and Feeding in Fish. Academic Press, London, pp. 1–19.

Taheri, A., Anvar, S.A.A., Ahari, H., Fogliano, V., 2012. Comparison the functional properties of protein hydrolysates from poultry by-products and rainbow trout (*Onchorhynchus mykiss*) viscera. Iranian Journal of Fisheries Sciences 12 (1), 154–169.

Tang, H.G., Wu, T.X., Zhao, Z.Y., Pan, X.D., 2008. Effects of fish protein hydrolysate on growth performance and humoral immune response in large yellow croaker (*Pseudosciaena crocea* R.). Journal of Zhejiang University. Science. B 9 (9), 684–690.

Terzioğlu, P., Ogut, H., Kalemtas, A., 2018. Natural calcium phosphates from fish bones and their potential biomedical application. Materials Science and Engineering C 91, 899–911.

Teves, J.F.C., Ragaza, J.A., 2016. The quest for indigenous aquafeed ingredients: a review. Reviews in Aquaculture 8, 154–171.

Tocher, D.R., Betancor, M.B., Sprague, M., Olsen, R.E., Napier, J.A., 2019. Omega-3 long-chain polyunsaturated fatty acids, EPA and DHA: bridging the gap between supply and demand. Nutrients 11 (1), 89.

Toppe, J., Albrektsen, S., Hope, B., Aksnes, A., 2007. Chemical composition, mineral content and amino acid and lipid profiles in bones from various fish species. Comparative Biochemistry and Physiology, Part B 146, 395–401.

Uyan, O., Koshio, S., Teshima, S., Ishikawa, M., Thu, M., Alam, M.S., et al., 2006. Growth and phosphorous loading by partially replacing fishmeal with tuna muscle by-product powder in the diet of juvenile Japanese flounder, *Paralichths olivaceus*. Aquaculture 257, 437–445.

Vazquez, J.A., Blanco, M., Massa, A.E., Amado, I.R., Perez-Martin, R.I., 2017. Production of fish protein hydrolysates from *Scyliorhinus canicula* discards with antihypertensive and antioxidant activities by enzymatic hydrolysis and mathematical optimization using response surface methodology. Marine Drugs 15 (10), 306.

Vidotti, R.M., Carneiro, D.J., Viegas, E., 2002. Growth rate of Pacu, *Piaractus mesopotamicus*, fingerlings fed diets containing co-dried fish silage as replacement of fish meal. Journal of Applied Aquaculture 12 (4), 77–88.

Vidotti, R.M., Viegas, E.M.M., Carneiro, D.J., 2003. Amino acid composition of processed fish silage using different raw materials. Animal Feed Science and Technology 105, 199–204.

Villamil, O., Vaquiro, H., Solanilla, J.F., 2017. Fish viscera protein hydrolysates: Production, potential applications and functional and bioactive properties. Food Chemistry 224, 160–171.

Villarreal, H., Hernandez-llamas, A., Rivera, M.C., Millan, A., Rocha, S., 2004. Effect of substitution of shrimp meal, fish meal and soy meal with red crab *Pleuroncodes planipes* (Stimpson) meal in pelleted diets for post larvae and juvenile *Farfantepenaeus californiensis* (Holmes). Aquaculture Research 35 (2), 178–183.

Wassef, E.A., 1991. Development of diets for gilthead seabream *Sparus aurata* L, cultured in Egypt. Journal King Abdul- Aziz University, Marine Science 2, 101–109.

Wassef, E.A., 2005. Alternative protein sources for fish feeds in Egypt. Cahier Option Méditerranéennes. Mediterranean Fish Nutrition 63, 127–142.

Wassef, E.A., Sweilam, M.A., Attala, R.F., 2001. Growth performance of Nile tilapia (*Oreochromis niloticus*) as affected with dietary incorporation of fermented fish silage or boilled fulfat soybean as protein replacers. Egyptian Journal of Aquatic Biology and Fisheries 5 (3), 137–150.

Wassef, E.A., Sweilam, M.A., Attalah, R.F., 2003. The use of fermented fish silage as a replacement for fish meal in Nile tilapia (*Oreochromis niloticus*) diets. Egyptian Journal of Nutrition and Feeds 6, 357–370.

Wasswa, J., Tang, J., Gu, X., 2007. Influence of the extent of enzymatic hydrolysis on the functional properties of protein hydrolysate from grass carp (*ctenopharyngodon idella*) skin. Food Chemistry 104 (4), 1698–1704.

Wei, Y., Liang, M., Mu, Y., Zheng, K., Xu, H., 2016. The effect of ultrafiltered fish protein hydrolysate level on growth performance, protein digestibility and mRNA expression of PepT1 in juvenile turbot (*Scophthalmus maximus* L.). Aquaculture Nutrition 22, 1006–1017.

Wells, G.A., Scott, A.C., Johnson, C.T., Gunning, R.F., Hancock, R.D., Jeffrey, M., et al., 1987. A novel progressive spongiform encephalopathy in cattle. The Veterinary Record 31, 419–420.

Westendorf, M.L., Dong, Z.C., Schoknecht, P.A., 1998. Recycled cafeteria food waste as a feed for swine: nutrient content digestibility, growth, and meat quality. Journal of Animal Science 76, 2976–2983.

Westendorf, M.L., 2000. Food waste as animal feed: an introduction. In: Westendorf, M.L. (Ed.), Food Waste to Animal Feed. Iowa State University Press, Ames, IA, pp. 3–16.

Wu, D., Zhou, L., Gao, M., Wang, M., Wang, B., He, J., et al., 2018. Effects of stick water hydrolysates on growth performance for yellow catfish (*Pelteobagrus fulvidraco*). Aquaculture 488, 161–173.

Yadav, M., Goswami, P., Paritosh, K., et al., 2019. Seafood waste: a source for preparation of commercially employable chitin/chitosan materials. Bioresources and Bioprocessing. 6, 8.

Yaghoubzadeh, Z., Peyravii Ghadikolaii, F., Kaboosi, H., Safari, R., Fattahi, E., 2020. Antioxidant activity and anticancer effect of bioactive peptides from rainbow trout (*Oncorhynchus mykiss*) skin hydrolysate. International Journal of Peptide Research and Therapeutics 26, 625–632.

Yan, J., Guo, C., Dawood, M.A.O., Gao, J., 2017. Effects of dietary chitosan on growth, lipid metabolism, immune response and antioxidant-related gene expression in *Misgurnus anguillicaudatus*. Beneficial Microbes 8, 1–12.

Yin, H., Pu, J., Wan, Y., Xiang, B., Bechtel, P.J., Sathivel, S., 2010. Rheological and functional properties of catfish skin protein hydrolysates. Journal of Food Science 75, 11–17.

Zaki, M.A., Salem, M. El-S., Gaber, M.M., Nour, A.M., 2015. Effect of chitosan supplemented diet on survival, growth, feed utilization, body composition and histology of sea bass (*Dicentrarchus labrax*). World Journal of Engineering and Technology 3, 38–47.

Zamora-Sillero, J., Gharsallaoui, A., Prentice, C., 2018. Peptides from fish by-product protein hydrolysates and its functional properties: an overview. Marine Biotechnology 20, 118–130.

Zhang, L., Zhao, G.X., Zhao, Y.Q., Qiu, Y.T., Chi, C.F., Wang, B., 2019. Identification and active evaluation of antioxidant peptides from protein hydrolysates of skipjack tuna (*Katsuwonus pelamis*) Head. Antioxidants 8, 318.

Zhao, Y.M., de Alba, M., Sun, D.W., Tiwari, B., 2019. Principles and recent applications of novel non-thermal processing technologies for the fish industry—a review. Critical Reviews in Food Science and Nutrition 59, 728–742.

Zheng, K., Xu, T., Qian, C., Liang, M., Wang, X., 2013. Effect of low molecular weight fish protein hydrolysate on growth performance and IGF-I expression in Japanese flounder (*Paralichthys olivaceus*) fed high plant protein diets. Aquaculture Nutrition 20, 372–380.

Zhou, P., Regenstein, J.M., 2005. Effects of alkaline and acid pretreatments on *Alaska pollo*ck skin gelatin extraction. Journal of Food Science 70 (6), 392–396.

Zhou, L., Budge, S.M., Ghaly, A.E., Brooks, M.S., Dave, D., 2011. Extraction, purification and characterization of fish chymotrypsin. A review. American Journal of Biochemistry and Biotechnology 7, 104–123.

FURTHER READING

Vázquez, J.A., Durán, A.I., Menduíña, A., Nogueira, M., Fraguas, J., Mirón, J., et al., 2019. Tailor-made process to recover high added value compounds from fishery byproducts. In: Barba, F.J., Roselló-Soto, E., Brncic, M., Lorenzo, J.M. (Eds.), Green Extraction and Valorization of Byproducts from Food Processing. CRC Press, Boca Raton, FL, pp. 91–140.

Chapter | ten

Lipid Extraction From Fish Processing Residues for Sustainable Biofuel Production

Krishna Kumar Jaiswal[1,2], Swapnamoy Dutta[3], Ishita Banerjee[4], Mayookha V.P.[5] and Mayank Bhushan[6]

[1]Department of Chemistry, Uttaranchal University, Dehradun, India, [2]Institute for Water and Wastewater Technology, Durban University of Technology, Durban, South Africa, [3]Department of Green Energy Technology, Pondicherry University, Puducherry, India, [4]Department of Biochemistry and Molecular Biology, Pondicherry University, Puducherry, India [5]Department of Food Science and Technology, Pondicherry University, Puducherry, India, [6]Department of Nanotechnology, North-Eastern Hill University, Shillong, India

10.1 INTRODUCTION

The fisheries sector is an emerging sector around the world. It grows at a faster rate than any other agricultural and livestock industry and plays a vital role in the development of the livelihoods of developing countries. This prevails the way to generate employment and income, as it influences various subsidiary industries (Kumar et al., 2006; Kumar et al., 2010). Nearly 50 million people worldwide depend on fish farming for their family income, and another 150 million depend on the fish processing industry. More than 10 million people embark on 2.5 million small-scale fishing vehicles and account for 50% of global catches. India ranks second in fish production and produces 5.43% of world production. More than 50% of the total catch of fishes are not consumed for food and generates nearly 32 MT (million tons)

Sustainable Fish Production and Processing. DOI: https://doi.org/10.1016/B978-0-12-824296-4.00009-8

of wastes. India generates >2 MT of wastes from fish processing industries (Kumar et al., 2010; Yuvaraj et al., 2019).

Approximately >170 MT of fishes (including capture and aquaculture), crustaceans, and others are produced worldwide every year. Only about 50%–60% of this total production is used for human consumption. Worldwide the annual production of fish discards is estimated at 25% of the catch, and they are called "nontarget" species and by-products of processing (FAO, 2018a,b; Ormanci et al., 2019; Ferraro et al., 2010). One billion tons of fish have been used for human consumption each year and the waste is discarded from the inedible parts, including head, viscera, fins, tail, skin, and liver. Discarded parts have been considered as valueless waste and are usually dumped on land at dumping sites or into the ocean. Organic components in discarded fish waste have harmful effects on the environment, as they decompose with the production of H_2S, amplified numbers of scavengers in the dumping sites and the production of bad odors (Arvanitoyannis and Kassaveti, 2008).

Therefore the management of fish farming waste has become a serious concern for all. To address this contamination and promote proper utilization of fishery waste, these waste products can be potentially transformed into multiple value-added products. Discarded fish waste is a potential source to produce various bioactive compounds including astaxanthin, chitin, protein hydrolysates, collagen, and lipids, which have pharmaceutical and biotechnological significance (Šilovs, 2018; Ghaly et al., 2013). The hydrolyzing of fish wastes can be used to feed animals and also as manure components. Fish wastes/by-products can be applied to generate mainly proteins and oils. Fish oils are rich in polyunsaturated fatty acids (PUFAs), especially eicosapentaenoic acids (EPA) and docosahexaenoic acids (DHA), which ensure a high commercial value. These two compositions have several essential bioactivities. These can prevent atherosclerosis, lower blood pressure, and protect against arrhythmias (Kim and Wijesekara, 2010; Ngo et al., 2011; Harnedy and FitzGerald, 2012; Rincón-Cervera et al., 2017).

Fish oils can be extracted from the whole fishes, skins, viscera, or livers. Recently the most important, environmentally friendly, and cost-effective way to use fish waste or fish processing wastes have been attracted in the field of biofuels production. Several investigations have been described for the prospective of biodiesel production from fish oils and biogas production from the discarded fish wastes (Jaiswal et al., 2014). This study has incited the renewable and sustainable utilization of fishery waste and fish processing residues for lipid extraction and to apply for sustainable biodiesel production. It has been represented as a clean source of bioenergy and biofuels production.

10.2 WORLDWIDE FISHERIES

Fisheries and aquaculture products have gained worldwide importance for the resource of high-quality aquatic animal protein, employment provider, cash income, and foreign exchange (Bostock et al., 2010). Fish products are the world's most traded food dominated by developing countries. Fish products are the main source of the nutritional proteins for ~950 million people around the world and are an important part of the diet (Porter, 1999). Globally, fisheries provide ~16% of the animal proteins for human consumptions. It is an appreciated source to be used for minerals and essential fatty acids in the diet. Fishes are also important for the omega-3 fatty acids in the diet. Omega-3 fatty acid is a critical nutrient for the typical development of the brain and eyes of babies and has preventive functions in several human diseases, such as cardiac disease, depression, and other cerebral complaints (Pradeepkiran, 2019).

Asia dominates in the capture fisheries and aquaculture production with China being the leading nation in the world. Fish used as raw material to produce animal feeds (i.e., fish meals and fish oil) (Cao et al., 2015). The exponential development in this region has been motivated due to the multiple features, which includes preexisting aquaculture practices, population and economic development, the comfortable governing environment, and escalating trade prospects (Bostock et al., 2010). The nutritional role of aquaculture differes extensively, which depends not only on the fish species and feeds but also on the ecological, social, and cost-effective framework of the production and delivery system (Tacon et al., 2020). Therefore assessing nutritional contributions needs a systematic way to comprehend the delivery, as well as the economic feasibility resulting throughout the supply chains. Both the ecological and nutritional management of farmed fisheries and aquaculture essentially be considered in the framework of the comprehensive food system and the local perspective of both production and utilization (Little et al., 2016). Most of the progression in the fishery sector is expected to ensure in developing country, which may accounts for ~79% of fisheries production for human consumption in 2020 (Gephart et al., 2020).

The maximum relative growth rates in 2006–2007 have been recorded in countries with relatively low productivity (e.g. 6450% in Lesotho, 909.5% in Rwanda, and 590.8% in Ukraine (USAID SPARE Fisheries & Aquaculture Panel, 2010). While these can be an appropriate sign of new ingenuity, a lower percentage of development in countries with previously significant productivity has a better influence. For example, the 5.2% growth in China accounted for 52.3% of the entire upsurge in world aquaculture export in 2007 (Msangi et al., 2013). The 2nd most notable country in this regard was Vietnam, which rewarded 16.7% of surplus aquaculture production with a productivity rate of ~30%. Few countries with substantial productivity experienced shrinkage in 2007, particularly, Thailand, Spain, and Canada. The causes for this were mostly related to the marketplace and affordability, while diseases and unique ecological measures can also influence the annual figures. Overall, these reductions equaled 1.6% of world supply (i.e., more than counterbalance by growth elsewhere) (Mathiesen, 2015).

Fish and other freshwater and marine aquatic foods (i.e., seafood) are key point to achieve food and nutritional security goal and potentially provide more environmentally sustainable animal feed (Belton et al., 2020; Thilsted et al., 2016; Tacon and Metian, 2013). Globally the supply of seafood per capita improved from 9.0 kg in 1961 to 20.2 kg in 2015, and the increase in costs in capture fisheries indicates an even higher demand (Guillen et al., 2019). All future forecasts ecpect the necessity for an enhanced supply of fish protein to meet the health requirements and overall desire of people. Also, it should be at reasonable levels relative to income. As with terrestrial animal proteins, fish protein productivity is ecologically more expensive than plant protein production due to the higher trophic level, while the systems (e.g., enriched polyculture ponds) compares remarkably (Béné et al., 2015).

10.3 INDIAN FISHERIES

Indian fisheries have enormous potential for the growth of fish-based industries and currently one of the focused sectors for fish production, marketing, and consumption. India is the second largest fish producer after China (Bais, 2018). Indonesia ranks third in aquaculture production. In India, this sector constitutes about 5% of world fish production, and 3% of world fish trade. The world supply of fish for human consumption per capita enhanced from an average of 9.9 kg in the 1960s–18.7 kg in 2011 and preliminary estimates for 2012 point to a further increase in consumption of fish to 19.2 kg. The availability of fish per capita in India has increased from 3 to 9 kg. India is one of the world's leading fish-producing countries, employing more than 7 million people in fishing and related industries and contributing Rs. 60 million a year to the national income (Bais, 2018). In addition to providing direct employment, this sector also supports canneries and processing establishments, gear, equipment manufacturers, shipyards, refrigeration, ice making plants, and transportation services. The shortage of human dietary protein can be supplemented by fish protein, especially in developing countries where protein shortages are severe. Aquaculture can play a significant role to meet the future nutritive needs of the human population in terms of providing quality protein. Fish protein is easily digestible and has proven its value as a growth promoter for human consumption. Fish protein comprises the 10 essential amino acids with desirable strength for body metabolism. In addition to protein, fish meat also offers minerals, vitamins, and fats (Jobling, 2016).

Analysis of the growth rate of fish production in various states from 1990–1981 to 2003–2004 showed a significant growth in all states except Karnataka and Madhya Pradesh (Kumar et al., 2010; Bais, 2018). In some of the states, namely, Andhra Pradesh, Karnataka, Kerala, Maharashtra, Orissa, and West Bengal, the growth of inland fisheries was found to be greater than the marine fisheries. But in Gujarat and Tamil Nadu the growth of marine fisheries has been observed to be greater than the inland fisheries. In 1990–91 Kerala, Gujarat, Maharashtra, Tamil Nadu, and Karnataka together contributed about 81% of total marine fish production. In the last 15 years, Gujarat was reaching a plateau and its contribution has been declining recently. Kerala has shown static on its rise while the ratio of Maharashtra and Tamil Nadu did not change. The only state that has been increasing at a rapid rate in marine fish production is Andhra Pradesh, which replaced Karnataka in its share of total fish production in the last 15 years. Among the other states the main contributors to total marine fish production are West Bengal and Orissa. Among the Union Territories (UT) the main contributors are the Puducherry and Andaman Nicobar Islands. Although their current fish production has stagnated in the last years. In the years 1990–1991 the main producing states were West Bengal, Bihar, Andhra Pradesh, Uttar Pradesh, and Orissa. Together they contributed about 68% to total inland fish production. In the next 15 years, Andhra Pradesh and Uttar Pradesh had made tremendous progress in their share of total national production, overtaking West Bengal, which remained static in its rise and therefore lost to these emerging states. Bihar and Orissa maintained their participation in national production, although they have reached stagnation in production. Apart from these the states that are reaching peak production are Assam, Kerala, Maharashtra, and Tamil Nadu (Duarah and Mall, 2020; Gopalakrishnan et al., 2017; Ayyappan and Krishnan, 2004).

10.4 FISH PROCESSING INDUSTRIAL WASTES AND USES

In the current food generation scenario, waste has been observed at various levels, such as in the initial stage of processing, at the time of distribution, during use by the consumer, etc. In all this waste generation a more significant portion is contributed by food processing industries. The fish processing industries are the one of the most critical sectors that generate an excessive amount of wastes, including liquid and solid waste. In general, waste is generated in stages such as handling by retailers and consumers, transport, storage, etc. During processing activities, mainly muscle segments are used, and the remaining parts are discarded. In the fishing industry the meat portion is mainly considered consumable. The residual pieces, such as head, trimmings, skin, viscera, scales, bones, etc., are considered waste segments and generally discarded without any further usage. In the fishing sector the harvest site generates many wastes as around 20%–35% of the catches are lost in different phases of the value chain in the postharvest period. Also it was found that ~18–40 MT of entire fishes has been discarded each year in industrial fisheries operations. Therefore this significant amount that is conventionally assigned a lower value can be used for various biofuels, pharmaceutical, bioactive compounds, biotechnological, etc.

Also products extracted from fish have higher nutritional value, which is considered a useful resource for the bioprocess industry to produce several valuable products. In a study, it was reported that carotenoid pigments, aromas, and intestinal enzymes generated from fish waste can be considered for valuable applications. Some of the primary uses suggested by current researchers based on profitability and value-added applications are fertilizers, aquaculture feed, livestock, domestic animals, the food industry, biotechnology, and pharmaceutical sectors.

The application of fish wastes has been reported to be used as a fertilizer for agricultural purposes, and fish-based manures have been applied for several decades in the agricultural sector. Another, the fish meal production, has been intended for animal feed, and it can cover up to ~14% of the world's animal protein requirements (Venugopal et al., 1995). The viscera of fish, mainly digestive tissues, can be used to produce fish silage, which is a liquefied product composed of any type of fish with the addition of acids and enzymes present in the fish itself. Fish fins, bones, and skins are the main by-products of the fishing industries, containing a high level of collagen as a structural protein. The collagen extracted from the fish wastes can be utilized in various pharmaceutical and biomedical fields. Collagen can be used to produce wound dressings, a vehicle for drug delivery. Also, gelatin, a collagen breakdown product, can be obtained from the fish waste to be utilized in the food/cosmetic industry. Fishbone and teeth contain a high level of calcium phosphate, which is natural hydroxyapatite. Therefore hydroxyapatite extracted from fish waste has promising biomimetic and biocompatible characteristics, which can be used in the biomedical field, toothpaste production, and in bone cement (Ormanci et al.). Several studies have demonstrated that marine fish waste comprises a higher level of carotenoids, especially the head and shell of shrimp, eggs, and fish scales. These carotene-based extracted pigments, such as astaxanthin, b-carotene, lutein, astacene, canthaxanthin, and zeaxanthin, are valuable in medicinal applications. The omega-3 fatty acid of fish oil has been best known for preventing atherosclerosis, the reduction of blood pressure, and the protection against arrhythmias. Furthermore several researchers are using oils extracted from fish waste segments for biodiesel production, showing some promising results (Dutta et al., 2019).

10.5 FISH OIL FROM FISHERIES INDUSTRIES AND USES

From the fisheries industries, out of the total amount of fish produced, 60%–70% is used for human consumption, fish meal, and oil production (Eslick et al., 2009; Sharma et al., 2014). Fish oil has been found to have more than 60 catagories of fatty acids in it. Moreover, almost 80%–85% are classified into four fatty acid groups, such as (1) C14:0 and C16:0, (2) C16:1 and C18:1, (3) C20:1 and C22:1, and (4) C20:5, C22:5 and C22:6. EPA and DHA are enriched in fish oil, representing over 90% of all PUFAs (Girish et al., 2017).

In the fish processing industries, not all fish parts are consumed, certain parts are discarded. In the discarded parts the backbone, skin, head, tail, and stomach are not edible and considered as waste. The seafood industry generates lots of wastes that can be used to extract fish oils. Nutraceutical and food industries are utilizing an enormous quantity of lipid extracted from gut, head, fin, scale, and fish skin for their product formation. Coldwater fish are the primary source of the above mentioned fatty acids (Chowdhury et al., 2012). For instance, we can mention sardines, mackerels and anchovies, which are the rich sources of EPA and DHA. Fishes are rich in essential fatty acid, including long-chain fatty acids (C20 and C22), named PUFAs. These components are known as vital since the animals cannot synthesize these, and their precursors should be included in the food (Bell, 1995). Furthermore, PUFA is commonly utilized by the food industries to manufacture fortified foods since in addition to the fact that these compositions are in the ester form; these fatty acids are members of the omega-3 family. This fact illustrates the need to acquire more usable quantities of natural sources for the human body. Omega-3 fatty acids are alternate sources of development for functional foods to assist in protecting cell membranes and facilitate the improved functioning of the brain and eyes, among other benefits (Rubio-Rodríguez et al., 2010; Swanson et al., 2012). In matured, DHA participates in preserving usual brain and nervous system activity (Kuratko et al., 2013; Domenichiello et al., 2014).

Fish oil fractionation experiments were performed on EPA and DHA, and fortified foods were associated with health benefits, with cardiac protecting properties, including antiarrhythmic, antithrombotic, antiinflammatory, antihypertensive, and antihyperlipidemic effects (González-Sarrías et al., 2013). They can also prevent carcinogenesis and slow down the growth of the tumor. To attain the essential fatty acids, EPA and DHA for the exploitation of their prospective use in the pharmaceutical and food industries, there is a need and industrial interest to invest in further research given that these fatty acids perform a significant role to prevent human disorders. Also the usage of such fatty acid is considered as leftover from manufacturing operations, but concerning the use of fish oil, they may be used as by-products as an inexpensive solution to reduce disposal complications and ecological contamination (Rai et al., 2013).

Numerous techniques were applied to manufacture PUFAs concentrate, including urea complexation, extraction using supercritical fluids, crystallization at low temperatures, a distillation of molecules, concentration of lipase and chemo-alcoholysis (Adeoti and Hawboldt, 2014; Homayooni et al., 2014). While the current fish oils manufacture scenario is constant, upcoming forecasts indicate that the wild marine resources available are becoming increasingly restricted while demands for PUFA continue to increase. This not only indicates but also emphasizes the necessity for alternate outlets to be explored and exploited (Ivanovs and Blumberga, 2017).

The remainder of the fish processing industry's raw materials, commonly defined as by-products, fluctuate between the species and procedures of processing but typically comprise parts that persist after the removal of eatable portions of the fishes: head, viscera, bone, attached flesh, skin, fin, trims, blood, etc., which are not used as food. The remaining discards should not be considered as wastes or as less important than the main products (Ivanovs and

Blumberga, 2017). The high nutritional significance of fish discards are due to the quality of precious mineral, vitamin, protein, and lipid fraction, so they are applicable for a multiple purposes, which includes the manufacturing of fish cakes, gelatine, sauces, and other products for human ingestion to the sustainable biodiesel manufacture (Kara et al.,2017). It is also possible to derive bioactive compounds such as peptides, enzymes, or collagens being utilized in pharmaceuticals, biomedicals, and biotechnological productions from fish discards (Arason et al., 2009; FAO, 2018a,b). Products of low value (e.g. fish meal) have been substituted by-products of higher value with the concern and technical expertise in the area of by-product application and characteristics of some nutrients such as proteins and oils that can be derived from fish discards (Arason et al., 2009; Rustad et al., 2011).

Distinct species of fishes possess specific processing yields, resulting in varied amount of by-products being produced. Furthermore the nature and chemical nature of marine by-product differs as per the species, seasons and catches, as well as management and postcapture processes. By-product components, while not specifically used for human consumption, have been utilized for feeding cattles and aquaculture industries and contribute indirectly to human food assembly. Potential drawbacks to the use of by-products relate to the inferior quality of the fish meals and oil compared to product achieved from entire fishes and to the fast enzymatic activity which makes them extremely perishable. Numerous investigations have shown that by-product oils can be efficiently extracted and processed to eliminate certain unwanted components such as moisture, pigment, free fatty acid, phospholipid, mineral, off-flavor, etc. that affect the stability of the oils, the general quality and the suitability of consumers, devoid of the loss of PUFA that allow its further applicability (Menegazzo et al., 2014; Huang and Sathivel, 2010; Ferdosh et al., 2015; Kuo et al., 2017; Nazir et al., 2017).

10.6 EXTRACTION OF OIL FROM FISH BY-PRODUCTS AND RESIDUES

The extraction of fish oil has been carried out with different conventional and neoteric approaches. In recent years, various physicochemical techniques have been developed to extract fish oils efficiently from fisheries and fish processing residues. The extraction techniques have been described as follows:

10.6.1 Wet Rendering Technique

The wet rendering extraction technique for the extraction of fish oil using a potential feedstock of fish processing wastes is known as a traditional technique. Typically, this process is used in a conventional fish reduction plant. This particular method is beneficial when extracting oil from specific biomass categories that are made up of higher oil content (Table 10.1). This procedure follows several sets of steps, which are cooking, dehydrating, separating, and purifying.

TABLE 10.1 Fish Oil Extraction Through Wet Pressing

Fish Species	Temperature (°C)	Time (min)	Yield (%)	References
Scomber scombrus	95–100	10–20	18.7	Bako et al. (2014)
Sufflamen capistratus	35 ± 2	30	40	Immanuel et al. (2009)
Acipenseridae	85	60	52.51	Hao et al. (2015)
Sardinella longiceps	75	30	8.3	Chakraborty and Joseph (2015)
Fish by-products	90	30	2.13	Silva-Aguiar and Soares-Goulart (2013)
Theragra chalcogramma	50–60	15–30	~52	Ribeiro et al. (2013)

10.6.1.1 COOKING

Cooking is the process of beginning the wet rendering process. In this procedure the fish is cooked to obtain water and oil segments from the fish waste. Generally, to carry out the cooking procedure, two models of fish cookers are used: direct and indirect cookers. In general, direct cookers are used to load fish into a hopper on the internal screw conveyor. After that a flow of steam is induced through a set of smaller inlet pipes into the lower section of the pot. Especially in the case of direct cooking, the need for steam is much higher. It has been reported that for one ton of waste of fish, the amount of steam required would be about ~270,000 Btu or about 300 pounds with a temperature of about ~195°F (Yuvaraj et al., 2019). Cooked fish is discharged via a small choke screw.

10.6.1.2 DEWATERING

After the cooking procedure the cooked fish was dewatered to convert the biomass into a pumpable suspension. The dewatering procedure has been performed using a horizontal bucket centrifuge. The cooked fish biomass has been passed through a disintegrator to form a pumpable suspension. Mainly the slurry is pumped into the centrifuge and then pressed. By this procedure the liquor and concrete segments have been adequately recovered. In some of the cases, it has been observed that the dewatering procedure can be avoided, but still, this method is quite efficient, and more oil segments can be successfully recovered (Stansby, 1967; Yuvaraj et al., 2019).

10.6.1.3 SEPARATION

Once the dehydration process has been completed, some residual elements such as proteins, bone fragments, and scales remain in the pressed liquor, which are generally eradicated through small openings. A screening and centrifugation procedure usually removes these remaining particles. When making an initial selection the food is placed in the disk bowl centrifuge for an additional process. Subsequently the solids are recovered after centrifugation that has a higher amount of oil and then fed back through the presses. The pressing operation is done separately, or in some cases, they are added to the press cake and dried for later use. Since the solid elements in suspension are eradicated the press liquor is sent for isolation purposes where the liquid segments must be isolated by using channels in the set of disks. Also the rising channels used here are determined by the proportion of the liquid phases. Finally the liquid segments with a light phase such as oil pass into the bowl head's inner section and then released through a particular partition of the hood.

10.6.1.4 PURIFICATION

Purification is the essential step to finally obtain the pure extracted lipid that is used for different applications. For the most part, purification is done using the older sedimentation tank process. This is where oil oxidation and degradation occurs when the oil press liquor can have more prolonged contact between the oil, water, and flour. Although recent studies find that the novel centrifuges generate a more significant amount of oil since the oil separation takes place just after the separation procedure is performed. Therefore the use of these new centrifuges and their instantaneous operation improves the amount of oil produced in several gallons. Finally, to eradicate the resulting oil's moisture content, some heating procedures are required with the help of the oven (Oliveira et al., 2013; Yuvaraj et al., 2019).

10.6.2 Dry Rendering Technique

The dry rendering extraction procedure is used mainly in the case of feedstock composed of a smaller quantity of oil and also accessible in smaller amounts. Furthermore, this process is classified as a batch process that comprises both cooking and drying processes. Here the feedstock has been processed using rotary type kitchen dryer equipment where the dryer used is a type of vacuum dryer. Once the feedstock is dried under vacuum conditions the biomass segments containing the oil have been sending to the hydraulic batch type press. The hydraulic press recovers most of the oil. The oil segment retrieved by the hydraulic press is typically much darker in color than the oil recovered by the wet extraction process. Generally, to obtain an oil of the right color, procedures such as alkaline refining, clay bleaching process has been employed. The dry rendering extraction process also requires the separation of the oil with other impurities and wastewater. Here the residual fish biomass has been collected and divided into small segments. These segments are then put into boiling water for almost ~20–25 minutes and then cooled for ~35–40 minutes. The resulting stew is then repeatedly strained at a fast flow rate, and in some cases, a fine sieve has been used in the final filtration. This filtration and the filtration procedure are performed repeatedly for almost ~2–3 times. And finally the filtration result is stored in an isolation funnel to separate the oil from the wastewater.

10.6.3 Solvent Extraction Technique

In addition to the wet and dry extraction process, several solvent extraction methods have been described to extract oil from fish processing wastes. These extraction procedures using different types of solvents are mainly used for analytical purposes. This method is avoided for food industry production purposes due to some restrictions on the harmful effects of other solvents (Rubio-Rodríguez et al., 2010). Solvent extraction procedures are generally based on the solubility of lipids in organic solvents. Using these methods, soluble components such as protein, carbohydrates, and minerals have been extracted and isolated from the water medium. However, some of the disadvantages observed in this process are that a relatively dry sample is required for this process and a long duration of the process and, in some cases, an excessive amount of residual solvent has generated adverse effect on the extracted oil (Adeoti and Hawboldt, 2014; Sahena et al., 2009).

Different solvent extraction approaches vary depending on the type of solvent or the process methodology used during the extraction process. The most commonly used solvent extraction processes are the method of Soxhlet, Bligh-Dyer, and Folch (Fiori et al., 2012; Immanuel et al., 2009; Rincon Cervera et al., 2015). Various solvents are used in different types of approaches. For example, in the Bligh-Dyer method, the solvents chloroform, methanol, and water have been used. In the Soxhlet extraction method, n-hexane, chloroform, methanol, petroleum ether, etc., have been used. In the given table the different solvents have been evaluated to extract the lipids from various species of fishes (Table 10.2). The Randall and Folch methods are not among the most widely used procedures, but some researchers have still explored these methods. The Folch method was used using chloroform and methanol solvents using the Solomon fish species. In a different study by Fiori et al. (2012), rainbow trout fish species (*Oncorhynchus mykiss*) were used to observe the extraction result using the Randall procedure, where n-hexane was used as a solvent (Rincon Cervera et al., 2015).

10.6.4 Supercritical Fluid Extraction Technique

Recently, there have been some advances in extraction technologies. Among the more recent methods established the supercritical fluid extraction procedure is a potential method that can have a significant impact on the extraction of oil. This extraction procedure uses different solvents for extraction purposes. This method has certain benefits: the extraction process that can be carried out at a moderate temperature, an oxygen-free environment, and the extraction of low polarity lipids. With this method the extraction of impurities with the oil can be avoided (Rubio-Rodríguez et al., 2012). These benefits make the process attractive enough for future research. Generally, supercritical fluids are less viscous, and

TABLE 10.2 Lipid Extraction From the Different Species of Fishes

Fish Species	Extraction Methods	Solvents	Reference
Oncorhynchus mykiss	Bligh and dryer	Chloroform, methanol, water	Kołakowska et al. (2006)
Theragra chalcogramma	Bligh and dryer	Chloroform, methanol, water	Silva-Aguiar and Soares-Goulart (2013)
Sufflamen capistratus	Bligh and dryer	Chloroform, methanol, water	Immanuel et al. (2009)
Sardinella longiceps	Bligh and dryer	Chloroform, methanol, water	Chakraborty and Joseph (2015); Létisse et al. (2006)
Rastrelliger kanagurta	Soxhlet	Petroleum ether	Sahena et al. (2010)
Seriola lalandi, Genipterus chilensis	Soxhlet	n-hexane	Rincon Cervera et al. (2015)
Seriola nigrofasciata	Soxhlet	Chloroform	Shamsudin and Salimon (2006)
Scomber scombrus	Soxhlet	Methanol	Adeniyi and Bawa (2006)
Trachurus trachurus, Alosa fallax, Belone belone, Mugil auratus	Soxhlet	Petroleum ether	Boran et al. (2006)

their diffusivity is higher than traditionally used solvents. Therefore, due to the use of supercritical fluids, the mass transfer has been improved, and the duration (time) of extraction was also alleviated. Despite having these advantages, it has a significant drawback. This technology is not very economical for large-scale applications so far. But ongoing research may show some thoughtful way to use it for industrial-scale purposes. In this oil extraction procedure, CO_2 is the primary fluid used. This supercritical fluid is used as an inert solvent, and no risk has been observed when using this solvent for oil extraction. The benefit of this supercritical fluid solvent is that it turns into a gaseous state and evaporates as soon as extraction is performed. Ambient temperature and pressure are sufficient to vaporize this solvent. Because it does not remain in the product; therefore, impurities are generally not found in the extracted oil (Rubio-Rodríguez et al., 2010, 2012).

In general, it is observed in the literature that researchers mainly tried to estimate the nature of various parameters such as pressure, temperature, CO_2 flux, and the length of the extraction period. Besides, in some specific investigations, some other characteristics such as the impact of humidity, flow direction, solvent-material ratio, and particle size have been investigated using fish oil (Rubio-Rodríguez et al., 2008). On the other hand, some researchers tried to find the optimal conditions using this process to produce fish oils rich in PUFAs (Ferdosh et al., 2013; Sarker et al., 2012). Here in the table, we have shown some parameters of the process with their specific fish species investigated by different researchers (Table 10.3).

10.7 BIODIESEL PRODUCTION FROM THE LIPIDS OF FISH PROCESSING WASTES

Different transesterification processes have investigated biodiesel from the lipids of fish processing wastes. Here, we have deliberated the developments in transesterification processes, including microwave approaches (Jaiswal et al., 2014; Aidos et al., 2000; Sathivel et al., 2008; Sharma et al., 2014).

10.7.1 Transesterification of Fish Lipids

Biodiesel is generally generated by the transesterification reaction. The lipids extracted from fish processing wastes undergo the response with the methyl alcohol (methanol, ethanol, etc.) in the presence of catalysts (Fig. 10.1). The resulting products are to be fatty acids, alkyl esters, and glycerin. In transesterification reactions a different range of alcohol to oil molar ratio can be employed to achieve a satisfactory yield of biodiesel. Theoretically the stoichiometric ratio of 3:1 (alcohol to oil molar ratio) has been established as a

TABLE 10.3 Supercritical Fluid Extraction of Fish Oils

Fish Species	Pressure (MPa)	Temperature (°C)	References
Rastrelliger kanagurta	20–35	45–75	Sahena et al. (2010)
Oncorhynchus mykiss	50	60	Fiori et al. (2012)
Salmo salar	25	45	Haq et al. (2017)
Thunnus tonggol, Euthynnus affinis, Auxis thazard	40	65	Ferdosh et al. (2014)
Sardinella lemuru	20–40	40–70	Gedi et al. (2015)

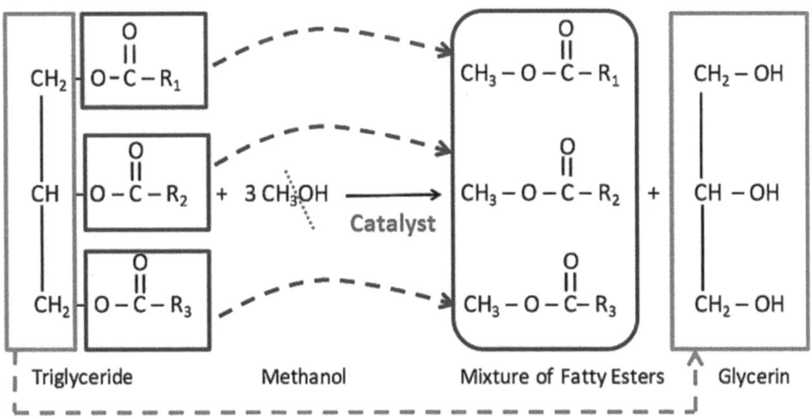

FIGURE 10.1 Typical mechanism of transesterification reaction for biodiesel synthesis.

significant ratio for the transesterification process. But in some experimental studies, it has been observed that a molar ratio can range from about 6:1 to 12:1 (alcohol to oil molar ratio) to provide an optimized yield of fatty acid methyl esters (Sharma et al., 2014). Furthermore, with the optimized alcohol to oil molar ratio, the use of appropriate catalysts also plays a vital role in the transesterification reaction. Here, we conferred the different catalysts that can be used in the transesterification reaction with particular advantages and disadvantages.

10.7.1.1 ACID-CATALYZED TRANSESTERIFICATION

Several researchers have been investigated the transesterification reaction with acid catalysts. It has been observed that in some cases the free fatty acid present in the oil reacts with a base catalyst and generates soap. The soap formation creates complications to separate the esters, glycerin, and wash water. Also, it affects the quality and quantity of biodiesel (Talha and Sulaiman, 2016; Canakci and Van Gerpen, 1999). To combat this complication the use of homogeneous acid catalysts has been suggested for the transesterification to produce biodiesel. This has been found to work mainly to mitigate these challenging situations caused by essential catalysts. The most commonly used homogeneous acid catalysts in transesterification reactions are sulfonic acid, hydrochloric acid, sulfuric acid, organic sulfonic acid, and ferric sulfate (Atadashi et al., 2013). In most studies the uses of hydrochloric acid and sulfuric acid have been investigated (Lam et al., 2010). Also the two-step process of esterification and transesterification has been performed using the acid catalyst, to improve the yield of the biodiesel (Helwani et al., 2009; Arzamendi et al., 2007; Kouzu et al., 2008). Furthermore, with the acid catalyst, the ratio of alcohol to oil (molar) was also the main parameter that impacted the reaction. Using an additional amount of alcohol can accelerate the reaction rate and also have a significant effect on the performance of biodiesel. After having such benefits, homogeneous acid catalysts also have certain drawbacks, such as contamination complications due to acid catalysts that act as inseparable from the final product. To resolve this problem, additional separation and purification steps are required that undesirable add costs (Lotero et al., 2005).

So to ease the complications the use of heterogeneous acid catalysts has been investigated and found to work more efficiently than homogeneous acid catalysts. The benefits of heterogeneous acid catalysts have been observed, as they are insensitive to the content of free fatty acids and manage esterification and transesterification at the same time, without the need for an additional washing and separation procedure to remove the catalysts from the resulting product. It also minimizes corrosion and the catalyst can be recycled for reuse (Helwani et al., 2009; Lopez et al., 2005; Di Serio et al., 2008). Furthermore, heterogeneous acid catalysts are suggested over homogeneous acid catalysts because they are composed of several sites with different Lewis and Bronsted acid strengths (Atadashi et al., 2013; Helwani et al., 2009;

Santacesaria et al., 2012). However, several concerns have been observed for the use of a heterogeneous acid catalyst. Lower reaction rates and adverse side reactions have been observed during the transesterification reaction (Helwani et al., 2009). These problems possibly deteriorate both the quantity and the quality of the biodiesel. Still, research is going to find an efficient way to use acid catalysts with maximum efficiency.

10.7.1.2 BASE-CATALYZED TRANSESTERIFICATION

Base catalysts are frequently used in the transesterification reaction. It is classified mainly as a homogeneous and heterogeneous category, and both have specific benefits and drawbacks that have been described by various researchers in their respective investigations. In some cases, it has been observed that to combat the slower rate of the acidic counterpart of the transesterification reaction; different homogeneous base catalysts have been used, which have also shown positive results (Lotero et al., 2005). Potassium hydroxide, potassium methoxide, sodium hydroxide, sodium methoxide, and sodium ethoxide are frequently used homogeneous base catalysts applicable in transesterification reaction (Lam et al., 2010; Narasimharao et al., 2007; Atadashi et al., 2013). Several benefits that have been observed through the use of these homogeneous base catalysts include comparatively lower reaction temperature, more downward pressure requirement, the higher conversion rate in less time, and economically feasible (Lam et al., 2010). It has been observed that sodium methoxide (CH_3NaO) and potassium methoxide (CH_3KO) catalysts showed a better performance than sodium hydroxide (NaOH) and potassium hydroxide (KOH) due to their ability to disintegration into CH_3O^- and Na^+ or CH_3O^- and K^+, respectively, and the clear impacts have been observed in the yield of the resulting biodiesel (Atadashi et al., 2013). Base catalysts are most often used for the commercial purpose of producing biodiesel because much less water is generated during the transesterification reaction (Sharma and Singh, 2008).

Not only homogeneous base catalysts but also heterogeneous base catalysts have been established for their working ability to verify the comparative analysis of both types and to find which one has more efficiency. The largest surface area has been observed in many heterogeneous base catalysts (Helwani et al., 2009). These catalysts are more active than the heterogeneous acid catalyst and their homogeneous counterparts (Arzamendi et al., 2007). In addition, heterogeneous-based catalysts, such as calcium oxide, have shown several benefits, such as higher operational performance, longer catalyst life (specifically recyclable and reusable), and able to maintain a moderate reaction environment (Kouzu et al., 2008). However, according to some other research, it has been observed that the calcium oxide catalyst harms the reaction rate (reducing the reaction rate) during the generation of biodiesel (Chew and Bhatia, 2008). Also, calcium oxide reacts with glycerol to form calcium diglyceroxide during the reaction of transesterification process, which requires additional purification steps (e.g., ion exchange resin) to remove soluble elements from biodiesel (Gryglewicz, 1999; Kouzu et al., 2008, 2009). In some of the reports, it has been found that researchers have tried different natural resources or waste materials to derive certain types of heterogeneous base catalysts. In a study the researchers have derived heterogeneous base catalysts from eggshells (Viriya-Empikul et al., 2012; Wei et al., 2009). Thus the use of this type of waste material can reduce the cost of the biodiesel production process. In addition, the amounts of heterogeneous base catalysts of 10%–25% have been reported to have a biodiesel yield of ~95%–97% (Nakatani et al., 2009; Viriya-Empikul et al., 2012). While in some other studies, it has been recognized that a smaller amount (3%–5%) of these catalysts can produce a noble amount (93%–95%) of biodiesel (Wei et al., 2009).

10.7.1.3 ENZYME-CATALYZED TRANSESTERIFICATION

The enzymatic catalysis of transesterification reaction has attracted a lot of attention due to the postprocessing problem that arises when using the chemical method of transesterification with acidic and basic catalysts, such as generating an excessive amount of wastewater and the recovery of glycerol. Therefore enzyme catalysts are considered to mitigate these problems. Using enzymatic catalysts, certain benefits have been observed, such as the generation of very few by-products or harmful elements, easy product recovery, moderate reaction condition, insensitive to the higher content of free fatty acids of the oil, and the reusable catalyst (Kulkarni and Dalai, 2006). Thus these advantages of enzyme catalysts make it a potential option and a promising substitute for the chemical transesterification to generate biodiesel in an environmentally friendly way. Although there are some limitations to employ enzyme catalysts for the transesterification, which are the higher cost and slower reaction rate (Bajaj et al., 2010).

10.7.1.4 ACID AND ALKALINE ESTERIFICATION PROCESS

This is a two-stage reaction process in which acid and alkali esterification processes have been carried out one after the other to obtain the final biodiesel product from fish waste. The extracted oil has been used for esterification with acids and alkalis to obtain biodiesel. Initially the acid esterification procedure needs to be carried out using an H_2SO_4 catalyst (2 wt. %). During acidic esterification, methanol has been added with the extracted oil maintaining the molar ratio of 6:1 (methanol to oil) with the acid catalyst (H_2SO_4). After acid esterification the solution has been separated into two layers when suspended in a segregated funnel: oil (top layer) and a liquid layer. Then, in the next step, alkaline esterification has been carried out. Here in this step a solution of NaOH (2 wt. %) dissolved in methanol used to mix with the resulting oil to perform the alkaline esterification process. At the end of the reaction the final product resulted in two layers, that is, the upper layer of biodiesel and the lower layer of glycerol. Further the purification process has been carried out using distilled water to eradicate the impurities from the product. Then, to make the biodiesel free of moisture, it is kept at a temperature of about ~90°C in a vacuum oven.

10.7.1.5 MICROWAVE-ASSISTED FISH LIPID EXTRACTION AND TRANSESTERIFICATION

Recently the microwave-assisted process has been investigated for efficient lipid extraction and transesterification. In general the dried and powdered biomass of fish processing wastes has been used to extract lipids. The microwave irradiation assists the extraction of lipids by shortening the time duration. The extracted lipids can be separated from the used solvent by filtration and rotatory evaporation. Another microwave-assisted transesterification has been performed using the extracted lipids to transform into fatty acid methyl esters (Yang et al., 2009). The researchers have attempted to estimate optimized parameters for microwave-assisted transesterification. It has been reported that the 6:1 methanol to oil (lipids) ratio is appropriate for efficient performance. The optimization of temperatures and time duration indicates the efficient transesterification at 65°C for a reaction period of 10 minutes. In addition the use of catalysts (e.g., KOH at 1.5 wt. %) has also been suggested to offer an adequate impact on the process. The by-product glycerol formed during the transesterification reaction can be separated from the biodiesel by centrifugation. The purification steps can be followed to remove any impurities.

10.8 CHARACTERISTICS OF BIODIESEL

Biodiesel production and characterization are the necessary elements to maintain the fuel standard to control the quality of standard specifications. The fuel standards have been developed to ensure that there are no engine performance issues by using biodiesel produced from different types of feedstocks. The properties of biodiesel have been carefully compared with the standards of national and international authorities. Some of the standard specifications such as ASTM (6751–3), EN 14214 (EU), DIN 51606 (Germany), ON (Austria), CSN (Czech Republic) have been considered by researchers to maintain the quality of biodiesel in production (Mishra and Goswami, 2018). Here in this section, we have discussed several essential fuel properties necessary to be maintained during the production of biodiesel.

10.8.1 Viscosity

The viscosity of the fuel generally defines its flow and atomization. Highly viscous fuel generally indicates poor fuel atomization (Lin and Li, 2009). For the most part the viscosity of biodiesel is much higher (10–15 times) than traditionally used diesel due to its more massive chemical structure, as well as its larger molecular mass (Mishra and Goswami, 2018). Therefore the problem with biodiesel is a higher viscosity, which also creates issues such as ring sticking, poor atomization, carbon deposits, and fuel pump failures

(Kuepethkaew et al., 2017). In literature, it has been observed that biodiesel derived from marine fish oil has a low atomization performance (Lin and Li, 2009). According to ASTM 445 standards the acceptable limit range of viscosity is 1.9–6.0 mm^2/second. Most biodiesel produced from different freshwater and marine fish discards has shown its viscosity limit of 2.90–7.2 mm^2/second. According to recent studies, it has been observed that the viscosity of biodiesel derived from used fish oil was maintained almost or was higher to a lesser degree than the international ASTM standards. However, most researchers stated that the viscosity of the oil derived from fish discards is based primarily on the fatty acid content of the fish oil.

10.8.2 Density

The density of biodiesel has a brief impact on fuel efficiency and engine performance. Mainly the effect of density has been observed during the injection and combustion of the derived fuel. The problems caused by higher density are the deterioration of the atomization efficiency, a slower injection, and a slower combustion rate. In studies, it has been observed that biodiesel has a higher density value than diesel. It has been found that biodiesel derived from freshwater fish discards has a density value between 0.87 and 0.88 (Fadhil and Ali, 2013; Fadhil et al., 2015). In another study, it was stated that biodiesel derived from fish processing waste has a density value of $\sim$0.87 (Fadhil and Ahmed, 2016). Therefore, from the investigations, it can be determined that the lower density biodiesel would possibly not affect the injection and combustion process of the engine.

10.8.3 Stability to Oxidation

Oxidation stability is an essential characteristic in determining the storage life of the biodiesel. The quality of biodiesel can vary due to the oxidation of fatty acids, which occurs during the most extended storage period of biodiesel. During the combustion of oxidized biodiesel, a certain amount of carbon has been deposited in the combustion chamber, and a large amount of nitrogen oxide emissions have been observed, which are often much more than nonoxidized fuels (Kim et al., 2018). Generally the degree of oxidation depends on the amount of unsaturated fatty acids contained in the fuel. Therefore the degree of oxidation increases if the amount of unsaturated fatty acids is more generous. Biodiesel made from fish oil has been found to have a higher amount of saturated fatty acids compared to vegetable oil. As a result, biodiesel derived from fish oil has slightly higher oxidation stability than biodiesel derived from oilseeds. However, it has been studies that the addition of specific antioxidants can improve oxidation stability. It has been reported that the oxidation stability of fish biodiesel can be improved using synthetic antioxidants, for example, butylated hydroxytoluene, butylated hydroxyanisole, and propyl gallate for about 1 month (Kudre et al., 2017).

TABLE 10.4 Acid Value of Fish Biodiesel

Fish Species	Ester Type	Acid Value (mg KOH/g)	References
Cyprinus carpio	Methyl esters	0.007	Fadhil et al. (2015)
Silurus triostegus Heckel	Methyl esters	0.39	Fadhil et al. (2013)
Labeo rohita	Methyl esters	0.25	Kudre et al. (2017)
Fish processing discard	Ethyl esters	0.14	Fadhil and Ahmed (2016)

10.8.4 Acidity Index

The acidity or neutralization number can be determined from the free fatty acid content of the fuel (Mishra and Goswami, 2018). According to international standards such as ASTM D664 and EN 14 104 the acid number of biodiesel must be around 0.50 mg KOH/g. Table 10.4 shows the acidity values of the biodiesel derived from different fish discards, which shows lower than the ASTM and EN standards, further representing the effectiveness of the fish oil biodiesel.

10.8.5 Flashpoint

The flashpoint is a specific temperature at which fuel ignites when the fuel is exposed to the source of ignition. The flashpoint temperature is related to the volatility of the fuel. It has been observed that biodiesel has a higher flash point temperature range than diesel (Kim et al., 2018). According to ASTM D93 standards the flashpoint limit is 93°C, and according to EN ISO standard the flashpoint extension is up to 120°C (Mishra and Goswami, 2018). In a study the flashpoint of biodiesel produced from fish oil of *Cyprinus carpio* and *Silurus triostegus* was observed to be ~83°C and ~87°C, respectively (Fadhil and Ahmed, 2016; Fadhil et al., 2015). These observed lower values of biodiesel derived from fish oil than international standards (ASTM and EN) further indicate the viability.

10.8.6 Cloud Point and Pour Point

Cloud point is defined as the temperature below which biodiesel has a cloudy appearance when cooled. The lowest temperature at which the filter has been tested occurs due to the crystallization of fuel components. It is the plugging point of the cold filter (Lin and Li, 2009). The International standards ASTM D97 and ASTM D2500 are also used to measure cloud points. To determine the cold flow properties the pour point and cloud point of biodiesel are used to monitor (Kim et al., 2018). The determination of these values is crucial as they help to observe the solidification of the fuel, otherwise, it can partially or entirely clog the fuel lines and filters and create problems while starting the engine.

10.8.7 Cetane Number

The compression ignition quality of the fuel is determined by using a cetane number. The cetane number is inversely related to the ignition and combustion delay time. Generally, higher cetane numbers result in relief of detonation events, and the formation of nitrogen oxide compounds can also be reduced. The international standards ASTM D613 and EN ISO 5165 are generally used to monitor the cetane number for biodiesel. In recent studies the cetane number of biodiesel derived from fish discards has been observed to be analogous to the cetane number standardized by international standards (ASTM D613 and EN ISO 5165) (Mishra and Goswami, 2018). Therefore it indicates that the biodiesel from fish oil can be used for combustion engines.

To evaluate fish oil with greater precision, properties such as calorific value, auto ignition temperature, carbon residues, moisture, elemental oxygen, and sediment residues are also monitored. Thus these properties are a comprehensive approach to understand the various advantages and disadvantages of biodiesel derived from the fish processing wastes and the correlation between them.

10.9 FUTURE PROSPECTS AND INDUSTRIAL PERSPECTIVE FOR THE PRODUCTION OF BIODIESEL FROM FISH PROCESSING WASTES

Several studies have reported that fish oil could be easily converted to biodiesel. But due to the dispersed market and the communication barrier, it has challenging to accumulate the fish processing wastes from different places. As a result, it has been disposed of as waste material without any other use. If biodiesel plants manage to accumulate and deposit fish waste materials in a particular place, then these feedstocks can easily be transformed into high-quality biodiesel with efficient technologies available. Although some of the commercial-scale projects are being executed successfully in different parts of the world. The extraction and conversion process is energy-intensive, but in the long run, it would be one of the most important and environmentally friendly sources of energy. Researchers have suggested that instead of many smaller production plants, only one large production plant will be economically useful, increasing the quantity and production potential (Yuvaraj et al., 2019). In addition, some researchers indicated that not only to generate biodiesel, it can also be used to produce biogas. By maintaining process stability, fish waste can be used as a cosubstrate with various agricultural waste materials such as manure and sewage sludge to improve the productivity rate and profitability for commercial biogas industries (Ward and Løes, 2011). Some of the enterprises and projects around the world that have taken promising approaches to using fish oil for industrial purposes, such as ENERFISH (European project initiated by the European Commission), AQUAFINCA (established in 1994, located in northwestern Honduras), PETROBRAS (Brazilian company, founded in 1953). Aspects of fish oil extraction and conversion technologies are still being investigated to make it a more reliable option, and efforts are also being made to transition to the pilot-scale generation of various laboratory scales.

10.10 CONCLUSION

The fishery industry developed throughout the world, primarily for human consumption. In recent years, fish production rate has increased across the globe with various productivity technologies and approaches. Besides the waste from fish farming and fish processing industries has also generated wastes. Every year almost 90 million tonnes of fish discards have been reported to generate. However, various by-products have been used in different applications such as biofertilizer, animal feed, and in biopharmaceuticals. But the remaining waste is simply dumped into the ocean or the ground without modifying it. Finally, these products have several negative impacts on the environment and cause environmental pollution. To solve these problems, fish waste can be channeled to produce fish oil, and this oil is a very cheap and sustainable source of biodiesel production. The production of biodiesel from fish farm waste can overcome the production defects of other feedstocks, such as used cooking oil and vegetable oil. However, the rate of fish biodiesel production varies from season to season. The production rate is not constant. It depends on several factors, such as the content and composition of the fat, some physiological properties of the fish species. This biodiesel produced from fish oil could be used to power machinery. It can be used as an alternative source of fossil fuels and free of greenhouse gas emissions. Commercial biodiesel production from the combination of fish oil with fish farming and processing industries could be a beneficial integrated system by generating more income and employment. Various by-products generated at the time of biodiesel production could be of great value in other fields. Glycerol produced from fish biodiesel plants can be widely used in the food and cosmetic industries. This whole approach to producing biodiesel from fish discards could stimulate the reduction of environmental pollution and the energy crisis. There is a great need for more research work on the production of biodiesel from waste materials and its proper use.

REFERENCES

Adeniyi, O.D., Bawa, A.A., 2006. Mackerel (*Scomber scrombrus*) oil extraction and evaluation as raw materials for industrial utilization. Leonardo Journal of Sciences 5 (8), 33–42.

Adeoti, I.A., Hawboldt, K., 2014. A review of lipid extraction from fish processing by-product for use as a biofuel. Biomass and Bioenergy 63, 330–340.

Aidos, I., Luten, J.B., Boonman, M., van der Padt, A., Boom, R.M., 2000. Production of fish oil from maatjes herring by-products. In: Proceedings of 29th WEFTA Meeting: 29th WEFTA Meeting, Leptocarya, Pieria, Greece, 1999 (pp. 352–360).

Arason, S., Karlsdottir, M., Valsdottir, T., Slizyte, R., Rustad, T., Falch, E., et al., 2009. Maximum Resource Utilisation—Value Added Fish By-Products. Nordic Innovation Centre, Oslo.

Arvanitoyannis, I.S., Kassaveti, A., 2008. Fish industry waste: treatments, environmental impacts, current and potential uses. International Journal of Food Science & Technology 43 (4), 726–745.

Arzamendi, G., Campo, I., Arguinarena, E., Sánchez, M., Montes, M., Gandia, L.M., 2007. Synthesis of biodiesel with heterogeneous NaOH/alumina catalysts: comparison with homogeneous NaOH. Chemical Engineering Journal 134 (1–3), 123–130.

Atadashi, I.M., Aroua, M.K., Aziz, A.A., Sulaiman, N.M.N., 2013. The effects of catalysts in biodiesel production: a review. Journal of Industrial and Engineering Chemistry 19 (1), 14–26.

Ayyappan, S., Krishnan, M., 2004. Fisheries sector in India: dimensions of development. Indian Journal of Agricultural Economics 59 (3), 392–412.

Bais, B., 2018. Fish scenario in India with emphasis on Indian major carps. International Journal of Avian &Wildlife Biology 3 (6).

Bajaj, A., Lohan, P., Jha, P.N., Mehrotra, R., 2010. Biodiesel production through lipase catalyzed transesterification: an overview. Journal of Molecular Catalysis B: Enzymatic 62 (1), 9–14.

Bako, T., Umogbai, V. I., Obetta, S. E., 2014. Extraction and characterizatin of Mackery (Scomber scombrus) oil for industrial use. Researcher 6, 80–85.

Bell, A.W., 1995. Regulation of organic nutrient metabolism during transition from late pregnancy to early lactation. Journal Animal Science 73, 2804–2819.

Belton, B., Little, D.C., Zhang, W., Edwards, P., Skladany, M., Thilsted, S.H., 2020. Farming fish in the sea will not nourish the world. Nature Communications 11 (1), 1–8.

Béné, C., Barange, M., Subasinghe, R., Pinstrup-Andersen, P., Merino, G., Hemre, G.I., et al., 2015. Feeding 9 billion by 2050–Putting fish back on the menu. Food Security 7 (2), 261–274.

Boran, G., Karaçam, H., Boran, M., 2006. Changes in the quality of fish oils due to storage temperature and time. Food Chemistry 98 (4), 693–698.

Bostock, J., McAndrew, B., Richards, R., Jauncey, K., Telfer, T., Lorenzen, K., et al., 2010. Aquaculture: global status and trends. Philosophical Transactions of the Royal Society of London. Series B, Biological Sciences 365 (1554), 2897–2912.

Canakci, M., Van Gerpen, J., 1999. Biodiesel production viaacid catalysis. Transactions of the ASAE 42 (5), 1203.

Cao, L., Naylor, R., Henriksson, P., Leadbitter, D., Metian, M., Troell, M., et al., 2015. China's aquaculture and the world's wild fisheries. Science 347 (6218), 133–135.

Chakraborty, K., Joseph, D., 2015. Cooking and pressing is an effective and eco-friendly technique for obtaining high quality oil from Sardinella longiceps. European Journal of Lipid Science and Technology 117 (6), 837–850.

Chew, T.L., Bhatia, S., 2008. Catalytic processes towards the production of biofuels in a palm oil and oil palm biomass-based biorefinery. Bioresource Technology 99 (17), 7911–7922.

Chowdhury, R., Stevens, S., Gorman, D., Pan, A., Warnakula, S., Chowdhury, S., et al., 2012. Association between fish consumption, long chain omega 3 fatty acids, and risk of cerebrovascular disease: systematic review and *meta*-analysis. British Medical Journal 345, e6698.

Di Serio, M., Tesser, R., Pengmei, L., Santacesaria, E., 2008. Heterogeneous catalysts for biodiesel production. Energy & Fuels 22 (1), 207–217.

Domenichiello, A.F., Chen, C.T., Trepanier, M.O., Stavro, P.M., Bazinet, R.P., 2014. Whole body synthesis rates of DHA from α-linolenic acid are greater than brain DHA accretion and uptake rates in adult rats. Journal of Lipid Research 55 (1), 62–74.

Duarah, J.P., Mall, M., 2020. Diversified fish farming for sustainable livelihood: A case-based study on small and marginal fish farmers in Cachar district of Assam, India. Aquaculture 529, 735569.

Dutta, S., Jaiswal, K.K., Verma, R., Basavaraju, D.M., Ramaswamy, A.P., 2019. Green synthesis of zinc oxide catalyst under microwave irradiation using banana (*Musa* spp.) corm (rhizome) extract for biodiesel synthesis from fish waste lipid. Biocatalysis and Agricultural Biotechnology 22, 101390.

Eslick, G.D., Howe, P.R.C., Smith, C., Priest, R., Bensoussan, A., 2009. Benefits of fish oil supplementation in hyperlipidemia: a systematic review and meta analysis. International Journal of Cardiology 136, 4–16.

Fadhil, A.B., Ahmed, A.I., 2016. Ethanolysis of fish oil via optimized protocol and purification by dry washing of crude ethyl esters. Journal of the Taiwan Institute of Chemical Engineers 58, 71–83.

Fadhil, A.B., Ali, L.H., 2013. Alkaline-catalyzed transesterification of Silurus triostegus Heckel fish oil: optimization of transesterification parameters. Renewable Energy 60, 481–488.

Fadhil, A.B., Al-Tikrity, E.T., Albadree, M.A., 2015. Transesterification of a novel feedstock, *Cyprinus carpio* fish oil: influence of co-solvent and characterization of biodiesel. Fuel 162, 215–223.

FAO, 2018a. The State of World Fisheries and Aquaculture 2018—Meeting the Sustainable Development Goals. FAO, Rome, p. 227.

(FAO), 2018b. The State of World Fisheries and Aquaculture 2018—Meeting the Sustainable Development Goals. Rome. License: CC BY-NC-SA 3.0 IGO.

Ferdosh, S., Sarker, M.Z.I., Ab Rahman, N.N.N., Akand, M.J.H., Ghafoor, K., Awang, M. B., et al., 2013. Supercritical carbon dioxide extraction of oil from Thunnus tonggol head by optimization of process parameters using response surface methodology. Korean Journal of Chemical Engineering 30 (7), 1466–1472.

Ferdosh, S., Sarker, Z. I., Norulaini, N., Oliveira, A., Yunus, K., Chowdury, A. J., Akanda, J., Omar, M., 2015. Quality of tuna fish oils extracted from processing the by-products of three species of neritic tuna using supercritical carbon dioxide. Journal of Food Processing and Preservation 39, 432–441.

Ferdosh, S., Sarker, Z.I., Norulaini, N., Oliviera, A., Yunus, K., Choedury, A.J., et al., 2015. Quality of tuna fish oils extracted from processing the by- products of three species of neritic tuna using supercritical carbon dioxide. Journal of Food Processing and Preservation 39, 432–441.

Ferraro, V., Cruz, I.B., Jorge, R.F., Malcata, F.X., Pintado, M.E., Castro, P.M., 2010. Valorisation of natural extracts from marine source focused on marine by-products: a review. Food Research International 43 (9), 2221–2233.

Fiori, L., Solana, M., Tosi, P., Manfrini, M., Strim, C., Guella, G., 2012. Lipid profiles of oil from trout (*Oncorhynchus mykiss*) heads, spines and viscera: trout by-products as a possible source of omega-3 lipids? Food Chemistry 134 (2), 1088–1095.

Gedi, M.A., Bakar, J., Mariod, A.A., 2015. Optimization of supercritical carbon dioxide (CO_2) extraction of sardine (*Sardinella lemuru* Bleeker) oil using response surface methodology (RSM). Grasas y Aceites 66 (2), 074.

Gephart, J.A., Golden, C.D., Asche, F., Belton, B., Brugere, C., Froehlich, H.E., et al., 2020. Scenarios for global aquaculture and its role in human nutrition. Reviews in Fisheries Science & Aquaculture 1–17.

Ghaly, A.E., Ramakrishnan, V.V., Brooks, M.S., Budge, S.M., Dave, D., 2013. Fish processing wastes as a potential source of proteins. Amino acids and oils: a critical review. Journal of Microbial and Biochemical Technology 5 (4), 107–129.

Girish, C.R., Gambhir, M.M., Deshmukh, T., 2017. Production of biodiesel from fish waste and characterization of produced biodiesel. International Journal of Civil Engineering and Technology 8, 1–6.

González-Sarrías, A., Larrosa, M., García-Conesa, M.T., Tomás-Barberán, F.A., Espín, J.C., 2013. Nutraceuticals for older people: facts, fictions and gaps in knowledge. Maturitas 75 (4), 313–334.

Gopalakrishnan, A., George, G., Ignatius, B., 2017. Supporting the Livelihood of four million fisher folk in Indian marine fisheries-ICAR-CMFRI's recent initiatives.

Gryglewicz, S., 1999. Rapeseed oil methyl esters preparation using heterogeneous catalysts. Bioresource Technology 70 (3), 249–253.

Guillen, J., Natale, F., Carvalho, N., Casey, J., Hofherr, J., Druon, J.N., et al., 2019. Global seafood consumption footprint. Ambio 48 (2), 111–122. Available from: https://doi.org/10.1007/s13280-018-1060-9.

Hao, S., Wei, Y., Li, L., Yang, X., Cen, J., Huang, H., Lin, W., Yuan, X., 2015. The effects of different extraction methods on composition and storage stability of sturgeon oil. Food Chemistry 173, 274–282.

Haq, M., Ahmed, R., Cho, Y.J., Chun, B.S., 2017. Quality properties and bio-potentiality of edible oils from atlantic salmon by-products extracted by supercritial carbon dioxide and conventional methods. Waste and Biomass Valorization 8 (6), 1953–1967.

Harnedy, P.A., FitzGerald, R.J., 2012. Bioactive peptides from marine processing waste and shellfish: a review. Journal of functional foods 4 (1), 6–24.

Helwani, Z., Othman, M.R., Aziz, N., Kim, J., Fernando, W.J.N., 2009. Solid heterogeneous catalysts for transesterification of triglycerides with methanol: a review. Applied Catalysis A: General 363 (1–2), 1–10.

Homayooni, B., Sahari, M.A., Barzegar, M., 2014. Concentrations of omega-3 fatty acids from rainbow sardine fish oil by various methods. International Food Research Journal 21, 743–748.

Huang, J., Sathivel, S., 2010. Purifying salmon oil using adsorption, neutralization, and a combined neutralization and adsorption process. Journal of Food Engineering 96, 51–58.

Immanuel, G., Sathasivan, S., Shankar, V.S., Peter, M.P., Palavesam, A., 2009. Processing and characterisation of low cost Balistid fish *Sufflamen capistratus* liver oil for edible purpose. Food Chemistry 115 (2), 430–435.

Ivanovs, K., Blumberga, D., 2017. Extraction of fish oil using green extraction methods: a short review. Energy Procedia 128, 477–483.

Jaiswal, K.K., Jha, B., Prasath, R.A., 2014. Biodiesel production from discarded fish waste for sustainable clean energy development. Journal of Chemical and Pharmaceutical Sciences (4), 113–114.

Jobling, M., 2016. Fish nutrition research: past, present and future. Aquaculture International 24 (3), 767–786.

Kara, K., Ouanji, F., El Lotfia, M., El Mahia, M., Kacimi, M., Ziyad, M., 2017. Biodiesel production from waste fish oil with high free fatty acid content from Moroccan fish-processing industries. Egyptian Journal of Petroleum 27, 249–255.

Kim, S.K., Wijesekara, I., 2010. Development and biological activities of marine-derived bioactive peptides: a review. Journal of Functional foods 2 (1), 1–9.

Kim, D.S., Hanifzadeh, M., Kumar, A., 2018. Trend of biodiesel feedstock and its impact on biodiesel emission characteristics. Environmental Progress & Sustainable Energy 37 (1), 7–19.

Kołakowska, A., Domiszewski, Z., Kozłowski, D., Gajowniczek, M., 2006. Effects of rainbow trout freshness on n-3 polyunsaturated fatty acids in fish offal. European Journal of Lipid Science and Technology 108 (9), 723–729.

Kouzu, M., Kasuno, T., Tajika, M., Sugimoto, Y., Yamanaka, S., Hidaka, J., 2008. Calcium oxide as a solid base catalyst for transesterification of soybean oil and its application to biodiesel production. Fuel 87 (12), 2798–2806.

Kouzu, M., Yamanaka, S.Y., Hidaka, J.S., Tsunomori, M., 2009. Heterogeneous catalysis of calcium oxide used for transesterification of soybean oil with refluxing methanol. Applied Catalysis A: General 355 (1–2), 94–99.

Kudre, T.G., Bhaskar, N., Sakhare, P.Z., 2017. Optimization and characterization of biodiesel production from rohu (*Labeo rohita*) processing waste. Renewable Energy 113, 1408–1418.

Kuepethkaew, S., Sangkharak, K., Benjakul, S., Klomklao, S., 2017. Optimized synthesis of biodiesel using lipase from Pacific white shrimp (*Litopenaeus vannamei*) hepatopancreas. Renewable Energy 104, 139–147.

Kulkarni, M.G., Dalai, A.K., 2006. Waste cooking oil an economical source for biodiesel: a review. Industrial & Engineering Chemistry Research 45 (9), 2901–2913.

Kumar, P., Dey, M.M., Paraguas, F.J., 2006. Fish supply projections by production environments and species types in India. Agricultural Economics Research Review 19 (2), 327–352.

Kumar, G.B., Datta, K.K., Joshi, P.K., 2010. Growth of fisheries and aquaculture sector in India: Needed policy directions for future. World Aquaculture 41 (3), 45–51.

Kuo, C.H., Liao, H.Z., Wang, Y.H., Wang, H.M.D., Shieh, C.J., Tseng, C.Y., 2017. Highly efficient extraction of EPA/DHA-enriched oil from cobia liver using homogenization plus sonication. European Journal of Lipid Science and Technology 119, 1600466.

Kuratko, C.N., Barrett, E.C., Nelson, E.B., Salem Jr., N., 2013. The relationship of docosahexaenoic acid (DHA) with learning and behavior in healthy children: a review. Nutrients 5 (7), 2777–2810.

Lam, M.K., Lee, K.T., Mohamed, A.R., 2010. Homogeneous, heterogeneous and enzymatic catalysis for transesterification of high free fatty acid oil (waste cooking oil) to biodiesel: a review. Biotechnology Advances 28 (4), 500–518.

Létisse, M., Rozières, M., Hiol, A., Sergent, M., Comeau, L., 2006. Enrichment of EPA and DHA from sardine by supercritical fluid extraction without organic modifier: I. optimization of extraction conditions. The Journal of Supercritical Fluids 38 (1), 27–36.

Lin, C.Y., Li, R.J., 2009. Fuel properties of biodiesel produced from the crude fish oil from the soapstock of marine fish. Fuel Processing Technology 90 (1), 130–136.

Little, D.C., Newton, R.W., Beveridge, M.C.M., 2016. Aquaculture: a rapidly growing and significant source of sustainable food? Status, transitions and potential. Proceedings of the Nutrition Society 75 (3), 274–286.

Lopez, D.E., Goodwin Jr, J.G., Bruce, D.A., Lotero, E., 2005. Transesterification of triacetin with methanol on solid acid and base catalysts. Applied Catalysis A: General 295 (2), 97–105.

Lotero, E., Liu, Y., Lopez, D.E., Suwannakarn, K., Bruce, D.A., Goodwin, J.G., 2005. Synthesis of biodiesel via acid catalysis. Industrial & Engineering Chemistry Research 44 (14), 5353–5363.

Mathiesen, Á.M., 2015. The State of World Fisheries and Aquaculture 2012.

Menegazzo, M.L., Petenuci, M.E., Fonseca, G.G., 2014. Production and characterization of crude and refined oils obtained from the co-products of Nile tilapia and hybrid sorubim processing. Food Chemistry 157, 100–104.

Mishra, V.K., Goswami, R., 2018. A review of production, properties and advantages of biodiesel. Biofuels 9 (2), 273–289.

Msangi, S., Kobayashi, M., Batka, M., Vannuccini, S., Dey, M.M., Anderson, J.L., 2013. Fish to 2030: prospects for fisheries and aquaculture. World Bank Report 83177 (1), 102.

Nakatani, N., Takamori, H., Takeda, K., Sakugawa, H., 2009. Transesterification of soybean oil using combusted oyster shell waste as a catalyst. Bioresource Technology 100 (3), 1510–1513.

Narasimharao, K., Lee, A., Wilson, K., 2007. Catalysts in production of biodiesel: a review. Journal of Biobased Materials and Bioenergy 1 (1), 19–30.

Nazir, N., Diana, A., Sayuti, K., 2017. Physicochemical and fatty acid profile of fish oil from head of tuna (*Thunnus albacares*) extracted from various extraction method. International Journal on Advanced Science, Engineering and Information Technology 7, 709–715.

Ngo, D.H., Wijesekara, I., Vo, T.S., Van Ta, Q., Kim, S.K., 2011. Marine food-derived functional ingredients as potential antioxidants in the food industry: an overview. Food Research International 44 (2), 523–529.

Oliveira, L.E., Barboza, J.C.S., Da Silva, M.L.C.P., 2013. Production of ethylic biodiesel from Tilápia visceral oil. Renewable Energy and Power Quality Journal 13, 412.

Ormanci, H.B., Künili, İ.E., Colakoğlu, F.A., 2019. Fish processing wastes: potential source of byproducts. In: Proceedings of the International scientific and practical conference "Bulgaria of regions" (vol. 2, no. 1).

Porter, G., 1999. Fisheries subsidies and overfishing: towards a structured discussion. UN Environment Programme. Economics and Trade Branch.

Pradeepkiran, J.A., 2019. Aquaculture role in global food security with nutritional value: a review. Translational Animal Science 3 (2), 903–910.

Rai, A.K., Bhaskar, N., Baskaran, V., 2013. Bioefficacy of EPA-DHA from lipids recovered from fish processing wastes through biotechnological approaches. Food Chemistry 136 (1), 80–86.

Ribeiro, V., Oliveira, A., Bechtel, P., & Prentice, C., 2013. Avaliação do processamento a baixas temperaturas do óleo de fígado de Alaska pollock (Theragra chalcogramma). Brazilian Journal of Food Technology 16, 99–106.

Rincon Cervera, M.A., Venegas, E., Ramos Bueno, R.P., Suarez Medina, M.D., Guil Guerrero, J.L., 2015. Docosahexaenoic acid purification from fish processing industry by-products. European Journal of Lipid Science and Technology 117 (5), 724–729.

Rincón-Cervera, M.Á., Villarreal-Rubio, M.B., Valenzuela, R., Valenzuela, A., 2017. Comparison of fatty acid profiles of dried and raw by-products from cultured and wild fishes. European Journal of Lipid Science and Technology 119 (9), 1600516.

Rubio-Rodríguez, N., Sara, M., Beltrán, S., Jaime, I., Sanz, M.T., Rovira, J., 2008. Supercritical fluid extraction of the omega-3 rich oil contained in hake (*Merluccius capensis–Merluccius paradoxus*) by-products: study of the influence of process parameters on the extraction yield and oil quality. The Journal of Supercritical Fluids 47 (2), 215–226.

Rubio-Rodríguez, N., Beltrán, S., Jaime, I., Diego, S.M., Sanz, M.T., Rovira, C.J., 2010. Production of omega-3 polyunsaturated fatty acid concentrates: a review. Innovative Food Science & Emerging Technologies 11 (1), 1–12.

Rubio-Rodríguez, N., Sara, M., Beltrán, S., Jaime, I., Sanz, M.T., Rovira, J., 2012. Supercritical fluid extraction of fish oil from fish by-products: a comparison with other extraction methods. Journal of Food Engineering 109 (2), 238–248.

Rustad, T., Storro, I., Slizyte, R., 2011. Possibilities for the utilisation of marine by-products. International Journal of Food Science & Technology 46, 2001–2014.

Sahena, F., Zaidul, I.S.M., Jinap, S., Saari, N., Jahurul, H.A., Abbas, K.A., et al., 2009. PUFAs in fish: extraction, fractionation, importance in health. Comprehensive Reviews in Food Science and Food Safety 8 (2), 59–74.

Sahena, F., Zaidul, I.S.M., Jinap, S., Jahurul, M.H.A., Khatib, A., Norulaini, N.A.N., 2010. Extraction of fish oil from the skin of Indian mackerel using supercritical fluids. Journal of Food Engineering 99 (1), 63–69.

Santacesaria, E., Vicente, G.M., Di Serio, M., Tesser, R., 2012. Main technologies in biodiesel production: state of the art and future challenges. Catalysis Today 195 (1), 2–13.

Sarker, M.Z.I., Selamat, J., Habib, A.S.M., Ferdosh, S., Akanda, M.J.H., Jaffri, J.M., 2012. Optimization of supercritical CO_2 extraction of fish oil from viscera of African catfish (*Clarias gariepinus*). International Journal of Molecular Sciences 13 (9), 11312–11322.

Sathivel, S., Prinyawiwatkul, W., Negulescu, I.I., King, J.M., 2008. Determination of melting points, specific heat capacity and enthalpy of catfish visceral oil during the purification process. Journal of the American Oil Chemists' Society 85 (3), 291–296.

Shamsudin, S., Salimon, J., 2006. Physicochemical characteristics of aji-aji fish (*Seriola nigrofasciata*) lipids. Malaysia Journal of Analytical Sciences 10 (1), 55–58.

Sharma, Y.C., Singh, B., 2008. Development of biodiesel from karanja, a tree found in rural India. Fuel 87 (8–9), 1740–1742.

Sharma, Y.C., Singh, B., Madhu, D., Liu, Y., Yaakob, Z., 2014. Fast synthesis of high quality biodiesel from "waste fish oil" by single step transesterification. Biofuel Research Journal 1 (3), 78–80.

Šilovs, M. 2018. Fish Processing By-Products Exploitation and Innovative Fish-Based Food Production. Research for Rural Development, 2.

Silva-Aguiar, G.P., Soares-Goulart, G.A., 2013. Utilização de material residual da indústria de pescado para obtenção de óleo e farinha. Tecnologia & Ciência Agropecuária 7 (4), 55–60.

Swanson, D., Block, R., Mousa, S.A., 2012. Omega-3 fatty acids EPA and DHA: health benefits throughout life. Advances in Nutrition 3 (1), 1–7.

Tacon, A.G., Metian, M., 2013. Fish matters: importance of aquatic foods in human nutrition and global food supply. Reviews in Fisheries Science 21 (1), 22–38.

Tacon, A.G., Lemos, D., Metian, M., 2020. Fish for health: improved nutritional quality of cultured fish for human consumption. Reviews in Fisheries Science & Aquaculture 28 (2), 1–10.

Talha, N.S., Sulaiman, S., 2016. Overview of catalysts in biodiesel production. ARPN Journal of Engineering and Applied Sciences 11 (1), 439–448.

Thilsted, S.H., Thorne-Lyman, A., Webb, P., Bogard, J.R., Subasinghe, R., Phillips, M.J., et al., 2016. Sustaining healthy diets: The role of capture fisheries and aquaculture for improving nutrition in the post-2015 era. Food Policy 61, 126–131.

USAID SPARE Fisheries & Aquaculture Panel. 2010. Review of the Status, Trends and Issues in Global Fisheries and Aquaculture, with Recommendations for USAID Investments. Retrieved 9th November.

Venugopal, V., Shahidi, F., Lee, T.C., 1995. Value-added products from underutilized fish species. Critical Reviews in Food Science & Nutrition 35 (5), 431–453.

Viriya-Empikul, N., Krasae, P., Nualpaeng, W., Yoosuk, B., Faungnawakij, K., 2012. Biodiesel production over Ca-based solid catalysts derived from industrial wastes. Fuel 92 (1), 239–244.

Ward, A.J., Løes, A.K., 2011. The potential of fish and fish oil waste for bioenergy generation: Norway and beyond. Biofuels 2 (4), 375–387.

Wei, Z., Xu, C., Li, B., 2009. Application of waste eggshell as low-cost solid catalyst for biodiesel production. Bioresource Technology 100 (11), 2883–2885.

Yang, L., Zhang, A., Zheng, X., 2009. Shrimp shell catalyst for biodiesel production. Energy & Fuels 23 (8), 3859–3865.

Yuvaraj, D., Bharathiraja, B., Rithika, J., Dhanasree, S., Ezhilarasi, V., Lavanya, A., et al., 2019. Production of biofuels from fish wastes: an overview. Biofuels 10 (3), 301–307.

FURTHER READING

Bimbo, A.P., Crowther, J.B., 1992. Fish meal and oil: current uses. Journal of the American Oil Chemists Society 69 (3), 221–227.

Feeney, R.E., Yeh, Y., 1998. Antifreeze proteins: current status and possible food uses. Trends in Food Science & Technology 9 (3), 102–106.

Joseph, T.C., Remya, S., Renuka, V., Jha, A.K., 2019. Fishery Industry Waste: A Resource to Be Valorised. Veraval Research Centre of ICAR-CIFT, Veraval, Gujarat.

Karkal, S.S., Kudre, T.G., 2020. Valorization of fish discards for the sustainable production of renewable fuels. Journal of Cleaner Production 275, 122985.

Kim, H.J., Lee, J.H., Hur, Y.B., Lee, C.W., Park, S.H., Koo, B.W., 2017. Marine antifreeze proteins: structure, function, and application to cryopreservation as a potential cryoprotectant. Marine Drugs 15 (2), 27.

Kusmiyati, K., 2015. Fish Waste Oil Conversion For Biodiesel Production Using Two Stages Reaction.

Pigott, G.M., 1967. Production of Fish Oil, vol. 277. United States Department of the Interior, Fish and Wildlife Service, Bureau of Commercial Fisheries, Seattle.

Rizliya, V., Mendis, E., 2014. Biological, physical, and chemical properties of fish oil and industrial applications. Seafood Processing By-Products. Springer, New York, NY, pp. 285–313.

Rustad, T., 2003. Utilisation of marine by-products. Electronic Journal of Environmental, Agricultural and Food Chemistry 2 (4), 458–463.

Venketesh, S., Dayananda, C., 2008. Properties, potentials, and prospects of antifreeze proteins. Critical Reviews in Biotechnology 28 (1), 57–82.

Index

Note: Page numbers followed by "*f*" and "*t*" refer to figures and tables, respectively.

C

D

E

F

Q

R

S